T0188276

VISCOUS
FLUID FLOW

VISCOUS FLUID FLOW

by

Tasos C. Papanastasiou

Georgios C. Georgiou
Department of Mathematics and Statistics
University of Cyprus
Nicosia, Cyprus

Andreas N. Alexandrou
Department of Mechanical Engineering
Worcester Polytechnic Institute
Worcester, MA

CRC Press
Taylor & Francis Group
Boca Raton London New York

CRC Press is an imprint of the
Taylor & Francis Group, an **informa** business

CRC Press
Taylor & Francis Group
6000 Broken Sound Parkway NW, Suite 300
Boca Raton, FL 33487-2742

First issued in paperback 2019

© 2000 by Taylor & Francis Group, LLC
CRC Press is an imprint of Taylor & Francis Group, an Informa business

No claim to original U.S. Government works

ISBN-13: 978-0-8493-1606-7 (hbk)
ISBN-13: 978-0-367-39924-5 (pbk)

Library of Congress Card Number 99-042163

This book contains information obtained from authentic and highly regarded sources. Reasonable efforts have been made to publish reliable data and information, but the author and publisher cannot assume responsibility for the validity of all materials or the consequences of their use. The authors and publishers have attempted to trace the copyright holders of all material reproduced in this publication and apologize to copyright holders if permission to publish in this form has not been obtained. If any copyright material has not been acknowledged please write and let us know so we may rectify in any future reprint.

Except as permitted under U.S. Copyright Law, no part of this book may be reprinted, reproduced, transmitted, or utilized in any form by any electronic, mechanical, or other means, now known or hereafter invented, including photocopying, microfilming, and recording, or in any information storage or retrieval system, without written permission from the publishers.

For permission to photocopy or use material electronically from this work, please access www.copyright.com (http://www.copyright.com/) or contact the Copyright Clearance Center, Inc. (CCC), 222 Rosewood Drive, Danvers, MA 01923, 978-750-8400. CCC is a not-for-profit organization that provides licenses and registration for a variety of users. For organizations that have been granted a photocopy license by the CCC, a separate system of payment has been arranged.

Trademark Notice: Product or corporate names may be trademarks or registered trademarks, and are used only for identification and explanation without intent to infringe.

Library of Congress Cataloging-in-Publication Data

Papanastasiou, Tasos C.
    Viscous fluid flow / by Tasos C. Papanastasiou, Georgios C. Georgiou, Andreas N. Alexandrou.
        p. cm.
    Includes bibliographical references and index.
    ISBN 0-8493-1606-5
    1. Viscous flow. I. Georgiou, Georgios C. II. Alexandrou, Andreas N. III. Title.
QA929 .P35 1999
532′.0533—dc21   99-042163
                                                                                                          CIP

Visit the Taylor & Francis Web site at
http://www.taylorandfrancis.com

and the CRC Press Web site at
http://www.crcpress.com

To

Androula, Charis, and *Yiangos Papanastasiou*

and to

Dimitra, Lisa, and *Nadia*

Contents

Preface

The original draft of this textbook was prepared by the late Professor Tasos Papanastasiou. Following his unfortunate death in 1994, we assumed the responsibility of completing and publishing the manuscript. In editing and completing the final text, we made every effort to retain the original approach of Professor Papanastasiou. However, parts of the book have been revised and rewritten so that the material is consistent with the intent of the book. The book is intended for upper-level undergraduate and graduate courses.

The educational purpose of the book is twofold: (a) to develop and rationalize the mathematics of viscous fluid flow using basic principles, such as mass, momentum conservation, and constitutive equations; and (b) to exhibit the systematic application of these principles to flows occurring in fluid processing and other applications.

The mass conservation or continuity equation is the mathematical expression of the statement that *"mass cannot be produced nor can it be destroyed to zero."* The equation of momentum conservation is the mathematical expression of Newton's law of motion that *"action of forces results in change of momentum and therefore acceleration."* The constitutive equation is inherent to the molecular structure of the continuous medium and describes the state of the material under stress: in static equilibrium, this state is fully described by pressure; in flow, it is fully described by deformation and pressure.

This book examines, in detail, flows of *Newtonian fluids*, i.e., of fluids that follow Newton's law of *viscosity*: *"viscous stress is proportional to the velocity gradient,"* the constant of proportionality being the viscosity. Some aspects of non-Newtonian flow are discussed briefly in Chapters 2 and 4.

Chapter 1, on *"Vector and Tensor Calculus,"* builds the mathematical prerequisites for studying Fluid Mechanics, particularly the theory of vectors and tensors and their operations. In this chapter, we introduce important vectors and tensors encountered in Fluid Mechanics such as the position, velocity, acceleration, momentum, and vorticity vectors, and the stress, velocity gradient, rate of strain, and vorticity tensors. We also discuss the integral theorems of vector and tensor calculus, i.e., the Gauss, the Stokes, and the Reynolds transport theorems. These theorems are used in subsequent chapters to derive the conservation equations. It takes six to seven hourly lectures to cover the material of Chapter 1.

Chapter 2, on *"Introduction to the Continuum Fluid,"* introduces the approximation of a fluid as a *continuum*, rather than as a discontinuous molecular medium. Properties associated with the continuum character such as density, mass, volume, linear and angular momentum, viscosity, kinematic viscosity, body and contact forces, mechanical pressure, and surface tension are introduced and discussed. The control volume concept is introduced and combined with the integral theorems and the differential operators of Chapter 1 to derive both macroscopic and microscopic conservation equations. The motion of fluid particles is described by using both Lagrangian and Eulerian descriptions. The chapter concludes with the local kinematics around a fluid particle that are responsible for stress, strain, and rate of strain development and propagation. The decomposition of the instantaneous velocity of a fluid particle into four elementary motions, i.e., rigid-body translation, rigid-body rotation, isotropic expansion, and pure straining motion without change of volume, is also demonstrated. It takes two to three hourly lectures to cover Chapter 2.

Chapter 3, on *"Conservation Laws,"* utilizes differential operators of Chapter 1 and conservation and control volume principles of Chapter 2, to develop the general integral conservation equation. This equation is first turned into differential form, by means of the Gauss theorem. It is then specialized to express mass, momentum, energy, and heat conservation equation. The conservation of momentum equations are expressed in terms of the stresses, which implies that they hold for any fluid. (The specialization of these equations to *incompressible Newtonian* fluids, the primary target of this book, is done in Chapter 5.) It takes two to three hourly lectures to cover Chapter 3.

Chapter 4, on *"Static Equilibrium of Fluids and Interfaces,"* deals with the application of conservation principles in the absence of relative flow. The general hydrostatics equation under rigid-body translation and rigid-body rotation for a single fluid in gravity and centrifugal fields is derived. It is then applied to barotropic and other fluids, yielding Bernoulli-like equations and the Archimedes' principle of buoyancy in fluids and across interfaces. The second part of the chapter deals with immiscible liquids across interfaces at static equilibrium. Normal and shear stress interface boundary conditions are derived in terms of bulk properties of fluids and the interface tension and curvature. The Young–Laplace equation is used to compute interface configurations at static equilibrium. It takes four to five lectures to cover

Chapter 4.

Chapter 5, on *"The Navier–Stokes Equations,"* starts with the concept of constitutive equations based on continuum mechanics. We then focus on Newtonian fluids by reducing the general Stokes constitutive equation for compressible Newtonian fluid to Newton's law of viscosity for incompressible Newtonian fluid. Alternative forms of the Navier–Stokes equations are also discussed. The dynamics of generation, intensification, convection, and diffusion of vorticity, which are directly related to the physics of flow, are projected and discussed along with the concepts of irrotationality, potentiality, local rigid-body rotation, and circulation that may be formulated and related by means of Bernoulli's and Euler's inviscid flow equations, the Stokes circulation theorem, and Kelvin's circulation conservation. Initial and boundary conditions necessary to solve the Navier–Stokes and related equations are also discussed. Chapter 5 concludes the part of the book that develops and discusses basic principles. It takes three to four lectures to cover Chapter 5.

The application part of the book starts with Chapter 6, on *"Unidirectional Flows,"* where steady-state and transient *unidirectional* flows amenable to analytical solution are studied. We first analyze five classes of steady, unidirectional, incompressible Newtonian flow in which the unknown velocity component is a function of just one spatial dependent variable: (a) steady, one-dimensional rectilinear flows; (b) steady, axisymmetric rectilinear flows; (c) steady, axisymmetric torsional flows; (d) steady, axisymmetric radial flows; and (e) steady, spherically symmetric radial flows. In all the above classes, the flow problem is reduced to an ordinary differential equation (ODE) subject to appropriate boundary conditions. This ODE results from the conservation of momentum (in the first three classes) or from the conservation of mass (in the last two classes). Next, we study two classes of unidirectional flow in which the unknown velocity component is a function of two independent variables: (a) transient one-dimensional unidirectional flows; and (b) steady two-dimensional rectilinear flows. In these two classes, conservation of momentum results in a partial differential equation (PDE) which must be solved together with appropriate boundary and initial conditions. For this purpose, techniques like the separation of variables and the similarity method are employed. Representative examples are provided throughout the chapter: steady and transient Poiseuille and Couette flows; film flow down an inclined plane or a vertical cylinder; flow between rotating cylinders; bubble growth; flow near a plate suddenly set in motion; steady Poiseuille flows in tubes of elliptical, rectangular, and triangular cross sections; and others. It takes six to seven lectures to cover Chapter 6.

Chapter 7, on *"Approximate Methods,"* introduces dimensional and order of magnitude analyses. It then focuses on the use of regular and singular perturbation methods in approximately solving flow problems in extreme limits of key parameters such as the Reynolds, Stokes, and capillary numbers, and inclination and geometrical aspect ratios. The chapter concludes with a brief discussion of the most important applications of perturbation methods in fluid mechanics, which are the subject of the subsequent chapters. It takes three to four hourly lectures to cover Chapter 7.

In Chapter 8, on *"Laminar Boundary Layer Flows,"* we examine laminar, high-Reynolds-number flows in irregular geometries and over submerged bodies. Flows are characterized as potential flows away from solid boundaries, and as boundary-layer flows in the vicinity of solid boundaries. Following the development of the boundary-layer equations by means of the stream function, exact solutions are examined by means of the Blasius' and Sakiades' analyses, and approximate, yet accurate enough, solutions are constructed along the lines of von Karman's analysis. The stagnation-point and rotating boundary-layer flows are also covered. It takes three to four hourly lectures to cover Chapter 8.

Chapter 9, on *"Nearly Unidirectional Flows,"* addresses lubrication and thin-film flows. Typical lubrication-flow applications considered are piston-cylinder and piston-ring lubrication of engines, journal-bearing system lubrication, and flows in nearly rectilinear channel or pipe. Flows of thin films under the combined action of viscosity, gravity, and surface tension are also analyzed. The integral mass and momentum equations lead to the celebrated Reynolds lubrication equation that relates the conduit width or film thickness to the pressure distribution in terms of the capillary and Stokes numbers and aspect ratios. The solution of the Reynolds equation in confined flows yields the pressure and shear stress distributions, which are directly responsible for load capacity, friction, and wear. The solution of the Reynolds equation in film flows - where the pressure gradient is related to the external pressure, the surface tension and the surface curvature - yields the configuration of the free surface and the final film thickness. Stretching flows, such as spinning of fibers, casting of sheets, and blowing of films, are also analyzed by means of the thin-beam approximation to yield the free surface profile and the final film thickness or fiber diameter, and the required tensions to achieve target fiber diameter and film thickness, depending on the spinnability of the involved liquid. It takes three to four hourly lectures to cover Chapter 9.

Chapter 10, on *"Creeping Bidirectional Flows,"* examines *slow* flows dominated by viscous forces or, equivalently, small Reynolds number flows. In the limit of zero Reynolds number, the equations of flow are simplified to the so-called Stokes equations. Stokes flow is conveniently studied with the introduction of the *stream function*, by means of which the system of the governing conservation equations is reduced to a much-easier-to-handle single, fourth-order PDE. Representative creeping flow examples, such as the flow near a corner and the flow past a sphere, are discussed in detail. It takes two to three hourly lectures to cover Chapter 10.

All chapters are accompanied by problems, which are often open-ended. The student is expected to spend time understanding the physical problem, developing the mathematical formulation, identifying assumptions and approximations, solving the problem, and evaluating the results by comparison to intuition, data, and other analyses.

We would like to express our gratitude to our colleagues and friends who read early drafts of chapters and provided useful suggestions: Dr. N. Adoniades (Greek Telecommunications Organization), Prof. A. Boudouvis, (NTU, Athens), Dr. M. Fyrillas (University of California, San Diego), Prof. A. Karageorghis (University of Cyprus), Dr. P. Papanastasiou (Schlumberger Cambridge Research), Dr. A. Poullikkas (Electricity Authority of Cyprus), Dr. M. Syrimis (University of Cyprus), and Prof. J. Tsamopoulos (University of Patras). We thank them all.

GG and AA
Worcester
July 1999

Below is the original acknowledgement text written by the late Professor Tasos Papanastasiou.

Several environments and individuals contributed directly or indirectly to the realization of this book, whom I would like to greatly acknowledge: my primary school teacher, George Maratheftis; my high school physics teacher, Andreas Stylianidis; my undergraduate fluid mechanics professor, Nikolaos Koumoutsos; and my graduate fluid mechanics professors, Prof. L.E. Scriven and C.W. Macosko of Minnesota. From the University of Michigan, my first school as assistant professor, I would like to thank the 1987-89 graduate fluid mechanics students and my research students; Prof. Andreas Alexandrou of Worcester Polytechnic Institute; Prof. Rose Wesson of LSU; Dr. Zhao Chen of Eastern Michigan University; Mr. Joe Greene of General Motors; Dr. Nick Malamataris from Greece; Dr. Kevin Ellwood of Ford Motor Company; Dr. N. Anturkar of Ford Motor Company; and Dr. Mehdi Alaie from Iran. Many thanks go to Mrs. Paula Bousley of Dixboro Designs for her prompt completion of both text and illustrations, and to the unknown reviewers of the book who suggested significant improvements.

Tasos C. Papanastasiou
Thessaloniki
March 1994

Chapter 1

VECTOR AND TENSOR CALCULUS

The physical quantities encountered in fluid mechanics can be divided into three classes: (a) *scalars*, such as pressure, density, viscosity, temperature, length, mass, volume, and time; (b) *vectors*, such as velocity, acceleration, displacement, linear momentum, and force; and (c) *tensors*, such as stress, rate of strain, and vorticity tensors.

Scalars are completely described by their *magnitude* or absolute value, and they do not require direction in space for their specification. In most cases, we shall denote scalars by lower case, lightface italic type, such as p for pressure and ρ for density. Operations with scalars, i.e., addition and multiplication, follow the rules of elementary algebra. A *scalar field* is a real-valued function that associates a scalar (i.e., a real number) with each point of a given region in space. Let us consider, for example, the *right-handed Cartesian coordinate system* of Fig. 1.1 and a closed three-dimensional region V occupied by a certain amount of a moving fluid at a given time instance t. The density ρ of the fluid at any point (x, y, z) of V defines a scalar field denoted by $\rho(x, y, z)$. If the density is, in addition, time-dependent, one may write $\rho = \rho(x, y, z, t)$.

Vectors are specified by their *magnitude* and their *direction* with respect to a given frame of reference. They are often denoted by lower case, boldface type, such as $\mathbf{u}$ for the velocity vector. A *vector field* is a vector-valued function that associates a vector with each point of a given region in space. For example, the velocity of the fluid in the region V of Fig. 1.1 defines a vector field denoted by $\mathbf{u}(x, y, z, t)$. A vector field which is independent of time is called a *steady-state* or *stationary* vector field. The magnitude of a vector $\mathbf{u}$ is designated by $|\mathbf{u}|$ or simply by u.

Vectors can be represented geometrically as arrows; the direction of the arrow specifies the direction of the vector and the length of the arrow, compared to some chosen scale, describes its magnitude. Vectors having the same length and the same

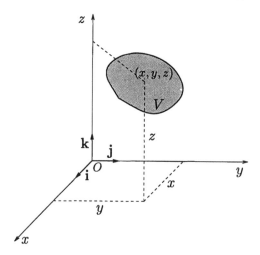

Figure 1.1. *Cartesian system of coordinates.*

direction, regardless of the position of their initial points, are said to be *equal*. A vector having the same length but the opposite direction to that of the vector **u** is denoted by −**u** and is called the *negative* of **u**.

The *sum* (or the *resultant*) **u**+**v** of two vectors **u** and **v** can be found using the *parallelogram law* for vector addition, as shown in Fig. 1.2a. Extensions to sums of more than two vectors are immediate. The difference **u**-**v** is defined as the sum **u**+(−**v**); its geometrical construction is shown in Fig. 1.2b.

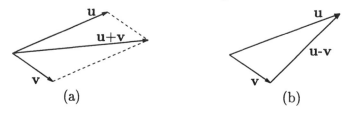

(a) (b)

Figure 1.2. *Addition and subtraction of vectors.*

The vector of length zero is called the *zero vector* and is denoted by **0**. Obviously, there is no natural direction for the zero vector. However, depending on the problem, a direction can be assigned for convenience. For any vector **u**,

$$\mathbf{u} + \mathbf{0} = \mathbf{0} + \mathbf{u} = \mathbf{u}$$

and

$$\mathbf{u} + (-\mathbf{u}) = \mathbf{0}.$$

Vector addition obeys the *commutative* and *associative* laws. If **u**, **v**, and **w** are vectors, then

$\mathbf{u} + \mathbf{v} = \mathbf{v} + \mathbf{u}$	Commutative law
$(\mathbf{u} + \mathbf{v}) + \mathbf{w} = \mathbf{u} + (\mathbf{v} + \mathbf{w})$	Associative law

If **u** is a nonzero vector and m is a nonzero scalar, then the *product* $m\mathbf{u}$ is defined as the vector whose length is $|m|$ times the length of **u**, and whose direction is the same as that of **u** if $m > 0$, and opposite to that of **u** if $m < 0$. If $m{=}0$ or $\mathbf{u}{=}\mathbf{0}$, then $m\mathbf{u}{=}\mathbf{0}$. If **u** and **v** are vectors and m and n are scalars, then

$m\mathbf{u} = \mathbf{u}m$	Commutative law
$m(n\mathbf{u}) = (mn)\mathbf{u}$	Associative law
$(m+n)\mathbf{u} = m\mathbf{u} + n\mathbf{u}$	Distributive law
$m(\mathbf{u}+\mathbf{v}) = m\mathbf{u} + m\mathbf{v}$	Distributive law

Note also that $(-1)\mathbf{u}$ is just the negative of **u**,

$$(-1)\mathbf{u} = -\mathbf{u}.$$

A *unit vector* is a vector having unit magnitude. The three vectors **i**, **j**, and **k** which have the directions of the positive x, y, and z axes, respectively, in the Cartesian coordinate system of Fig. 1.1 are unit vectors.

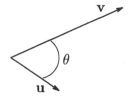

Figure 1.3. *Angle between vectors* **u** *and* **v**.

Let **u** and **v** be two nonzero vectors in a two- or three-dimensional space positioned so that their initial points coincide (Fig. 1.3). The *angle* θ between **u** and **v** is the angle determined by **u** and **v** that satisfies $0 \leq \theta \leq \pi$. The *dot product* (or *scalar product*) of **u** and **v** is a scalar quantity defined by

$$\mathbf{u} \cdot \mathbf{v} \equiv uv \, \cos\theta. \tag{1.1}$$

If **u**, **v**, and **w** are vectors and m is a scalar, then

$\mathbf{u} \cdot \mathbf{v} = \mathbf{v} \cdot \mathbf{u}$	Commutative law
$\mathbf{u} \cdot (\mathbf{v}+\mathbf{w}) = \mathbf{u} \cdot \mathbf{v} + \mathbf{u} \cdot \mathbf{w}$	Distributive law
$m(\mathbf{u} \cdot \mathbf{v}) = (m\mathbf{u}) \cdot \mathbf{v} = \mathbf{u} \cdot (m\mathbf{v})$	

Moreover, the dot product of a vector with itself is a positive number that is equal to the square of the length of the vector:

$$\mathbf{u} \cdot \mathbf{u} = u^2 \quad \Longleftrightarrow \quad u = \sqrt{\mathbf{u} \cdot \mathbf{u}}. \tag{1.2}$$

If $\mathbf{u}$ and $\mathbf{v}$ are nonzero vectors and

$$\mathbf{u} \cdot \mathbf{v} = 0\,,$$

then $\mathbf{u}$ and $\mathbf{v}$ are *orthogonal* or *perpendicular* to each other.

A vector set $\{\mathbf{u}_1, \mathbf{u}_2, \cdots, \mathbf{u}_n\}$ is said to be an *orthogonal set* or *orthogonal system* if every distinct pair of the set is orthogonal, i.e.,

$$\mathbf{u}_i \cdot \mathbf{u}_j = 0\,, \quad i \neq j\,.$$

If, in addition, all its members are unit vectors, then the set $\{\mathbf{u}_1, \mathbf{u}_2, \cdots, \mathbf{u}_n\}$ is said to be *orthonormal.* In such a case,

$$\mathbf{u}_i \cdot \mathbf{u}_j = \delta_{ij}\,, \tag{1.3}$$

where δ_{ij} is the *Kronecker delta,* defined as

$$\delta_{ij} \equiv \begin{cases} 1, & i = j \\ 0, & i \neq j \end{cases}. \tag{1.4}$$

The three unit vectors $\mathbf{i}, \mathbf{j}$, and $\mathbf{k}$ defining the Cartesian coordinate system of Fig. 1.1 form an orthonormal set:

$$\begin{aligned} \mathbf{i} \cdot \mathbf{i} = \mathbf{j} \cdot \mathbf{j} = \mathbf{k} \cdot \mathbf{k} &= 1 \\ \mathbf{i} \cdot \mathbf{j} = \mathbf{j} \cdot \mathbf{k} = \mathbf{k} \cdot \mathbf{i} &= 0 \end{aligned} \tag{1.5}$$

The *cross product* (or *vector product* or *outer product*) of two vectors $\mathbf{u}$ and $\mathbf{v}$ is a vector defined as

$$\mathbf{u} \times \mathbf{v} \equiv uv \sin\theta\, \mathbf{n}\,, \tag{1.6}$$

where $\mathbf{n}$ is the unit vector normal to the plane of $\mathbf{u}$ and $\mathbf{v}$, such that $\mathbf{u}$, $\mathbf{v}$ and $\mathbf{n}$ form a *right-handed* orthogonal system, as illustrated in Fig. 1.4. The magnitude of $\mathbf{u} \times \mathbf{v}$ is the same as that of the area of a parallelogram with sides $\mathbf{u}$ and $\mathbf{v}$. If $\mathbf{u}$ and $\mathbf{v}$ are parallel, then $\sin\theta=0$ and $\mathbf{u} \times \mathbf{v}=0$. For instance, $\mathbf{u} \times \mathbf{u}=0$.

If $\mathbf{u}$, $\mathbf{v}$, and $\mathbf{w}$ are vectors and m is a scalar, then

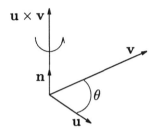

Figure 1.4. *The cross product* $\mathbf{u} \times \mathbf{v}$.

$\mathbf{u} \times \mathbf{v} = -\mathbf{v} \times \mathbf{u}$	Not commutative
$\mathbf{u} \times (\mathbf{v} + \mathbf{w}) = \mathbf{u} \times \mathbf{v} + \mathbf{u} \times \mathbf{w}$	Distributive law
$m(\mathbf{u} \times \mathbf{v}) = (m\mathbf{u}) \times \mathbf{v} = \mathbf{u} \times (m\mathbf{v}) = (\mathbf{u} \times \mathbf{v})m$	

For the three unit vectors $\mathbf{i}$, $\mathbf{j}$, and $\mathbf{k}$ one gets:

$$\mathbf{i} \times \mathbf{i} = \mathbf{j} \times \mathbf{j} = \mathbf{k} \times \mathbf{k} = \mathbf{0},$$

$$\mathbf{i} \times \mathbf{j} = \mathbf{k}, \quad \mathbf{j} \times \mathbf{k} = \mathbf{i}, \quad \mathbf{k} \times \mathbf{i} = \mathbf{j}, \tag{1.7}$$

$$\mathbf{j} \times \mathbf{i} = -\mathbf{k}, \quad \mathbf{k} \times \mathbf{j} = -\mathbf{i}, \quad \mathbf{i} \times \mathbf{k} = -\mathbf{j}.$$

Note that the cyclic order $(\mathbf{i}, \mathbf{j}, \mathbf{k}, \mathbf{i}, \mathbf{j}, \cdots)$, in which the cross product of any neighboring pair in order is the next vector, is consistent with the right-handed orientation of the axes as shown in Fig. 1.1.

The product $\mathbf{u} \cdot (\mathbf{v} \times \mathbf{w})$ is called the *scalar triple product* of $\mathbf{u}$, $\mathbf{v}$, and $\mathbf{w}$, and is a scalar representing the volume of a parallelepiped with $\mathbf{u}$, $\mathbf{v}$, and $\mathbf{w}$ as the edges. The product $\mathbf{u} \times (\mathbf{v} \times \mathbf{w})$ is a vector called the *vector triple product*. The following laws are valid:

$(\mathbf{u} \cdot \mathbf{v})\mathbf{w} \neq \mathbf{u}(\mathbf{v} \cdot \mathbf{w})$	Not associative
$\mathbf{u} \times (\mathbf{v} \times \mathbf{w}) \neq (\mathbf{u} \times \mathbf{v}) \times \mathbf{w}$	Not associative
$\mathbf{u} \times (\mathbf{v} \times \mathbf{w}) = (\mathbf{u} \cdot \mathbf{w})\mathbf{v} - (\mathbf{u} \cdot \mathbf{v})\mathbf{w}$	
$(\mathbf{u} \times \mathbf{v}) \times \mathbf{w} = (\mathbf{u} \cdot \mathbf{w})\mathbf{v} - (\mathbf{v} \cdot \mathbf{w})\mathbf{u}$	
$\mathbf{u} \cdot (\mathbf{v} \times \mathbf{w}) = \mathbf{v} \cdot (\mathbf{w} \times \mathbf{u}) = \mathbf{w} \cdot (\mathbf{u} \times \mathbf{v})$	

Thus far, we have presented vectors and vector operations from a geometrical viewpoint. These are treated analytically in Section 1.2.

Tensors may be viewed as *generalized vectors* being characterized by their magnitude and more than one *ordered* direction with respect to a given frame of reference.

Tensors encountered in fluid mechanics are of second order, i.e., they are characterized by an *ordered pair* of coordinate directions. Tensors are often denoted by uppercase, boldface type or lower case, boldface Greek letters, such as τ for the stress tensor. A *tensor field* is a tensor-valued function that associates a tensor with each point of a given region in space. Tensor addition and multiplication of a tensor by a scalar are commutative and associative. If $\mathbf{R}$, $\mathbf{S}$, and $\mathbf{T}$ are tensors of the same type, and m and n are scalars, then

$\mathbf{R} + \mathbf{S} = \mathbf{S} + \mathbf{R}$	Commutative law
$(\mathbf{R} + \mathbf{S}) + \mathbf{T} = \mathbf{S} + (\mathbf{R} + \mathbf{T})$	Associative law
$m\mathbf{R} = \mathbf{R}m$	Commutative law
$m(n\mathbf{R}) = (mn)\mathbf{R}$	Associative law
$(m+n)\mathbf{R} = m\mathbf{R} + n\mathbf{R}$	Distributive law
$m(\mathbf{R}+\mathbf{S}) = m\mathbf{R} + m\mathbf{S}$	Distributive law

Tensors and tensor operations are discussed in more detail in Section 1.3.

1.1 Systems of Coordinates

A coordinate system in the three-dimensional space is defined by choosing a set of three *linearly independent* vectors, $B=\{\mathbf{e}_1, \mathbf{e}_2, \mathbf{e}_3\}$, representing the three fundamental directions of the space. The set B is a *basis* of the three-dimensional space, i.e., each vector $\mathbf{v}$ of this space is uniquely written as a *linear combination* of $\mathbf{e}_1$, $\mathbf{e}_2$, and $\mathbf{e}_3$:

$$\mathbf{v} = v_1\,\mathbf{e}_1 + v_2\,\mathbf{e}_2 + v_3\,\mathbf{e}_3\,. \tag{1.8}$$

The scalars v_1, v_2, and v_3 are the *components* of $\mathbf{v}$ and represent the magnitudes of the *projections* of $\mathbf{v}$ onto each of the fundamental directions. The vector $\mathbf{v}$ is often denoted by $\mathbf{v}(v_1, v_2, v_3)$ or simply by (v_1, v_2, v_3).

In most cases, the vectors $\mathbf{e}_1$, $\mathbf{e}_2$, and $\mathbf{e}_3$ are *unit* vectors. In the three coordinate systems that are of interest in this book, i.e., *Cartesian, cylindrical*, and *spherical* coordinates, the three vectors are, in addition, orthogonal. Hence, in all these systems, the basis $B=\{\mathbf{e}_1, \mathbf{e}_2, \mathbf{e}_3\}$ is orthonormal:

$$\mathbf{e}_i \cdot \mathbf{e}_j = \delta_{ij}\,. \tag{1.9}$$

(In some cases, nonorthogonal systems are used for convenience; see, for example, [1].) For the cross products of $\mathbf{e}_1$, $\mathbf{e}_2$, and $\mathbf{e}_3$, one gets:

$$\mathbf{e}_i \times \mathbf{e}_j = \sum_{k=1}^{3} \epsilon_{ijk}\,\mathbf{e}_k\,, \tag{1.10}$$

where ϵ_{ijk} is the *permutation symbol* defined as

$$\epsilon_{ijk} \equiv \begin{cases} 1\,, & \text{if } ijk=123,\ 231,\ \text{or } 312 \text{ (i.e., an even permutation of 123)} \\ -1\,, & \text{if } ijk=321,\ 132,\ \text{or } 213 \text{ (i.e., an odd permutation of 123)}\,. \\ 0\,, & \text{if any two indices are equal} \end{cases}$$

$$(1.11)$$

A useful relation involving the permutation symbol is the following:

$$\begin{vmatrix} a_1 & a_2 & a_3 \\ b_1 & b_2 & b_3 \\ c_1 & c_2 & c_3 \end{vmatrix} = \sum_{i=1}^{3}\sum_{j=1}^{3}\sum_{k=1}^{3} \epsilon_{ijk}\, a_i b_j c_k\,. \qquad (1.12)$$

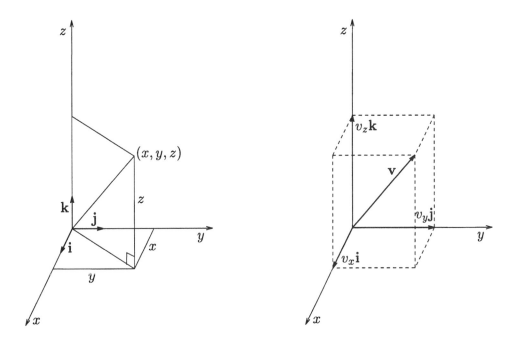

Figure 1.5. *Cartesian coordinates* (x, y, z) *with* $-\infty < x < \infty$, $-\infty < y < \infty$ *and* $-\infty < z < \infty$.

The Cartesian (or *rectangular*) system of coordinates (x, y, z), with

$$-\infty < x < \infty\,, \qquad -\infty < y < \infty \quad \text{and} \quad -\infty < z < \infty\,,$$

has already been introduced in previous examples. Its basis is often denoted by $\{\mathbf{i}, \mathbf{j}, \mathbf{k}\}$ or $\{\mathbf{e}_x, \mathbf{e}_y, \mathbf{e}_z\}$. The decomposition of a vector $\mathbf{v}$ into its three components

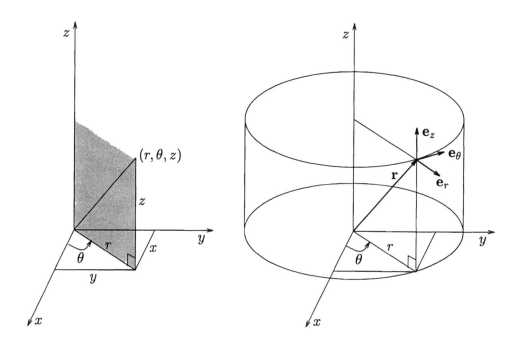

Figure 1.6. *Cylindrical polar coordinates* (r, θ, z) *with* $r \geq 0$, $0 \leq \theta < 2\pi$ *and* $-\infty < z < \infty$, *and the position vector* **r**.

$(r, \theta, z) \longrightarrow (x, y, z)$	$(x, y, z) \longrightarrow (r, \theta, z)$
<u>Coordinates</u>	
$x = r \cos \theta$	$r = \sqrt{x^2 + y^2}$
$y = r \sin \theta$	$\theta = \begin{cases} \arctan \frac{y}{x}, & x > 0,\ y \geq 0 \\ \pi + \arctan \frac{y}{x}, & x < 0 \\ 2\pi + \arctan \frac{y}{x}, & x > 0,\ y < 0 \end{cases}$
$z = z$	$z = z$
<u>Unit vectors</u>	
$\mathbf{i} = \cos \theta\, \mathbf{e}_r - \sin \theta\, \mathbf{e}_\theta$	$\mathbf{e}_r = \cos \theta\, \mathbf{i} + \sin \theta\, \mathbf{j}$
$\mathbf{j} = \sin \theta\, \mathbf{e}_r + \cos \theta\, \mathbf{e}_\theta$	$\mathbf{e}_\theta = -\sin \theta\, \mathbf{i} + \cos \theta\, \mathbf{j}$
$\mathbf{k} = \mathbf{e}_z$	$\mathbf{e}_z = \mathbf{k}$

Table 1.1. *Relations between Cartesian and cylindrical polar coordinates.*

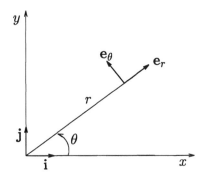

Figure 1.7. *Plane polar coordinates (r, θ).*

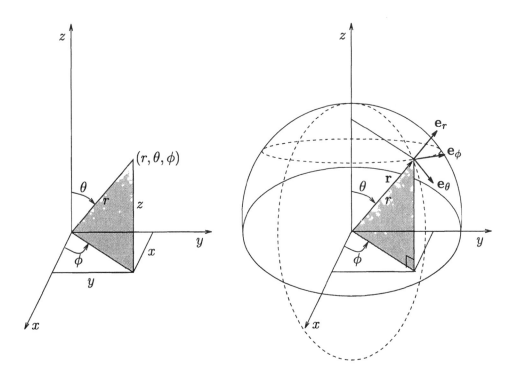

Figure 1.8. *Spherical polar coordinates (r, θ, ϕ) with $r \geq 0$, $0 \leq \theta \leq \pi$ and $0 \leq \phi \leq 2\pi$, and the position vector $\mathbf{r}$.*

$(r, \theta, \phi) \longrightarrow (x, y, z)$	$(x, y, z) \longrightarrow (r, \theta, \phi)$
Coordinates $x = r \sin\theta \cos\phi$ $y = r \sin\theta \sin\phi$ $z = r \cos\theta$	$r = \sqrt{x^2 + y^2 + z^2}$ $\theta = \begin{cases} \arctan \frac{\sqrt{x^2+y^2}}{z}, & z > 0 \\ \frac{\pi}{2}, & z = 0 \\ \pi + \arctan \frac{\sqrt{x^2+y^2}}{z}, & z < 0 \end{cases}$ $\phi = \begin{cases} \arctan \frac{y}{x}, & x > 0, \ y \geq 0 \\ \pi + \arctan \frac{y}{x}, & x < 0 \\ 2\pi + \arctan \frac{y}{x}, & x > 0, \ y < 0 \end{cases}$
Unit vectors $\mathbf{i} = \sin\theta \cos\phi \, \mathbf{e}_r + \cos\theta \cos\phi \, \mathbf{e}_\theta - \sin\phi \, \mathbf{e}_\phi$ $\mathbf{j} = \sin\theta \sin\phi \, \mathbf{e}_r + \cos\theta \sin\phi \, \mathbf{e}_\theta + \cos\phi \, \mathbf{e}_\phi$ $\mathbf{k} = \cos\theta \, \mathbf{e}_r - \sin\theta \, \mathbf{e}_\theta$	$\mathbf{e}_r = \sin\theta \cos\phi \, \mathbf{i} + \sin\theta \sin\phi \, \mathbf{j} + \cos\theta \, \mathbf{k}$ $\mathbf{e}_\theta = \cos\theta \cos\phi \, \mathbf{i} + \cos\theta \sin\phi \, \mathbf{j} - \sin\theta \, \mathbf{k}$ $\mathbf{e}_\phi = -\sin\phi \, \mathbf{i} + \cos\phi \, \mathbf{j}$

Table 1.2. *Relations between Cartesian and spherical polar coordinates.*

(v_x, v_y, v_z) is depicted in Fig. 1.5. It should be noted that, throughout this book, we use *right-handed* coordinate systems.

The cylindrical and spherical polar coordinates are the two most important orthogonal *curvilinear* coordinate systems. The cylindrical polar coordinates (r, θ, z), with

$$r \geq 0, \quad 0 \leq \theta < 2\pi \quad \text{and} \quad -\infty < z < \infty,$$

are shown in Fig. 1.6 together with the Cartesian coordinates sharing the same origin. The basis of the cylindrical coordinate system consists of three orthonormal vectors: the radial vector $\mathbf{e}_r$, the azimuthal vector $\mathbf{e}_\theta$, and the axial vector $\mathbf{e}_z$. Note that the azimuthal angle θ revolves around the z axis. Any vector $\mathbf{v}$ is decomposed into, and is fully defined by, its components $\mathbf{v}(v_r, v_\theta, v_z)$ with respect to the cylindrical system. By invoking simple trigonometric relations, any vector, including those of the bases, can be transformed from one system to another. Table 1.1 lists the formulas for making coordinate conversions from cylindrical to Cartesian coordinates and vice versa.

On the xy plane, i.e., if the z coordinate is ignored, the cylindrical polar coordinates are reduced to the familiar *plane polar coordinates* (r, θ) shown in Fig. 1.7.

The spherical polar coordinates (r, θ, ϕ), with

$$r \geq 0, \quad 0 \leq \theta \leq \pi \quad \text{and} \quad 0 \leq \phi < 2\pi,$$

together with the Cartesian coordinates with the same origin, are shown in Fig. 1.8. It should be emphasized that r and θ in cylindrical and spherical coordinates are not the same. The basis of the spherical coordinate system consists of three orthonormal vectors: the radial vector $\mathbf{e}_r$, the meridional vector $\mathbf{e}_\theta$, and the azimuthal vector $\mathbf{e}_\phi$. Any vector $\mathbf{v}$ can be decomposed into the three components, $\mathbf{v}(v_r, v_\theta, v_\phi)$, which are the scalar projections of $\mathbf{v}$ onto the three fundamental directions. The transformation of a vector from spherical to Cartesian coordinates (sharing the same origin) and vice versa obeys the relations of Table 1.2.

The choice of the appropriate coordinate system, when studying a fluid mechanics problem, depends on the geometry and symmetry of the flow. Flow between parallel plates is conveniently described by Cartesian coordinates. *Axisymmetric* (i.e., *axially symmetric*) flows, such as flow in an annulus, are naturally described using cylindrical coordinates, and flow around a sphere is expressed in spherical coordinates. In some cases, nonorthogonal systems might be employed too. More details on other coordinate systems and transformations can be found elsewhere [1].

Example 1.1.1. Basis of the cylindrical system
Show that the basis $B = \{\mathbf{e}_r, \mathbf{e}_\theta, \mathbf{e}_z\}$ of the cylindrical system is orthonormal.

Solution:

Since $\mathbf{i} \cdot \mathbf{i} = \mathbf{j} \cdot \mathbf{j} = \mathbf{k} \cdot \mathbf{k} = 1$ and $\mathbf{i} \cdot \mathbf{j} = \mathbf{j} \cdot \mathbf{k} = \mathbf{k} \cdot \mathbf{i} = 0$, we obtain:

$$
\begin{aligned}
\mathbf{e}_r \cdot \mathbf{e}_r &= (\cos\theta\,\mathbf{i} + \sin\theta\,\mathbf{j}) \cdot (\cos\theta\,\mathbf{i} + \sin\theta\,\mathbf{j}) = \cos^2\theta + \sin^2\theta = 1 \\
\mathbf{e}_\theta \cdot \mathbf{e}_\theta &= (-\sin\theta\,\mathbf{i} + \cos\theta\,\mathbf{j}) \cdot (-\sin\theta\,\mathbf{i} + \cos\theta\,\mathbf{j}) = \sin^2\theta + \cos^2\theta = 1 \\
\mathbf{e}_z \cdot \mathbf{e}_z &= \mathbf{k} \cdot \mathbf{k} = 1 \\
\mathbf{e}_r \cdot \mathbf{e}_\theta &= (\cos\theta\,\mathbf{i} + \sin\theta\,\mathbf{j}) \cdot (-\sin\theta\,\mathbf{i} + \cos\theta\,\mathbf{j}) = 0 \\
\mathbf{e}_r \cdot \mathbf{e}_z &= (\cos\theta\,\mathbf{i} + \sin\theta\,\mathbf{j}) \cdot \mathbf{k} = 0 \\
\mathbf{e}_\theta \cdot \mathbf{e}_z &= (-\sin\theta\,\mathbf{i} + \cos\theta\,\mathbf{j}) \cdot \mathbf{k} = 0
\end{aligned}
$$

$\square$

Example 1.1.2. The position vector
The *position vector* $\mathbf{r}$ defines the position of a point in space with respect to a coordinate system. In Cartesian coordinates,

$$\mathbf{r} = x\,\mathbf{i} + y\,\mathbf{j} + z\,\mathbf{k}, \tag{1.13}$$

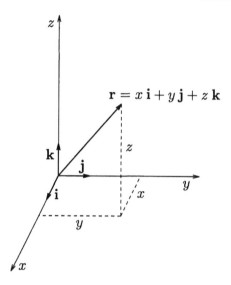

Figure 1.9. *The position vector,* **r**, *in Cartesian coordinates.*

and thus

$$|\mathbf{r}| = (\mathbf{r} \cdot \mathbf{r})^{\frac{1}{2}} = \sqrt{x^2 + y^2 + z^2} \,. \tag{1.14}$$

The decomposition of **r** into its three components (x, y, z) is illustrated in Fig. 1.9.
In cylindrical coordinates, the position vector is given by

$$\mathbf{r} = r\,\mathbf{e}_r + z\,\mathbf{e}_z \quad \text{with} \quad |\mathbf{r}| = \sqrt{r^2 + z^2} \,. \tag{1.15}$$

Note that the magnitude $|\mathbf{r}|$ of the position vector *is not* the same as the radial
cylindrical coordinate r. Finally, in spherical coordinates,

$$\mathbf{r} = r\,\mathbf{e}_r \quad \text{with} \quad |\mathbf{r}| = r \,, \tag{1.16}$$

that is, $|\mathbf{r}|$ is the radial spherical coordinate r. Even though expressions (1.15) and
(1.16) for the position vector are obvious (see Figs. 1.6 and 1.8, respectively), we will
derive both of them, starting from Eq. (1.13) and using coordinate transformations.
In cylindrical coordinates,

$$
\begin{aligned}
\mathbf{r} &= x\,\mathbf{i} + y\,\mathbf{j} + z\,\mathbf{k} \\
&= r\cos\theta\,(\cos\theta\,\mathbf{e}_r - \sin\theta\,\mathbf{e}_\theta) + r\sin\theta\,(\sin\theta\,\mathbf{e}_r + \cos\theta\,\mathbf{e}_\theta) + z\,\mathbf{e}_z \\
&= r\,(\cos^2\theta + \sin^2\theta)\,\mathbf{e}_r + r\,(-\sin\theta\cos\theta + \sin\theta\cos\theta)\,\mathbf{e}_\theta + z\,\mathbf{e}_z \\
&= r\,\mathbf{e}_r + z\,\mathbf{e}_z \,.
\end{aligned}
$$

In spherical coordinates,

$$
\begin{aligned}
\mathbf{r} &= x\,\mathbf{i} + y\,\mathbf{j} + z\,\mathbf{k} \\
&= r\sin\theta\cos\phi\,(\sin\theta\cos\phi\,\mathbf{e}_r + \cos\theta\cos\phi\,\mathbf{e}_\theta - \sin\phi\,\mathbf{e}_\phi) \\
&\quad + r\sin\theta\sin\phi\,(\sin\theta\sin\phi\,\mathbf{e}_r + \cos\theta\sin\phi\,\mathbf{e}_\theta + \cos\phi\,\mathbf{e}_\phi) \\
&\quad + r\cos\theta\,(\cos\theta\,\mathbf{e}_r - \sin\theta\,\mathbf{e}_\theta) \\
&= r\,[\sin^2\theta\,(\cos^2\phi + \sin^2\phi)\,\cos^2\theta]\,\mathbf{e}_r \\
&\quad + r\sin\theta\cos\theta\,[(\cos^2\phi + \sin^2\phi) - 1]\,\mathbf{e}_\theta \\
&\quad + r\sin\theta\,(-\sin\phi\cos\phi + \sin\phi\cos\phi)\,\mathbf{e}_\phi \\
&= r\,\mathbf{e}_r\,.
\end{aligned}
$$

$\square$

Example 1.1.3. Derivatives of the basis vectors

The basis vectors $\mathbf{i}$, $\mathbf{j}$, and $\mathbf{k}$ of the Cartesian coordinates are fixed and do not change with position. This is not true for the basis vectors in curvilinear coordinate systems. From Table 1.1, we observe that, in cylindrical coordinates,

$$
\mathbf{e}_r = \cos\theta\,\mathbf{i} + \sin\theta\,\mathbf{j} \quad\text{and}\quad \mathbf{e}_\theta = -\sin\theta\,\mathbf{i} + \cos\theta\,\mathbf{j}\,;
$$

therefore, $\mathbf{e}_r$ and $\mathbf{e}_\theta$ change with θ. Taking the derivatives with respect to θ, we obtain:

$$
\frac{\partial \mathbf{e}_r}{\partial\theta} = -\sin\theta\,\mathbf{i} + \cos\theta\,\mathbf{j} = \mathbf{e}_\theta
$$

and

$$
\frac{\partial \mathbf{e}_\theta}{\partial\theta} = -\cos\theta\,\mathbf{i} - \sin\theta\,\mathbf{j} = -\mathbf{e}_r\,.
$$

All the other spatial derivatives of $\mathbf{e}_r$, $\mathbf{e}_\theta$, and $\mathbf{e}_z$ are zero. Hence,

$$
\begin{array}{lll}
\dfrac{\partial \mathbf{e}_r}{\partial r} = 0 & \dfrac{\partial \mathbf{e}_\theta}{\partial r} = 0 & \dfrac{\partial \mathbf{e}_z}{\partial r} = 0 \\[2mm]
\dfrac{\partial \mathbf{e}_r}{\partial\theta} = \mathbf{e}_\theta & \dfrac{\partial \mathbf{e}_\theta}{\partial\theta} = -\mathbf{e}_r & \dfrac{\partial \mathbf{e}_z}{\partial\theta} = 0 \\[2mm]
\dfrac{\partial \mathbf{e}_r}{\partial z} = 0 & \dfrac{\partial \mathbf{e}_\theta}{\partial z} = 0 & \dfrac{\partial \mathbf{e}_z}{\partial z} = 0
\end{array}
\tag{1.17}
$$

Similarly, for the spatial derivatives of the unit vectors in spherical coordinates, we obtain:

$$
\begin{aligned}
&\frac{\partial \mathbf{e}_r}{\partial r} = 0 &\quad &\frac{\partial \mathbf{e}_\theta}{\partial r} = 0 &\quad &\frac{\partial \mathbf{e}_\phi}{\partial r} = 0 \\[2mm]
&\frac{\partial \mathbf{e}_r}{\partial \theta} = \mathbf{e}_\theta &\quad &\frac{\partial \mathbf{e}_\theta}{\partial \theta} = -\mathbf{e}_r &\quad &\frac{\partial \mathbf{e}_\phi}{\partial \theta} = 0 \\[2mm]
&\frac{\partial \mathbf{e}_r}{\partial \phi} = \sin\theta\,\mathbf{e}_\phi &\quad &\frac{\partial \mathbf{e}_\theta}{\partial \phi} = \cos\theta\,\mathbf{e}_\phi &\quad &\frac{\partial \mathbf{e}_\phi}{\partial \phi} = -\sin\theta\,\mathbf{e}_r - \cos\theta\,\mathbf{e}_\theta
\end{aligned}
\tag{1.18}
$$

Equations (1.17) and (1.18) are very useful in converting differential operators from Cartesian to orthogonal curvilinear coordinates.

$\square$

1.2   Vectors

In this section, vector operations are considered from an analytical point of view. Let $B=\{\mathbf{e}_1, \mathbf{e}_2, \mathbf{e}_3\}$ be an *orthonormal* basis of the three-dimensional space, which implies that

$$
\mathbf{e}_i \cdot \mathbf{e}_j = \delta_{ij},
\tag{1.19}
$$

and

$$
\mathbf{e}_i \times \mathbf{e}_j = \sum_{k=1}^{3} \epsilon_{ijk}\, \mathbf{e}_k.
\tag{1.20}
$$

Any vector $\mathbf{v}$ can be expanded in terms of its components (v_1, v_2, v_3):

$$
\mathbf{v} = v_1\,\mathbf{e}_1 + v_2\,\mathbf{e}_2 + v_3\,\mathbf{e}_3 = \sum_{i=1}^{3} v_i\,\mathbf{e_i}.
\tag{1.21}
$$

Any operation between two or more vectors is easily performed by first decomposing each vector into its components and then invoking the basis relations (1.19) and (1.20). If $\mathbf{u}$ and $\mathbf{v}$ are vectors, then

$$
\mathbf{u} \pm \mathbf{v} = (u_1 \pm v_1)\,\mathbf{e}_1 + (u_2 \pm v_2)\,\mathbf{e}_2 + (u_3 \pm v_3)\,\mathbf{e}_3 = \sum_{i=1}^{3} (u_i \pm v_i)\,\mathbf{e_i},
\tag{1.22}
$$

i.e., addition (or subtraction) of two vectors corresponds to adding (or subtracting) their corresponding components. If m is a scalar, then

$$
m\mathbf{v} = m\left(\sum_{i=1}^{3} v_i\,\mathbf{e_i}\right) = \sum_{i=1}^{3} m v_i\,\mathbf{e_i},
\tag{1.23}
$$

i.e., multiplication of a vector by a scalar corresponds to multiplying each of its components by the scalar.

For the dot product of **u** and **v**, we obtain:

$$\mathbf{u} \cdot \mathbf{v} = \left(\sum_{i=1}^{3} u_i \, \mathbf{e_i} \right) \cdot \left(\sum_{i=1}^{3} v_i \, \mathbf{e_i} \right) \quad \Longrightarrow$$

$$\mathbf{u} \cdot \mathbf{v} = u_1 v_1 + u_2 v_2 + u_3 v_3 = \sum_{i=1}^{3} u_i v_i \, . \tag{1.24}$$

The magnitude of **v** is thus given by

$$v = (\mathbf{v} \cdot \mathbf{v})^{\frac{1}{2}} = \sqrt{v_1^2 + v_2^2 + v_3^2} \, . \tag{1.25}$$

Finally, for the cross product of **u** and **v**, we get

$$\mathbf{u} \times \mathbf{v} = \left(\sum_{i=1}^{3} u_i \, \mathbf{e_i} \right) \times \left(\sum_{j=1}^{3} v_j \, \mathbf{e_j} \right) = \sum_{i=1}^{3} \sum_{j=1}^{3} u_i v_j \, \mathbf{e_i} \times \mathbf{e_j} \quad \Longrightarrow$$

$$\mathbf{u} \times \mathbf{v} = \sum_{i=1}^{3} \sum_{j=1}^{3} \sum_{k=1}^{3} \epsilon_{ijk} \, u_i v_j \, \mathbf{e_k} \tag{1.26}$$

or

$$\mathbf{u} \times \mathbf{v} = \begin{vmatrix} \mathbf{e_1} & \mathbf{e_2} & \mathbf{e_3} \\ u_1 & u_2 & u_3 \\ v_1 & v_2 & v_3 \end{vmatrix} = (u_2 v_3 - u_3 v_2)\mathbf{e_1} - (u_1 v_3 - u_3 v_1)\mathbf{e_2} + (u_1 v_2 - u_2 v_1)\mathbf{e_3} . \tag{1.27}$$

Example 1.2.1. The scalar triple product

For the scalar triple product $(\mathbf{u} \times \mathbf{v}) \cdot \mathbf{w}$, we have:

$$(\mathbf{u} \times \mathbf{v}) \cdot \mathbf{w} = \left(\sum_{i=1}^{3} \sum_{j=1}^{3} \sum_{k=1}^{3} \epsilon_{ijk} \, u_i v_j \, \mathbf{e_k} \right) \cdot \left(\sum_{k=1}^{3} w_k \, \mathbf{e_k} \right) \quad \Longrightarrow$$

$$(\mathbf{u} \times \mathbf{v}) \cdot \mathbf{w} = \sum_{i=1}^{3} \sum_{j=1}^{3} \sum_{k=1}^{3} \epsilon_{ijk} \, u_i v_j w_k \tag{1.28}$$

or

$$(\mathbf{u} \times \mathbf{v}) \cdot \mathbf{w} = \begin{vmatrix} u_1 & u_2 & u_3 \\ v_1 & v_2 & v_3 \\ w_1 & w_2 & w_3 \end{vmatrix}. \tag{1.29}$$

Using basic properties of determinants, one can easily show the following identity:

$$(\mathbf{u} \times \mathbf{v}) \cdot \mathbf{w} = (\mathbf{w} \times \mathbf{u}) \cdot \mathbf{v} = (\mathbf{v} \times \mathbf{w}) \cdot \mathbf{u}. \tag{1.30}$$

$\square$

In the following subsections, we will make use of the vector differential operator *nabla* (or *del*), ∇. In Cartesian coordinates, ∇ is defined by

$$\nabla \equiv \frac{\partial}{\partial x} \mathbf{i} + \frac{\partial}{\partial y} \mathbf{j} + \frac{\partial}{\partial z} \mathbf{k}. \tag{1.31}$$

The *gradient* of a scalar field $f(x, y, z)$ is a vector field defined by

$$\nabla f = \frac{\partial f}{\partial x} \mathbf{i} + \frac{\partial f}{\partial y} \mathbf{j} + \frac{\partial f}{\partial z} \mathbf{k}. \tag{1.32}$$

The *divergence* of a vector field $\mathbf{v}(x, y, z)$ is a scalar field defined by

$$\nabla \cdot \mathbf{v} = \frac{\partial v_x}{\partial x} + \frac{\partial v_y}{\partial y} + \frac{\partial v_z}{\partial z}. \tag{1.33}$$

More details about ∇ and its forms in curvilinear coordinates are given in Section 1.4.

1.2.1   Vectors in Fluid Mechanics

As already mentioned, the *position* vector, $\mathbf{r}$, defines the position of a point with respect to a coordinate system. The *separation* or *displacement vector* between two points A and B (see Figure 1.10) is commonly denoted by $\Delta \mathbf{r}$, and is defined as

$$\Delta \mathbf{r}_{AB} \equiv \mathbf{r}_A - \mathbf{r}_B. \tag{1.34}$$

The *velocity vector*, $\mathbf{u}$, is defined as the *total time derivative* of the position vector:

$$\mathbf{u} \equiv \frac{d\mathbf{r}}{dt}. \tag{1.35}$$

Geometrically, the velocity vector is tangent to the curve C defined by the motion of the position vector $\mathbf{r}$ (Fig. 1.11). The *relative velocity* of a particle A, with respect to another particle B, is defined accordingly by

$$\mathbf{u}_{AB} \equiv \frac{d\Delta \mathbf{r}_{AB}}{dt} = \frac{d\mathbf{r}_A}{dt} - \frac{d\mathbf{r}_B}{dt} = \mathbf{u}_A - \mathbf{u}_B. \tag{1.36}$$

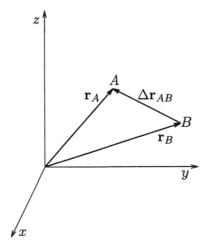

Figure 1.10. *Position and separation vectors.*

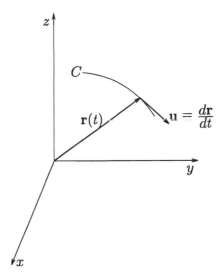

Figure 1.11. *Position and velocity vectors.*

The *acceleration vector*, **a**, is defined by

$$\mathbf{a} \equiv \frac{d\mathbf{u}}{dt} = \frac{d^2\mathbf{r}}{dt^2} \,. \tag{1.37}$$

The acceleration of gravity, **g**, is a vector directed towards the center of the earth. In problems where gravity is important, it is convenient to choose one of the axes, usually the z axis, to be collinear with **g**. In such a case, $\mathbf{g}=-g\mathbf{e}_z$ or $g\mathbf{e}_z$.

Example 1.2.2. Velocity components

In Cartesian coordinates, the basis vectors are fixed and thus time independent. So,

$$\mathbf{u} \equiv \frac{d}{dt}(x\mathbf{i} + y\mathbf{j} + z\mathbf{k}) = \frac{dx}{dt}\,\mathbf{i} + \frac{dy}{dt}\,\mathbf{j} + \frac{dz}{dt}\,\mathbf{k} \,.$$

Hence, the velocity components (u_x, u_y, u_z) are given by:

$$u_x = \frac{dx}{dt} \,, \qquad u_y = \frac{dy}{dt} \,, \qquad u_z = \frac{dz}{dt} \,. \tag{1.38}$$

In cylindrical coordinates, the position vector is given by $\mathbf{r} = r\,\mathbf{e_r} + z\,\mathbf{e_z}$, where $\mathbf{e_r}$ is time dependent:

$$\mathbf{u} \equiv \frac{d}{dt}(r\,\mathbf{e}_r + z\,\mathbf{e}_z) = \frac{dr}{dt}\mathbf{e}_r + r\frac{d\mathbf{e}_r}{dt} + \frac{dz}{dt}\mathbf{e}_z = \frac{dr}{dt}\mathbf{e}_r + r\frac{d\mathbf{e}_r}{d\theta}\frac{d\theta}{dt} + \frac{dz}{dt}\mathbf{e}_z \quad \Longrightarrow$$

$$\mathbf{u} = \frac{dr}{dt}\,\mathbf{e}_r + r\Omega\,\mathbf{e}_\theta + \frac{dz}{dt}\,\mathbf{e}_z \,,$$

where $\Omega \equiv d\theta/dt$ is the *angular velocity*. The velocity components (u_r, u_θ, u_z) are given by:

$$u_r = \frac{dr}{dt} \,, \qquad u_\theta = r\frac{d\theta}{dt} = r\Omega \,, \qquad u_z = \frac{dz}{dt} \,. \tag{1.39}$$

In spherical coordinates, all the basis vectors are time dependent. The velocity components (u_r, u_θ, u_ϕ) are easily found to be:

$$u_r = \frac{dr}{dt} \,, \qquad u_\theta = r\frac{d\theta}{dt} \,, \qquad u_\phi = r\sin\theta\,\frac{d\phi}{dt} \,. \tag{1.40}$$

$\square$

Example 1.2.3. Circular motion

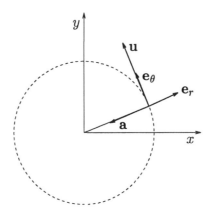

Figure 1.12. *Velocity and acceleration vectors in circular motion.*

Consider plane polar coordinates and suppose that a small solid sphere rotates at a constant distance, R, with constant angular velocity, Ω, around the origin (*uniform rotation*). The position vector of the sphere is $\mathbf{r} = R \, \mathbf{e}_r$ and, therefore,

$$\mathbf{u} \equiv \frac{d\mathbf{r}}{dt} = \frac{d}{dt}(R \, \mathbf{e}_r) = R \frac{d\mathbf{e}_r}{dt} = R \frac{d\mathbf{e}_r}{d\theta} \frac{d\theta}{dt} \qquad \Longrightarrow \qquad \mathbf{u} = R\Omega \, \mathbf{e}_\theta \, .$$

The acceleration of the sphere is:

$$\mathbf{a} \equiv \frac{d\mathbf{u}}{dt} = \frac{d}{dt}(R\Omega \, \mathbf{e}_\theta) = R\Omega \frac{d\mathbf{e}_\theta}{d\theta} \frac{d\theta}{dt} \qquad \Longrightarrow \qquad \mathbf{a} = - R\Omega^2 \, \mathbf{e}_r \, .$$

This is the familiar *centripetal acceleration* $R\Omega^2$ directed towards the axis of rotation.

□

The *force vector*, $\mathbf{F}$, is combined with other vectors to yield:

$$\text{Work}: \ W = \mathbf{F} \cdot \mathbf{r}_{AB} \ ; \tag{1.41}$$

$$\text{Power}: \ P = \frac{dW}{dt} = \mathbf{F} \cdot \frac{d\mathbf{r}_{AB}}{dt} \ ; \tag{1.42}$$

$$\text{Moment}: \ \mathbf{M} \equiv \mathbf{r} \times \mathbf{F} \ . \tag{1.43}$$

In the first two expressions, the force vector, $\mathbf{F}$, is considered constant.

Example 1.2.4. Linear and angular momentum

The *linear momentum*, $\mathbf{J}$, of a body of mass m moving with velocity $\mathbf{u}$ is defined by

$\mathbf{J} \equiv m\mathbf{u}$. The net force $\mathbf{F}$ acting on the body is given by *Newton's law of motion*,

$$\mathbf{F} = \frac{d\mathbf{J}}{dt} = \frac{d}{dt}(m\mathbf{u}) \,. \tag{1.44}$$

If m is constant, then

$$\mathbf{F} = m\frac{d\mathbf{u}}{dt} = m\mathbf{a} \,, \tag{1.45}$$

where $\mathbf{a}$ is the linear acceleration of the body.

The *angular momentum* (or *moment of momentum*) is defined as

$$\mathbf{J}_\theta \equiv \mathbf{r} \times \mathbf{J} \,. \tag{1.46}$$

Therefore,

$$\frac{d\mathbf{J}_\theta}{dt} = \frac{d}{dt}(\mathbf{r} \times \mathbf{J}) = \frac{d\mathbf{r}}{dt} \times \mathbf{J} + \mathbf{r} \times \frac{d\mathbf{J}}{dt} = \mathbf{u} \times (m\mathbf{u}) + \mathbf{r} \times \mathbf{F} = \mathbf{0} + \mathbf{r} \times \mathbf{F} \implies$$

$$\frac{d\mathbf{J}_\theta}{dt} = \mathbf{r} \times \mathbf{F} = \mathbf{M} \,, \tag{1.47}$$

where the identity $\mathbf{u} \times \mathbf{u} = \mathbf{0}$ has been used. □

1.2.2 Unit Tangent and Normal Vectors

Consider a *smooth* surface S, i.e., a surface at each point of which a *tangent plane* can be defined. At each point of S, one can then define an orthonormal set consisting of two *unit tangent vectors*, $\mathbf{t}_1$ and $\mathbf{t}_2$, lying on the tangent plane, and a *unit normal vector*, $\mathbf{n}$, which is perpendicular to the tangent plane:

$$\mathbf{n} \cdot \mathbf{n} = \mathbf{t}_1 \cdot \mathbf{t}_1 = \mathbf{t}_2 \cdot \mathbf{t}_2 = 1 \quad \text{and} \quad \mathbf{n} \cdot \mathbf{t}_1 = \mathbf{t}_1 \cdot \mathbf{t}_2 = \mathbf{t}_2 \cdot \mathbf{n} = 0 \,.$$

Obviously, there are two choices for $\mathbf{n}$; the first is the vector

$$\frac{\mathbf{t}_1 \times \mathbf{t}_2}{|\mathbf{t}_1 \times \mathbf{t}_2|} \,,$$

and the second one is just its opposite. Once one of these two vectors is chosen as the unit normal vector $\mathbf{n}$, the surface is said to be *oriented*; $\mathbf{n}$ is then called the *orientation* of the surface. In general, if the surface is the boundary of a control volume, we assume that $\mathbf{n}$ is positive when it points away from the system bounded by the surface.

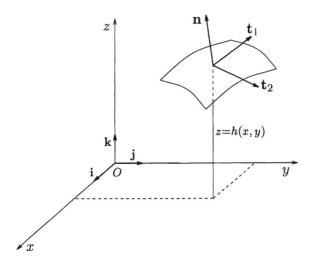

Figure 1.13. *Unit normal and tangent vectors to a surface defined by $z=h(x,y)$.*

The unit normal to a surface represented by

$$f(x,y,z) \; = \; z \; - \; h(x,y) \; = \; 0 \tag{1.48}$$

is given by

$$\mathbf{n} \; = \; \frac{\nabla f}{|\nabla f|} \qquad \Longrightarrow \tag{1.49}$$

$$\mathbf{n} \; = \; \frac{-\dfrac{\partial h}{\partial x}\,\mathbf{i} \; - \; \dfrac{\partial h}{\partial y}\,\mathbf{j} \; + \; \mathbf{k}}{\left[1 \; + \; \left(\dfrac{\partial h}{\partial x}\right)^2 \; + \; \left(\dfrac{\partial h}{\partial y}\right)^2\right]^{1/2}} \; . \tag{1.50}$$

Obviously, $\mathbf{n}$ is defined only if the gradient ∇f is defined and $|\nabla f| \neq 0$. Note that from Eq. (1.50) it follows that the unit normal vector is considered positive when it is upward, i.e., when its z component is positive, as in Fig. 1.13. One can easily choose two orthogonal unit tangent vectors, $\mathbf{t}_1$ and $\mathbf{t}_2$, so that the set $\{\mathbf{n}, \mathbf{t}_1, \mathbf{t}_2\}$ is orthonormal. Any vector field $\mathbf{u}$ can then be expanded as follows,

$$\mathbf{u} \; = \; u_n\mathbf{n} \; + \; u_{t1}\mathbf{t_1} \; + \; u_{t2}\mathbf{t_2} \tag{1.51}$$

where u_n is the *normal component*, and u_{t1} and u_{t2} are the *tangential components* of $\mathbf{u}$. The dot product $\mathbf{n} \cdot \mathbf{u}$ represents the normal component of $\mathbf{u}$, since

$$\mathbf{n} \cdot \mathbf{u} \; = \; \mathbf{n} \cdot (u_n\mathbf{n} \; + \; u_{t1}\mathbf{t_1} \; + \; u_{t2}\mathbf{t_2}) \; = \; u_n \; .$$

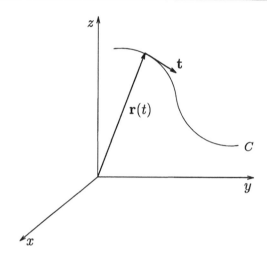

Figure 1.14. *The unit tangent vector to a curve.*

The integral of the normal component of $\mathbf{u}$ over the surface S,

$$Q \equiv \int_S \mathbf{n} \cdot \mathbf{u} \, dS \,, \tag{1.52}$$

is the *flux integral* or *flow rate* of $\mathbf{u}$ *across* S. In fluid mechanics, if $\mathbf{u}$ is the velocity vector, Q represents the *volumetric flow rate* across S. Setting $\mathbf{n}dS=d\mathbf{S}$, Eq. (1.52) takes the form

$$Q = \int_S \mathbf{u} \cdot d\mathbf{S} \,. \tag{1.53}$$

A curve C in the three-dimensional space can be defined as the graph of the position vector $\mathbf{r}(t)$, as depicted in Fig. 1.14. The motion of $\mathbf{r}(t)$ with parameter t indicates which one of the two possible directions has been chosen as the positive direction to trace C. We already know that the derivative $d\mathbf{r}/dt$ is tangent to the curve C. Therefore, the unit tangent vector to the curve C is given by

$$\mathbf{t} = \frac{\dfrac{d\mathbf{r}}{dt}}{\left|\dfrac{d\mathbf{r}}{dt}\right|} \,, \tag{1.54}$$

and is defined only at those points where the derivative $d\mathbf{r}/dt$ exists and is not zero.

As an example, consider the plane curve of Fig. 1.15, defined by

$$y = h(x) \,, \tag{1.55}$$

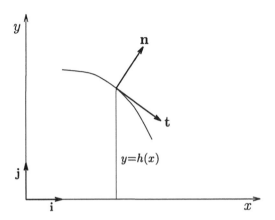

Figure 1.15. *Normal and tangent unit vectors to a plane curve defined by $y=h(x)$.*

or, equivalently, by $\mathbf{r}(t)=x\mathbf{i}+h(x)\mathbf{j}$. The unit tangent vector at a point of C is given by

$$\mathbf{t} = \frac{\frac{d\mathbf{r}}{dt}}{\left|\frac{d\mathbf{r}}{dt}\right|} = \frac{\mathbf{i} + \frac{\partial h}{\partial x}\mathbf{j}}{\left[1 + \left(\frac{\partial h}{\partial x}\right)^2\right]^{1/2}}. \tag{1.56}$$

By invoking the conditions $\mathbf{n} \cdot \mathbf{t}=0$ and $\mathbf{n} \cdot \mathbf{n}=1$, we find for the unit normal vector $\mathbf{n}$:

$$\mathbf{n} = \pm \frac{-\frac{\partial h}{\partial x}\mathbf{i} + \mathbf{j}}{\left[1 + \left(\frac{\partial h}{\partial x}\right)^2\right]^{1/2}}.$$

Choosing $\mathbf{n}$ to have positive y-component, as in Fig. 1.15, we get

$$\mathbf{n} = \frac{-\frac{\partial h}{\partial x}\mathbf{i} + \mathbf{j}}{\left[1 + \left(\frac{\partial h}{\partial x}\right)^2\right]^{1/2}}. \tag{1.57}$$

Note that the last expression for $\mathbf{n}$ can also be obtained from Eq. (1.50), as a degenerate case.

Let C be an arbitrary *closed* curve in the space, with the unit tangent vector $\mathbf{t}$ oriented in a specified direction, and let $\mathbf{u}$ be a vector field. The integral

$$\Gamma \equiv \oint_C \mathbf{t} \cdot \mathbf{u} \, d\ell, \tag{1.58}$$

where ℓ is the *arc length* around C, is called the *circulation* of $\mathbf{u}$ along C. If $\mathbf{r}$ is the position vector, then $\mathbf{t}d\ell = d\mathbf{r}$, and Equation (1.58) is written as follows

$$\Gamma \equiv \oint_C \mathbf{u} \cdot d\mathbf{r} \,. \qquad (1.59)$$

1.3 Tensors

Let $\{\mathbf{e}_1, \mathbf{e}_2, \mathbf{e}_3\}$ be an orthonormal basis of the three-dimensional space. This means that any vector $\mathbf{v}$ of this space can be uniquely expressed as a linear combination of the three coordinate directions $\mathbf{e}_1$, $\mathbf{e}_2$, and $\mathbf{e}_3$,

$$\mathbf{v} = \sum_{i=1}^{3} v_i \, \mathbf{e}_i \,, \qquad (1.60)$$

where the scalars v_i are the components of $\mathbf{v}$.

In the previous sections, two kinds of products that can be formed with any two unit basis vectors were defined, i.e., the *dot product*, $\mathbf{e}_i \cdot \mathbf{e}_j$, and the *cross product*, $\mathbf{e}_i \times \mathbf{e}_j$. A third kind of product is the *dyadic product*, $\mathbf{e}_i\mathbf{e}_j$, also referred to as a *unit dyad*. The unit dyad $\mathbf{e}_i\mathbf{e}_j$ represents *an ordered pair of coordinate directions*, and thus $\mathbf{e}_i\mathbf{e}_j \neq \mathbf{e}_j\mathbf{e}_i$ unless $i=j$. The nine possible unit dyads,

$$\{\mathbf{e}_1\mathbf{e}_1, \ \mathbf{e}_1\mathbf{e}_2, \ \mathbf{e}_1\mathbf{e}_3, \ \mathbf{e}_2\mathbf{e}_1, \ \mathbf{e}_2\mathbf{e}_2, \ \mathbf{e}_2\mathbf{e}_3, \ \mathbf{e}_3\mathbf{e}_1, \ \mathbf{e}_3\mathbf{e}_2, \ \mathbf{e}_3\mathbf{e}_3\} \,,$$

form the basis of the space of *second-order tensors*. A second-order tensor, $\boldsymbol{\tau}$, can thus be written as a linear combination of the unit dyads:

$$\boldsymbol{\tau} = \sum_{i=1}^{3} \sum_{j=1}^{3} \tau_{ij} \, \mathbf{e}_i\mathbf{e}_j \,, \qquad (1.61)$$

where the scalars τ_{ij} are referred to as the *components* of the tensor $\boldsymbol{\tau}$. Similarly, a third-order tensor can be defined as the linear combination of all possible *unit triads* $\mathbf{e}_i\mathbf{e}_j\mathbf{e}_k$, etc. Scalars can be viewed as zero-order tensors, and vectors as first-order tensors.

A tensor, $\boldsymbol{\tau}$, can be represented by means of a square matrix as

$$\boldsymbol{\tau} = (\mathbf{e}_1, \mathbf{e}_2, \mathbf{e}_3) \begin{bmatrix} \tau_{11} & \tau_{12} & \tau_{13} \\ \tau_{21} & \tau_{22} & \tau_{23} \\ \tau_{31} & \tau_{32} & \tau_{33} \end{bmatrix} \begin{pmatrix} \mathbf{e}_1 \\ \mathbf{e}_2 \\ \mathbf{e}_3 \end{pmatrix} \qquad (1.62)$$

and often simply by the matrix of its components,

$$\boldsymbol{\tau} = \begin{bmatrix} \tau_{11} & \tau_{12} & \tau_{13} \\ \tau_{21} & \tau_{22} & \tau_{23} \\ \tau_{31} & \tau_{32} & \tau_{33} \end{bmatrix} . \tag{1.63}$$

Note that the equality sign "=" is loosely used since $\boldsymbol{\tau}$ is a tensor and *not* a matrix. For a complete description of a tensor by means of Eq. (1.63), the basis $\{\mathbf{e}_1, \mathbf{e}_2, \mathbf{e}_3\}$ should be provided.

The *unit* (or *identity*) *tensor*, $\mathbf{I}$, is defined by

$$\mathbf{I} \equiv \sum_{i=1}^{3} \sum_{j=1}^{3} \delta_{ij} \, \mathbf{e}_i \mathbf{e}_j = \mathbf{e}_1 \mathbf{e}_1 + \mathbf{e}_2 \mathbf{e}_2 + \mathbf{e}_3 \mathbf{e}_3 . \tag{1.64}$$

Each diagonal component of the matrix form of $\mathbf{I}$ is unity and the nondiagonal components are zero:

$$\mathbf{I} = \begin{bmatrix} 1 & 0 & 0 \\ 0 & 1 & 0 \\ 0 & 0 & 1 \end{bmatrix} . \tag{1.65}$$

The sum of two tensors, $\boldsymbol{\sigma}$ and $\boldsymbol{\tau}$, is the tensor whose components are the sums of the corresponding components of the two tensors:

$$\boldsymbol{\sigma} + \boldsymbol{\tau} = \sum_{i=1}^{3} \sum_{j=1}^{3} \sigma_{ij} \, \mathbf{e}_i \mathbf{e}_j + \sum_{i=1}^{3} \sum_{j=1}^{3} \tau_{ij} \, \mathbf{e}_i \mathbf{e}_j = \sum_{i=1}^{3} \sum_{j=1}^{3} (\sigma_{ij} + \tau_{ij}) \, \mathbf{e}_i \mathbf{e}_j . \tag{1.66}$$

The product of a tensor, $\boldsymbol{\tau}$, and a scalar, m, is the tensor whose components are equal to the components of $\boldsymbol{\tau}$ multiplied by m:

$$m \, \boldsymbol{\tau} = m \left(\sum_{i=1}^{3} \sum_{j=1}^{3} \sigma_{ij} \, \mathbf{e}_i \mathbf{e}_j \right) = \sum_{i=1}^{3} \sum_{j=1}^{3} (m \tau_{ij}) \, \mathbf{e}_i \mathbf{e}_j . \tag{1.67}$$

The *transpose*, $\boldsymbol{\tau}^T$, of a tensor $\boldsymbol{\tau}$ is defined by

$$\boldsymbol{\tau}^T \equiv \sum_{i=1}^{3} \sum_{j=1}^{3} \tau_{ji} \, \mathbf{e}_i \mathbf{e}_j . \tag{1.68}$$

The matrix form of $\boldsymbol{\tau}^T$ is obtained by interchanging the rows and columns of the matrix form of $\boldsymbol{\tau}$:

$$\boldsymbol{\tau}^T = \begin{bmatrix} \tau_{11} & \tau_{21} & \tau_{31} \\ \tau_{12} & \tau_{22} & \tau_{32} \\ \tau_{13} & \tau_{23} & \tau_{33} \end{bmatrix} . \tag{1.69}$$

If $\tau^T = \tau$, i.e., if τ is equal to its transpose, the tensor τ is said to be *symmetric*. If $\tau^T = -\tau$, the tensor τ is said to be *antisymmetric* (or *skew symmetric*). Any tensor τ can be expressed as the sum of a symmetric, $\mathbf{S}$, and an antisymmetric tensor, $\mathbf{U}$,

$$\tau = \mathbf{S} + \mathbf{U}\,, \tag{1.70}$$

where

$$\mathbf{S} = \frac{1}{2}\left(\tau + \tau^T\right), \tag{1.71}$$

and

$$\mathbf{U} = \frac{1}{2}\left(\tau - \tau^T\right). \tag{1.72}$$

The dyadic product of two vectors $\mathbf{a}$ and $\mathbf{b}$ can easily be constructed as follows:

$$\mathbf{ab} = \left(\sum_{i=1}^{3} a_i\,\mathbf{e}_i\right)\left(\sum_{j=1}^{3} b_j\,\mathbf{e}_j\right) = \sum_{i=1}^{3}\sum_{j=1}^{3} a_i b_j\,\mathbf{e}_i\mathbf{e}_j\,. \tag{1.73}$$

Obviously, $\mathbf{ab}$ is a tensor, referred to as *dyad* or *dyadic tensor*. Its matrix form is

$$\mathbf{ab} = \begin{bmatrix} a_1 b_1 & a_1 b_2 & a_1 b_3 \\ a_2 b_1 & a_2 b_2 & a_2 b_3 \\ a_3 b_1 & a_3 b_2 & a_3 b_3 \end{bmatrix}. \tag{1.74}$$

Note that $\mathbf{ab} \neq \mathbf{ba}$ unless $\mathbf{ab}$ is symmetric. Given that $(\mathbf{ab})^T = \mathbf{ba}$, the dyadic product of a vector by itself, $\mathbf{aa}$, is symmetric.

The unit dyads $\mathbf{e}_i\mathbf{e}_j$ are dyadic tensors, the matrix form of which has only one unitary nonzero entry at the (i,j) position. For example,

$$\mathbf{e}_2\mathbf{e}_3 = \begin{bmatrix} 0 & 0 & 0 \\ 0 & 0 & 1 \\ 0 & 0 & 0 \end{bmatrix}.$$

The most important operations involving unit dyads are the following:

(i) The *single-dot product* (or *tensor product*) of two unit dyads is a tensor defined by

$$(\mathbf{e}_i\mathbf{e}_j)\cdot(\mathbf{e}_k\mathbf{e}_l) \equiv \mathbf{e}_i\,(\mathbf{e}_j\cdot\mathbf{e}_k)\,\mathbf{e}_l = \delta_{jk}\,\mathbf{e}_i\mathbf{e}_l\,. \tag{1.75}$$

This operation is *not* commutative.

(ii) The *double-dot product* (or *scalar product* or *inner product*) of two unit dyads is a scalar defined by

$$(\mathbf{e}_i\mathbf{e}_j) : (\mathbf{e}_k\mathbf{e}_l) \equiv (\mathbf{e}_i \cdot \mathbf{e}_l)(\mathbf{e}_j \cdot \mathbf{e}_k) = \delta_{il}\delta_{jk} \,. \tag{1.76}$$

It is easily seen that this operation is commutative.

(iii) The dot product of a unit dyad and a unit vector is a vector defined by

$$(\mathbf{e}_i\mathbf{e}_j) \cdot \mathbf{e}_k \equiv \mathbf{e}_i (\mathbf{e}_j \cdot \mathbf{e}_k) = \delta_{jk}\,\mathbf{e}_i \,, \tag{1.77}$$

or

$$\mathbf{e}_i \cdot (\mathbf{e}_j\mathbf{e}_k) \equiv (\mathbf{e}_i \cdot \mathbf{e}_j)\,\mathbf{e}_k = \delta_{ij}\,\mathbf{e}_k \,. \tag{1.78}$$

Obviously, this operation is *not* commutative.

Operations involving tensors are easily performed by expanding the tensors into components with respect to a given basis and using the elementary unit dyad operations defined in Eqs. (1.75)-(1.78). The most important operations involving tensors are summarized below.

The single-dot product of two tensors
If σ and τ are tensors, then

$$
\begin{aligned}
\boldsymbol{\sigma} \cdot \boldsymbol{\tau} &= \left(\sum_{i=1}^{3}\sum_{j=1}^{3}\sigma_{ij}\,\mathbf{e}_i\mathbf{e}_j\right) \cdot \left(\sum_{k=1}^{3}\sum_{l=1}^{3}\tau_{kl}\,\mathbf{e}_k\mathbf{e}_l\right) \\
&= \sum_{i=1}^{3}\sum_{j=1}^{3}\sum_{k=1}^{3}\sum_{l=1}^{3}\sigma_{ij}\tau_{kl}\,(\mathbf{e}_i\mathbf{e}_j)\cdot(\mathbf{e}_k\mathbf{e}_l) \\
&= \sum_{i=1}^{3}\sum_{j=1}^{3}\sum_{k=1}^{3}\sum_{l=1}^{3}\sigma_{ij}\tau_{kl}\,\delta_{jk}\,\mathbf{e}_i\mathbf{e}_l \\
&= \sum_{i=1}^{3}\sum_{j=1}^{3}\sum_{l=1}^{3}\sigma_{ij}\tau_{jl}\,\mathbf{e}_i\mathbf{e}_l \quad \Longrightarrow
\end{aligned}
$$

$$\boldsymbol{\sigma} \cdot \boldsymbol{\tau} = \sum_{i=1}^{3}\sum_{l=1}^{3}\left(\sum_{j=1}^{3}\sigma_{ij}\tau_{jl}\right)\mathbf{e}_i\mathbf{e}_l \,. \tag{1.79}$$

The operation is not commutative. It is easily verified that

$$\boldsymbol{\sigma} \cdot \mathbf{I} = \mathbf{I} \cdot \boldsymbol{\sigma} = \boldsymbol{\sigma} \,. \tag{1.80}$$

A tensor τ is said to be *invertible* if there exists a tensor τ^{-1} such that

$$\tau \cdot \tau^{-1} = \tau^{-1} \cdot \tau = \mathbf{I}. \tag{1.81}$$

If τ is invertible, then τ^{-1} is unique and is called the *inverse* of τ.

The double-dot product of two tensors

$$\boldsymbol{\sigma} : \boldsymbol{\tau} = \sum_{i=1}^{3} \sum_{j=1}^{3} \sigma_{ij} \tau_{ji} \, \mathbf{e}_i \mathbf{e}_j \,. \tag{1.82}$$

The dot product of a tensor and a vector
This is a very useful operation in fluid mechanics. If $\mathbf{a}$ is a vector, we have:

$$\begin{aligned}
\boldsymbol{\sigma} \cdot \mathbf{a} &= \left(\sum_{i=1}^{3} \sum_{j=1}^{3} \sigma_{ij} \, \mathbf{e}_i \mathbf{e}_j \right) \cdot \left(\sum_{k=1}^{3} a_k \, \mathbf{e}_k \right) \\
&= \sum_{i=1}^{3} \sum_{j=1}^{3} \sum_{k=1}^{3} \sigma_{ij} a_k \, (\mathbf{e}_i \mathbf{e}_j) \cdot \mathbf{e}_k \\
&= \sum_{i=1}^{3} \sum_{j=1}^{3} \sum_{k=1}^{3} \sigma_{ij} a_k \, \delta_{jk} \, \mathbf{e}_i \\
&= \sum_{i=1}^{3} \sum_{j=1}^{3} \sigma_{ij} a_j \, \delta_{jj} \, \mathbf{e}_i \quad \Longrightarrow
\end{aligned}$$

$$\boldsymbol{\sigma} \cdot \mathbf{a} = \sum_{i=1}^{3} \left(\sum_{j=1}^{3} \sigma_{ij} a_j \right) \mathbf{e}_i \,. \tag{1.83}$$

Similarly, we find that

$$\mathbf{a} \cdot \boldsymbol{\sigma} = \sum_{i=1}^{3} \left(\sum_{j=1}^{3} \sigma_{ji} a_j \right) \mathbf{e}_i \,. \tag{1.84}$$

The vectors $\boldsymbol{\sigma} \cdot \mathbf{a}$ and $\mathbf{a} \cdot \boldsymbol{\sigma}$ are not, in general, equal.

The following identities, in which $\mathbf{a}$, $\mathbf{b}$, $\mathbf{c}$, and $\mathbf{d}$ are vectors, $\boldsymbol{\sigma}$ and $\boldsymbol{\tau}$ are tensors, and $\mathbf{I}$ is the unit tensor, are easy to prove:

$$(\mathbf{ab}) \cdot (\mathbf{cd}) = (\mathbf{b} \cdot \mathbf{c}) \, \mathbf{ad} \,, \tag{1.85}$$

$$(\mathbf{ab}) : (\mathbf{cd}) = (\mathbf{cd}) : (\mathbf{ab}) = (\mathbf{a} \cdot \mathbf{d})(\mathbf{b} \cdot \mathbf{c}) \,, \tag{1.86}$$

$$(\mathbf{ab}) \cdot \mathbf{c} = (\mathbf{b} \cdot \mathbf{c}) \, \mathbf{a} \,, \tag{1.87}$$

$$\mathbf{c} \cdot (\mathbf{ab}) = (\mathbf{c} \cdot \mathbf{a})\,\mathbf{b}\,, \tag{1.88}$$

$$\mathbf{a} \cdot \mathbf{I} = \mathbf{I} \cdot \mathbf{a} = \mathbf{a}\,, \tag{1.89}$$

$$\boldsymbol{\sigma} : \mathbf{ab} = (\boldsymbol{\sigma} \cdot \mathbf{a}) \cdot \mathbf{b}\,, \tag{1.90}$$

$$\mathbf{ab} : \boldsymbol{\sigma} = \mathbf{a} \cdot (\mathbf{b} \cdot \boldsymbol{\sigma})\,. \tag{1.91}$$

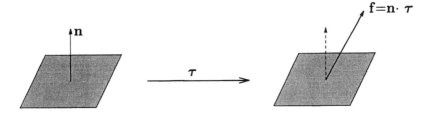

Figure 1.16. *The action of a tensor $\boldsymbol{\tau}$ on the normal vector* **n**.

The vectors forming an orthonormal basis of the three-dimensional space are normal to three mutually perpendicular plane surfaces. If $\{\mathbf{n}_1, \mathbf{n}_2, \mathbf{n}_3\}$ is such a basis and **v** is a vector, then

$$\mathbf{v} = \mathbf{n}_1 v_1 + \mathbf{n}_2 v_2 + \mathbf{n}_3 v_3\,, \tag{1.92}$$

where v_1, v_2 and v_3 are the components of **v** in the coordinate system defined by $\{\mathbf{n}_1, \mathbf{n}_2, \mathbf{n}_3\}$. Note that *a vector associates a scalar with each coordinate direction.* Since $\{\mathbf{n}_1, \mathbf{n}_2, \mathbf{n}_3\}$ is orthonormal,

$$v_1 = \mathbf{n}_1 \cdot \mathbf{v}\,, \quad v_2 = \mathbf{n}_2 \cdot \mathbf{v} \quad \text{and} \quad v_3 = \mathbf{n}_3 \cdot \mathbf{v}\,. \tag{1.93}$$

The component $v_i = \mathbf{n}_i \cdot \mathbf{v}$ might be viewed as *the result or flux produced by* **v** *through the surface* with unit normal $\mathbf{n}_i$, since the contributions of the other two components are tangent to that surface. Hence, the vector **v** is fully defined at a point by the *fluxes it produces through three mutually perpendicular infinitesimal surfaces.* We also note that a vector can be defined as an operator which produces a scalar flux when acting on an orientation vector.

Along these lines, a tensor can be conveniently defined as an operator of higher order that operates on an orientation vector and produces a *vector flux*. The action of a tensor $\boldsymbol{\tau}$ on the unit normal to a surface, **n**, is illustrated in Fig. 1.16. The dot product $\mathbf{f} = \mathbf{n} \cdot \boldsymbol{\tau}$ is a vector that differs from **n** in both length and direction. If the vectors

$$\mathbf{f}_1 = \mathbf{n}_1 \cdot \boldsymbol{\tau}\,, \quad \mathbf{f}_2 = \mathbf{n}_2 \cdot \boldsymbol{\tau} \quad \text{and} \quad \mathbf{f}_3 = \mathbf{n}_3 \cdot \boldsymbol{\tau}\,, \tag{1.94}$$

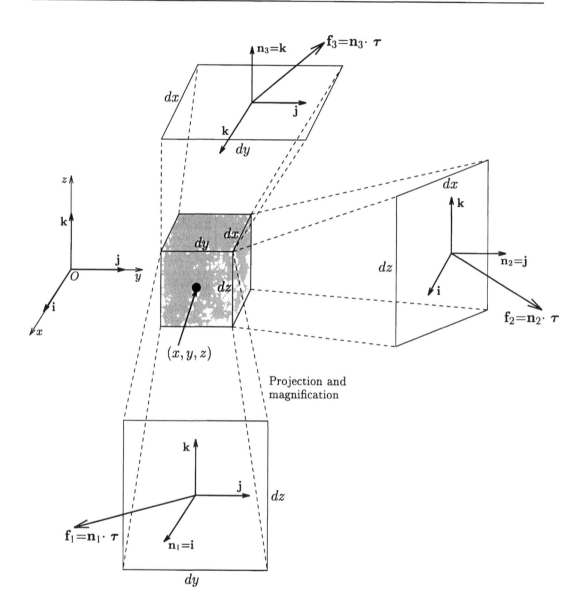

Figure 1.17. *Actions of a tensor τ on three mutually perpendicular infinitesimal plane surfaces.*

are the actions of a tensor τ on the unit normals $\mathbf{n}_1$, $\mathbf{n}_2$, and $\mathbf{n}_3$ of three mutually perpendicular infinitesimal plane surfaces, as illustrated in Fig. 1.17, then τ is given by

$$\tau = \mathbf{n}_1 \mathbf{f}_1 + \mathbf{n}_2 \mathbf{f}_2 + \mathbf{n}_3 \mathbf{f}_3 \,. \tag{1.95}$$

The tensor τ is thus represented by the sum of three dyadic products. Note that *a second-order tensor associates a vector with each coordinate direction*. The vectors $\mathbf{f}_1$, $\mathbf{f}_2$, and $\mathbf{f}_3$ can be further expanded into measurable scalar components,

$$
\begin{aligned}
\mathbf{f}_1 &= \tau_{11}\mathbf{n_1} + \tau_{12}\mathbf{n_2} + \tau_{13}\mathbf{n_3} \,, \\
\mathbf{f}_2 &= \tau_{21}\mathbf{n_1} + \tau_{22}\mathbf{n_2} + \tau_{23}\mathbf{n_3} \,, \\
\mathbf{f}_3 &= \tau_{31}\mathbf{n_1} + \tau_{32}\mathbf{n_2} + \tau_{33}\mathbf{n_3} \,.
\end{aligned}
\tag{1.96}
$$

The scalars that appear in Eq. (1.96) are obviously the components of τ with respect to the system of coordinates defined by $\{\mathbf{n}_1, \mathbf{n}_2, \mathbf{n}_3\}$:

$$\tau = \begin{bmatrix} \tau_{11} & \tau_{12} & \tau_{13} \\ \tau_{21} & \tau_{22} & \tau_{23} \\ \tau_{31} & \tau_{32} & \tau_{33} \end{bmatrix} . \tag{1.97}$$

The diagonal elements are the components of the *normal* on each of the three mutually perpendicular surfaces; the nondiagonal elements are the magnitudes of the two *tangential* or *shear* actions or fluxes on each of the three surfaces.

The most common tensor in fluid mechanics is the *stress tensor*, $\mathbf{T}$, which, when acting on a surface of unit normal $\mathbf{n}$, produces *surface stress* or *traction*,

$$\mathbf{f} = \mathbf{n} \cdot \mathbf{T} \,. \tag{1.98}$$

The traction $\mathbf{f}$ is the force per unit area exerted on an infinitesimal surface element. It can be decomposed into a normal component $\mathbf{f}_N$ that points in the direction of $\mathbf{n}$, and a tangential or shearing component $\mathbf{f}_T$:

$$\mathbf{f} = \mathbf{f}_N + \mathbf{f}_T \,. \tag{1.99}$$

The *normal traction*, $\mathbf{f}_N$, is given by

$$\mathbf{f}_N = (\mathbf{n} \cdot \mathbf{f}) \, \mathbf{n} = \mathbf{n} \cdot (\mathbf{n} \cdot \mathbf{T}) \, \mathbf{n} = (\mathbf{n}\mathbf{n} : \mathbf{T}) \, \mathbf{n} \,, \tag{1.100}$$

and, therefore, for the *tangetial traction* we obtain:

$$\mathbf{f}_T = \mathbf{f} - \mathbf{f}_N = \mathbf{n} \cdot \mathbf{T} - (\mathbf{n}\mathbf{n} : \mathbf{T}) \, \mathbf{n} \,. \tag{1.101}$$

It is left to the reader to show that the above equation is equivalent to

$$\mathbf{f}_T = \mathbf{n} \times (\mathbf{f} \times \mathbf{n}) = \mathbf{f} \cdot (\mathbf{I} - \mathbf{n}\mathbf{n}), \tag{1.102}$$

where $\mathbf{I}$ is the unit tensor.

Example 1.3.1. Vector-tensor operations[1]

Consider the Cartesian system of coordinates and the point $\mathbf{r} = \sqrt{3}\mathbf{j}$ m. Measurements of force per unit area on a small test surface give the following time-independent results:

Direction in which the test surface faces	Measured traction on the test surface (force/area)
i	$2\,\mathbf{i}+\mathbf{j}$
j	$\mathbf{i}+4\,\mathbf{j}+\mathbf{k}$
k	$\mathbf{j}+6\,\mathbf{k}$

(a) What is the state of stress at the point $\mathbf{r} = \sqrt{3}\,\mathbf{j}$?

(b) What is the traction on the test surface when it is oriented to face in the direction $\mathbf{n} = (1/\sqrt{3})(\mathbf{i}+\mathbf{j}+\mathbf{k})$?

(c) What is the moment of the traction found in Part (b)?

Solution:

(a) Let

$$\mathbf{n}_1 = \mathbf{i}, \quad \mathbf{n}_2 = \mathbf{j}, \quad \mathbf{n}_3 = \mathbf{k},$$

$$\mathbf{f}_1 = 2\mathbf{i}+\mathbf{j}, \quad \mathbf{f}_2 = \mathbf{i}+4\mathbf{j}+\mathbf{k} \quad \text{and} \quad \mathbf{f}_3 = \mathbf{j}+6\mathbf{k}.$$

The stress tensor, $\mathbf{T}$, is given by

$$
\begin{aligned}
\mathbf{T} &= \mathbf{n}_1\mathbf{f}_1 + \mathbf{n}_2\mathbf{f}_2 + \mathbf{n}_3\mathbf{f}_3 \\
&= \mathbf{i}(2\mathbf{i}+\mathbf{j}) + \mathbf{j}(\mathbf{i}+4\mathbf{j}+\mathbf{k}) + \mathbf{k}(\mathbf{j}+6\mathbf{k}) \\
&= 2\mathbf{ii} + \mathbf{ij} + 0\mathbf{ik} + \mathbf{ji} + 4\mathbf{jj} + \mathbf{jk} + 0\mathbf{ki} + \mathbf{kj} + 6\mathbf{kk}
\end{aligned}
$$

The matrix representation of $\mathbf{T}$ with respect to the basis $(\mathbf{i}, \mathbf{j}, \mathbf{k})$ is

$$
\mathbf{T} = \begin{bmatrix} 2 & 1 & 0 \\ 1 & 4 & 1 \\ 0 & 1 & 6 \end{bmatrix}.
$$

---

[1] Taken from Ref. [2].

Notice that **T** is symmetric.
(b) The traction **f** on the surface **n** is given by

$$\mathbf{f} = \mathbf{n} \cdot \mathbf{T} = \frac{1}{\sqrt{3}} (\mathbf{i} + \mathbf{j} + \mathbf{k}) \cdot (2\mathbf{ii} + \mathbf{ij} + \mathbf{ji} + 4\mathbf{jj} + \mathbf{jk} + \mathbf{kj} + 6\mathbf{kk}) = \frac{1}{\sqrt{3}} (3\mathbf{i} + 6\mathbf{j} + 7\mathbf{k}) .$$

(c) The moment of the traction at the point $\mathbf{r} = \sqrt{3}\,\mathbf{j}$ is a vector given by

$$\mathbf{M} = \mathbf{r} \times \mathbf{f} = \begin{vmatrix} \mathbf{i} & \mathbf{j} & \mathbf{k} \\ 0 & \sqrt{3} & 0 \\ \frac{3}{\sqrt{3}} & \frac{6}{\sqrt{3}} & \frac{7}{\sqrt{3}} \end{vmatrix} = 7\mathbf{i} - 3\mathbf{k} .$$

$\square$

Example 1.3.2. Normal and tangential tractions

Consider the state of stress given in Example 1.3.1. The normal and tangential components of the traction $\mathbf{f}_1$ are:

$$\mathbf{f}_{1N} = (\mathbf{n}_1 \cdot \mathbf{f}_1)\, \mathbf{n}_1 = \mathbf{i} \cdot (2\mathbf{i} + \mathbf{j})\, \mathbf{i} = 2\mathbf{i}$$

and

$$\mathbf{f}_{1T} = \mathbf{f} - \mathbf{f}_{1N} = (2\mathbf{i} + \mathbf{j}) - 2\mathbf{i} = \mathbf{j} ,$$

respectively. Similarly, for the tractions on the other two surfaces, we get:

$$\mathbf{f}_{2N} = 4\mathbf{j} , \quad \mathbf{f}_{2T} = \mathbf{i} + \mathbf{k} ;$$

$$\mathbf{f}_{3N} = 6\mathbf{k} , \quad \mathbf{f}_{3T} = \mathbf{j} .$$

Note that the normal tractions on the three surfaces are exactly the diagonal elements of the component matrix

$$\mathbf{T} = \begin{bmatrix} 2 & 1 & 0 \\ 1 & 4 & 1 \\ 0 & 1 & 6 \end{bmatrix} .$$

The nondiagonal elements of each line are the components of the corresponding tangential traction.

$\square$

1.3.1   Principal Directions and Invariants

Let $\{\mathbf{e}_1, \mathbf{e}_2, \mathbf{e}_3\}$ be an orthonormal basis of the three-dimensional space and $\boldsymbol{\tau}$ be a second-order tensor,

$$\boldsymbol{\tau} = \sum_{i=1}^{3} \sum_{j=1}^{3} \tau_{ij} \, \mathbf{e}_i \mathbf{e}_j \,, \tag{1.103}$$

or, in matrix notation,

$$\boldsymbol{\tau} = \begin{bmatrix} \tau_{11} & \tau_{12} & \tau_{13} \\ \tau_{21} & \tau_{22} & \tau_{23} \\ \tau_{31} & \tau_{32} & \tau_{33} \end{bmatrix} \,. \tag{1.104}$$

If certain conditions are satisfied, it is possible to identify an orthonormal basis $\{\mathbf{n}_1, \mathbf{n}_2, \mathbf{n}_3\}$ such that

$$\boldsymbol{\tau} = \lambda_1 \, \mathbf{n}_1 \mathbf{n}_1 + \lambda_2 \, \mathbf{n}_2 \mathbf{n}_2 + \lambda_3 \, \mathbf{n}_3 \mathbf{n}_3 \,, \tag{1.105}$$

which means that the matrix form of $\boldsymbol{\tau}$ in the coordinate system defined by the new basis is diagonal:

$$\boldsymbol{\tau} = \begin{bmatrix} \lambda_1 & 0 & 0 \\ 0 & \lambda_2 & 0 \\ 0 & 0 & \lambda_3 \end{bmatrix} \,. \tag{1.106}$$

The orthogonal vectors $\mathbf{n}_1$, $\mathbf{n}_2$, and $\mathbf{n}_3$ that *diagonalize* $\boldsymbol{\tau}$ are called the *principal directions*, and λ_1, λ_2, and λ_3 are called the *principal values* of $\boldsymbol{\tau}$. From Eq. (1.105), one observes that the vector fluxes through the surface of unit normal $\mathbf{n}_i$, i=1,2,3, satisfy the relation

$$\mathbf{f}_i = \mathbf{n}_i \cdot \boldsymbol{\tau} = \boldsymbol{\tau} \cdot \mathbf{n}_i = \lambda_i \mathbf{n}_i \,, \quad i = 1, 2, 3 \,. \tag{1.107}$$

What the above equation says is that the vector flux through the surface with unit normal $\mathbf{n}_i$ is collinear with $\mathbf{n}_i$, i.e., $\mathbf{n}_i \cdot \boldsymbol{\tau}$ is normal to that surface and its tangential component is zero. From Eq. (1.107) one gets:

$$(\boldsymbol{\tau} - \lambda_i \mathbf{I}) \cdot \mathbf{n}_i = \mathbf{0} \,, \tag{1.108}$$

where $\mathbf{I}$ is the unit tensor.

In mathematical terminology, Eq. (1.108) defines an *eigenvalue problem*. The principal directions and values of $\boldsymbol{\tau}$ are thus also called the *eigenvectors* and *eigenvalues* of $\boldsymbol{\tau}$, respectively. The eigenvalues are determined by solving the *characteristic equation*,

$$\det(\boldsymbol{\tau} - \lambda \mathbf{I}) = 0 \tag{1.109}$$

or

$$\begin{vmatrix} \tau_{11} - \lambda & \tau_{12} & \tau_{13} \\ \tau_{21} & \tau_{22} - \lambda & \tau_{23} \\ \tau_{31} & \tau_{32} & \tau_{33} - \lambda \end{vmatrix} = 0 \,, \tag{1.110}$$

which guarantees nonzero solutions to the homogeneous system (1.108). The characteristic equation is a cubic equation and, therefore, it has three roots, λ_i, $i=1,2,3$. After determining an eigenvalue λ_i, one can determine the eigenvectors, $\mathbf{n}_i$, associated with λ_i by solving the *characteristic system* (1.108). When the tensor (or matrix) τ is symmetric, all eigenvalues and the associated eigenvectors are real. This is the case with most tensors arising in fluid mechanics.

Example 1.3.3. Principal values and directions

(a) Find the principal values of the tensor

$$\tau = \begin{bmatrix} x & 0 & z \\ 0 & 2y & 0 \\ z & 0 & x \end{bmatrix} \,.$$

(b) Determine the principal directions $\mathbf{n}_1, \mathbf{n}_2, \mathbf{n}_3$ at the point $(0,1,1)$.

(c) Verify that the vector flux through a surface normal to a principal direction $\mathbf{n}_i$ is collinear with $\mathbf{n}_i$.

(d) What is the matrix form of the tensor τ in the coordinate system defined by $\{\mathbf{n}_1, \mathbf{n}_2, \mathbf{n}_3\}$?

Solution:
(a) The characteristic equation of τ is

$$0 = \det(\tau - \lambda \mathbf{I}) = \begin{vmatrix} x - \lambda & 0 & z \\ 0 & 2y - \lambda & 0 \\ z & 0 & x - \lambda \end{vmatrix} = (2y - \lambda) \begin{vmatrix} x - \lambda & z \\ z & x - \lambda \end{vmatrix} \implies$$

$$(2y - \lambda)(x - \lambda - z)(x - \lambda + z) = 0.$$

The eigenvalues of τ are $\lambda_1 = 2y$, $\lambda_2 = x - z$ and $\lambda_3 = x + z$.
(b) At the point $(0, 1, 1)$,

$$\tau = \begin{bmatrix} 0 & 0 & 1 \\ 0 & 2 & 0 \\ 1 & 0 & 0 \end{bmatrix} = \mathbf{ik} + 2\mathbf{jj} + \mathbf{ki} \,,$$

and $\lambda_1=2$, $\lambda_2=-1$ and $\lambda_3=1$. The associated eigenvectors are determined by solving the corresponding characteristic system:

$$(\boldsymbol{\tau} - \lambda_i \mathbf{I}) \cdot \mathbf{n}_i = \mathbf{0}, \quad i = 1, 2, 3.$$

For $\lambda_1=2$, one gets

$$\begin{bmatrix} 0-2 & 0 & 1 \\ 0 & 2-2 & 0 \\ 1 & 0 & 0-2 \end{bmatrix} \begin{bmatrix} n_{x1} \\ n_{y1} \\ n_{z1} \end{bmatrix} = \begin{bmatrix} 0 \\ 0 \\ 0 \end{bmatrix} \implies \left. \begin{array}{r} -2n_{x1} + n_{z1} = 0 \\ 0 = 0 \\ n_{x1} - 2n_{z1} = 0 \end{array} \right\} \implies$$

$$n_{x1} = n_{z1} = 0.$$

Therefore, the eigenvectors associated with λ_1 are of the form $(0, a, 0)$, where a is an arbitrary nonzero constant. For $a=1$, the eigenvector is *normalized*, i.e., it is of unit magnitude. We set

$$\mathbf{n}_1 = (0, 1, 0) = \mathbf{j}.$$

Similarly, solving the characteristic systems

$$\begin{bmatrix} 0+1 & 0 & 1 \\ 0 & 2+1 & 0 \\ 1 & 0 & 0+1 \end{bmatrix} \begin{bmatrix} n_{x2} \\ n_{y2} \\ n_{z2} \end{bmatrix} = \begin{bmatrix} 0 \\ 0 \\ 0 \end{bmatrix}$$

of $\lambda_2=-1$, and

$$\begin{bmatrix} 0-1 & 0 & 1 \\ 0 & 2-1 & 0 \\ 1 & 0 & 0-1 \end{bmatrix} \begin{bmatrix} n_{x3} \\ n_{y3} \\ n_{z3} \end{bmatrix} = \begin{bmatrix} 0 \\ 0 \\ 0 \end{bmatrix}$$

of $\lambda_3=1$, we find the normalized eigenvectors

$$\mathbf{n}_2 = \frac{1}{\sqrt{2}} (1, 0, -1) = \frac{1}{\sqrt{2}} (\mathbf{i} - \mathbf{k})$$

and

$$\mathbf{n}_3 = \frac{1}{\sqrt{2}} (1, 0, 1) = \frac{1}{\sqrt{2}} (\mathbf{i} + \mathbf{k}).$$

We observe that the three eigenvectors, $\mathbf{n}_1$, $\mathbf{n}_2$, and $\mathbf{n}_3$ are orthogonal:[2]

$$\mathbf{n}_1 \cdot \mathbf{n}_2 = \mathbf{n}_2 \cdot \mathbf{n}_3 = \mathbf{n}_3 \cdot \mathbf{n}_1 = 0.$$

---

[2] A well-known result of linear algebra is that the eigenvectors associated with distinct eigenvalues of a symmetric matrix are orthogonal. If two eigenvalues are the same, then the two linearly independent eigenvectors determined by solving the corresponding characteristic system may not be orthogonal. From these two eigenvectors, however, a pair of orthogonal eigenvectors can be obtained using the *Gram-Schmidt orthogonalization process*; see, for example, [3].

(c) The vector fluxes through the three surfaces normal to $\mathbf{n}_1$, $\mathbf{n}_2$, and $\mathbf{n}_3$ are:

$$\mathbf{n}_1 \cdot \boldsymbol{\tau} = \mathbf{j} \cdot (\mathbf{ik} + 2\mathbf{jj} + \mathbf{ki}) = 2\mathbf{j} = 2\,\mathbf{n}_1 \,,$$

$$\mathbf{n}_2 \cdot \boldsymbol{\tau} = \frac{1}{\sqrt{2}}(\mathbf{i} - \mathbf{k}) \cdot (\mathbf{ik} + 2\mathbf{jj} + \mathbf{ki}) = \frac{1}{\sqrt{2}}(\mathbf{k} - \mathbf{i}) = -\mathbf{n}_2 \,,$$

$$\mathbf{n}_3 \cdot \boldsymbol{\tau} = \frac{1}{\sqrt{2}}(\mathbf{i} + \mathbf{k}) \cdot (\mathbf{ik} + 2\mathbf{jj} + \mathbf{ki}) = \frac{1}{\sqrt{2}}(\mathbf{k} + \mathbf{i}) = \mathbf{n}_3 \,.$$

(d) The matrix form of $\boldsymbol{\tau}$ in the coordinate system defined by $\{\mathbf{n}_1, \mathbf{n}_2, \mathbf{n}_3\}$ is

$$\boldsymbol{\tau} = 2\mathbf{n}_1\mathbf{n}_1 - \mathbf{n}_2\mathbf{n}_2 + \mathbf{n}_3\mathbf{n}_3 = \begin{bmatrix} 2 & 0 & 0 \\ 0 & -1 & 0 \\ 0 & 0 & 1 \end{bmatrix}.$$

$\square$

The *trace*, $tr\boldsymbol{\tau}$, of a tensor $\boldsymbol{\tau}$ is defined by

$$tr\boldsymbol{\tau} \equiv \sum_{i=1}^{3} \tau_{ii} = \tau_{11} + \tau_{22} + \tau_{33} \,. \tag{1.111}$$

An interesting observation for the tensor $\boldsymbol{\tau}$ of Example 1.3.3 is that its trace is the same (equal to 2) in both coordinate systems defined by $\{\mathbf{i}, \mathbf{j}, \mathbf{k}\}$ and $\{\mathbf{n}_1, \mathbf{n}_2, \mathbf{n}_3\}$. Actually, it can be shown that the trace of a tensor is independent of the coordinate system to which its components are referred. Such quantities are called *invariants* of a tensor.[3] There are three independent invariants of a second-order tensor $\boldsymbol{\tau}$:

$$I \equiv tr\boldsymbol{\tau} = \sum_{i=1}^{3} \tau_{ii} \,, \tag{1.112}$$

$$II \equiv tr\boldsymbol{\tau}^2 = \sum_{i=1}^{3}\sum_{j=1}^{3} \tau_{ij}\tau_{ji} \,, \tag{1.113}$$

$$III \equiv tr\boldsymbol{\tau}^3 = \sum_{i=1}^{3}\sum_{j=1}^{3}\sum_{k=1}^{3} \tau_{ij}\tau_{jk}\tau_{ki} \,, \tag{1.114}$$

where $\boldsymbol{\tau}^2 = \boldsymbol{\tau} \cdot \boldsymbol{\tau}$ and $\boldsymbol{\tau}^3 = \boldsymbol{\tau} \cdot \boldsymbol{\tau}^2$. Other invariants can be formed by simply taking combinations of I, II, and III. Another common set of independent invariants is

---

[3]From a vector $\mathbf{v}$, only one independent invariant can be constructed. This is the magnitude $v = \sqrt{\mathbf{v} \cdot \mathbf{v}}$ of $\mathbf{v}$.

the following:

$$I_1 = I = tr\tau, \tag{1.115}$$

$$I_2 = \frac{1}{2}(I^2 - II) = \frac{1}{2}\left[(tr\tau)^2 - tr\tau^2\right], \tag{1.116}$$

$$I_3 = \frac{1}{6}(I^3 - 3I\ II + 2III) = \det\tau. \tag{1.117}$$

I_1, I_2, and I_3 are called *basic invariants* of τ. The characteristic equation of τ can be written as[4]

$$\lambda^3 - I_1\lambda^2 + I_2\lambda - I_3 = 0. \tag{1.118}$$

If λ_1, λ_2 and λ_3 are the eigenvalues of τ, the following identities hold:

$$I_1 = \lambda_1 + \lambda_2 + \lambda_3 = tr\tau, \tag{1.119}$$

$$I_2 = \lambda_1\lambda_2 + \lambda_2\lambda_3 + \lambda_3\lambda_1 = \frac{1}{2}\left[(tr\tau)^2 - tr\tau^2\right], \tag{1.120}$$

$$I_3 = \lambda_1\lambda_2\lambda_3 = \det\tau. \tag{1.121}$$

The theorem of Cayley-Hamilton states that a square matrix (or a tensor) is a root of its characteristic equation, i.e.,

$$\tau^3 - I_1\tau^2 + I_2\tau - I_3\,\mathbf{I} = \mathbf{O}. \tag{1.122}$$

Note that in the last equation, the boldface quantities $\mathbf{I}$ and $\mathbf{O}$ are the unit and zero tensors, respectively. As implied by its name, the *zero tensor* is the tensor whose components are all zero.

Example 1.3.4. The first invariant

Consider the tensor

$$\tau = \begin{bmatrix} 0 & 0 & 1 \\ 0 & 2 & 0 \\ 1 & 0 & 0 \end{bmatrix} = \mathbf{ik} + 2\mathbf{jj} + \mathbf{ki},$$

encountered in Example 1.3.3. Its first invariant is

$$I \equiv tr\tau = 0 + 2 + 0 = 2.$$

---

[4]The component matrices of a tensor in two different coordinate systems are *similar*. An important property of similar matrices is that they have the same characteristic polynomial; hence, the coefficients I_1, I_2, and I_3 and the eigenvalues λ_1, λ_2, and λ_3 are invariant under a change of coordinate system.

Verify that the value of I is the same in cylindrical coordinates.

Solution:
Using the relations of Table 1.1, we have

$$
\begin{aligned}
\boldsymbol{\tau} &= \mathbf{ik} + 2\mathbf{jj} + \mathbf{ki} \\
&= \left(\cos\theta\,\mathbf{e}_r - \sin\theta\,\mathbf{e}_\theta\right)\mathbf{e}_z + 2\left(\sin\theta\,\mathbf{e}_r + \cos\theta\,\mathbf{e}_\theta\right)\left(\sin\theta\,\mathbf{e}_r + \cos\theta\,\mathbf{e}_\theta\right) \\
&\quad + \mathbf{e}_z\left(\cos\theta\,\mathbf{e}_r - \sin\theta\,\mathbf{e}_\theta\right) \\
&= 2\sin^2\theta\,\mathbf{e}_r\mathbf{e}_r + 2\sin\theta\,\cos\theta\,\mathbf{e}_r\mathbf{e}_\theta + \cos\theta\,\mathbf{e}_r\mathbf{e}_z + \\
&\quad 2\sin\theta\,\cos\theta\,\mathbf{e}_\theta\mathbf{e}_r + 2\cos^2\theta\,\mathbf{e}_\theta\mathbf{e}_\theta - \sin\theta\,\mathbf{e}_\theta\mathbf{e}_z + \\
&\quad \cos\theta\,\mathbf{e}_z\mathbf{e}_r - \sin\theta\,\mathbf{e}_z\mathbf{e}_\theta + 0\,\mathbf{e}_z\mathbf{e}_z\,.
\end{aligned}
$$

Therefore, the component matrix of $\boldsymbol{\tau}$ in cylindrical coordinates $\{\mathbf{e}_r, \mathbf{e}_\theta, \mathbf{e}_z\}$ is

$$
\boldsymbol{\tau} = \begin{bmatrix} 2\sin^2\theta & 2\sin\theta\,\cos\theta & \cos\theta \\ 2\sin\theta\,\cos\theta & 2\cos^2\theta & -\sin\theta \\ \cos\theta & -\sin\theta & 0 \end{bmatrix}\,.
$$

Notice that $\boldsymbol{\tau}$ remains symmetric. Its first invariant is

$$
I = tr\boldsymbol{\tau} = 2\left(\sin^2\theta + \cos^2\theta\right) + 0 = 2\,,
$$

as it should be.   □

1.3.2   Index Notation and Summation Convention

So far, we have used three different ways for representing tensors and vectors:
(a) the compact *symbolic notation*, e.g., $\mathbf{u}$ for a vector and $\boldsymbol{\tau}$ for a tensor;
(b) the so-called *Gibbs' notation*, e.g.,

$$
\sum_{i=1}^{3} u_i\,\mathbf{e}_i \quad \text{and} \quad \sum_{i=1}^{3}\sum_{j=1}^{3} \tau_{ij}\,\mathbf{e}_i\mathbf{e}_j
$$

for $\mathbf{u}$ and $\boldsymbol{\tau}$, respectively; and
(c) the *matrix notation*, e.g.,

$$
\boldsymbol{\tau} = \begin{bmatrix} \tau_{11} & \tau_{12} & \tau_{13} \\ \tau_{21} & \tau_{22} & \tau_{23} \\ \tau_{31} & \tau_{32} & \tau_{33} \end{bmatrix}
$$

for $\boldsymbol{\tau}$.

Very frequently in the literature, use is made of the *index notation* and the so-called *Einstein's summation convention* in order to simplify expressions involving vector and tensor operations by omitting the summation symbols.

In index notation, a vector **v** is represented as

$$v_i \equiv \sum_{i=1}^{3} v_i \, \mathbf{e}_i = \mathbf{v} \,. \tag{1.123}$$

A tensor $\boldsymbol{\tau}$ is represented as

$$\tau_{ij} \equiv \sum_{i=1}^{3} \sum_{j=1}^{3} \tau_{ij} \, \mathbf{e}_i \, \mathbf{e}_j = \boldsymbol{\tau} \,. \tag{1.124}$$

The nabla operator, for example, is represented as

$$\frac{\partial}{\partial x_i} \equiv \sum_{i=1}^{3} \frac{\partial}{\partial x_i} \, \mathbf{e}_i = \frac{\partial}{\partial x} \mathbf{i} + \frac{\partial}{\partial y} \mathbf{j} + \frac{\partial}{\partial z} \mathbf{k} = \nabla \,, \tag{1.125}$$

where x_i is the general Cartesian coordinate taking on the values of x, y, and z. The unit tensor **I** is represented by Kronecker's delta:

$$\delta_{ij} \equiv \sum_{i=1}^{3} \sum_{j=1}^{3} \delta_{ij} \, \mathbf{e}_i \, \mathbf{e}_j = \mathbf{I} \,. \tag{1.126}$$

It is evident that an explicit statement must be made when the tensor τ_{ij} is to be distinguished from its (i, j) element.

With Einstein's summation convention, if an index appears twice in an expression, then summation is implied with respect to the *repeated* index over the range of that index. The number of the *free indices*, i.e., the indices that appear only once, is the number of directions associated with an expression; it thus determines whether an expression is a scalar, a vector, or a tensor. In the following expressions, there are no free indices, and thus these are scalars:

$$u_i v_i \equiv \sum_{i=1}^{3} u_i v_i = \mathbf{u} \cdot \mathbf{v} \,, \tag{1.127}$$

$$\tau_{ii} \equiv \sum_{i=1}^{3} \tau_{ii} = tr\boldsymbol{\tau} \,, \tag{1.128}$$

$$\frac{\partial u_i}{\partial x_i} \equiv \sum_{i=1}^{3} \frac{\partial u_i}{\partial x_i} = \frac{\partial u_x}{\partial x} + \frac{\partial u_y}{\partial y} + \frac{\partial u_z}{\partial z} = \nabla \cdot \mathbf{u} \,, \qquad (1.129)$$

$$\frac{\partial^2 f}{\partial x_i \partial x_i} \text{ or } \frac{\partial^2 f}{\partial x_i^2} \equiv \sum_{i=1}^{3} \frac{\partial^2 f}{\partial x_i^2} = \frac{\partial^2 f}{\partial x^2} + \frac{\partial^2 f}{\partial y^2} + \frac{\partial^2 f}{\partial z^2} = \nabla^2 f \,, \quad (1.130)$$

where ∇^2 is the *Laplacian operator* to be discussed in more detail in Section 1.4. In the following expression, there are two sets of double indices, and summation must be performed over both sets:

$$\sigma_{ij}\tau_{ji} \equiv \sum_{i=1}^{3}\sum_{j=1}^{3} \sigma_{ij}\tau_{ji} = \boldsymbol{\sigma} : \boldsymbol{\tau} \,. \qquad (1.131)$$

The following expressions, with one free index, are vectors:

$$\epsilon_{ijk}u_i v_j \equiv \sum_{k=1}^{3} \left(\sum_{i=1}^{3}\sum_{j=1}^{3} \epsilon_{ijk}u_i v_j \right) \mathbf{e}_k = \mathbf{u} \times \mathbf{v} \,, \qquad (1.132)$$

$$\frac{\partial f}{\partial x_i} \equiv \sum_{i=1}^{3} \frac{\partial f}{\partial x_i} \mathbf{e}_i = \frac{\partial f}{\partial x}\mathbf{i} + \frac{\partial f}{\partial y}\mathbf{j} + \frac{\partial f}{\partial z}\mathbf{k} = \nabla f \,, \qquad (1.133)$$

$$\tau_{ij}v_j \equiv \sum_{i=1}^{3} \left(\sum_{j=1}^{3} \tau_{ij}v_j \right) \mathbf{e}_i = \boldsymbol{\tau} \cdot \mathbf{v} \,. \qquad (1.134)$$

Finally, the following quantities, having two free indices, are tensors:

$$u_i v_j \equiv \sum_{i=1}^{3}\sum_{j=1}^{3} u_i v_j \, \mathbf{e}_i \mathbf{e}_j = \mathbf{uv} \,, \qquad (1.135)$$

$$\sigma_{ik}\tau_{kj} \equiv \sum_{i=1}^{3}\sum_{j=1}^{3} \left(\sum_{k=1}^{3} \sigma_{ik}\tau_{kj} \right) \mathbf{e}_i \mathbf{e}_j = \boldsymbol{\sigma} \cdot \boldsymbol{\tau} \,, \qquad (1.136)$$

$$\frac{\partial u_j}{\partial x_i} \equiv \sum_{i=1}^{3}\sum_{j=1}^{3} \frac{\partial u_j}{\partial x_i} \, \mathbf{e}_i \mathbf{e}_j = \nabla \mathbf{u} \,. \qquad (1.137)$$

Note that $\nabla \mathbf{u}$ in the last equation is a dyadic tensor.[5]

---

[5]Some authors use even simpler expressions for the nabla operator. For example, $\nabla \cdot \mathbf{u}$ is also represented as $\partial_i u_i$ or $u_{i,i}$, with a comma to indicate the derivative, and the dyadic ∇u is represented as $\partial_i u_j$ or $u_{i,j}$.

1.3.3 Tensors in Fluid Mechanics

Flows in the physical world are three-dimensional, and so are the tensors involved in the governing equations. Many flow problems, however, are often approximated as two- or even one-dimensional, in which cases the involved tensors and vectors degenerate to two- or one-dimensional forms. In this subsection, we give only a brief description of the most important tensors in fluid mechanics. More details are given in following chapters.

The *stress tensor*, $\mathbf{T}$, represents the state of the stress in a fluid. When operating on a surface, $\mathbf{T}$ produces a traction $\mathbf{f}=\mathbf{n}\cdot\mathbf{T}$, where $\mathbf{n}$ is the unit normal to the surface. In static equilibrium, the stress tensor is identical to the *hydrostatic pressure tensor*,

$$\mathbf{T}^{SE} = -p_H\,\mathbf{I}\,, \tag{1.138}$$

where p_H is the scalar hydrostatic pressure. The traction on any submerged surface is given by

$$\mathbf{f}^{SE} = \mathbf{n}\cdot\mathbf{T}^{SE} = \mathbf{n}\cdot(-p_H\,\mathbf{I}) = -p_H\,\mathbf{n}\,, \tag{1.139}$$

and is normal to the surface; its magnitude is identical to the hydrostatic pressure:

$$|\mathbf{f}^{SE}| = |-p_H\,\mathbf{n}| = p_H\,.$$

Since the resulting traction is independent of the orientation of the surface, the pressure tensor is *isotropic*, i.e., its components are unchanged by rotation of the frame of reference.

In flowing incompressible media, the stress tensor consists of an isotropic or pressure part which is, in general, different from the hydrostatic pressure tensor, and an *anisotropic* or *viscous part*, which resists relative motion:[6]

$$
\begin{array}{ccccc}
\mathbf{T} & = & -p\,\mathbf{I} & + & \tau \\[1em]
\begin{bmatrix} Total \\ Stress \end{bmatrix} & = & \begin{bmatrix} Isotropic \\ Pressure \\ Stress \end{bmatrix} & + & \begin{bmatrix} Anisotropic \\ Viscous \\ Stress \end{bmatrix}
\end{array} \tag{1.140}
$$

The viscous stress tensor τ is, of course, zero in static equilibrium. It is, in general, anisotropic, i.e., the viscous traction on a surface depends on its orientation: it can

---

[6]In some books (e.g., in [4] and [9]), a different sign convention is adopted for the total stress tensor $\mathbf{T}$, so that
$$\mathbf{T} = p\,\mathbf{I} - \tau\,.$$
An interesting discussion about the two sign conventions can be found in [9].

be normal, shear (i.e., tangential), or a mixture of the two. In matrix notation, Eq. (1.140) becomes

$$
\begin{bmatrix} T_{11} & T_{12} & T_{13} \\ T_{21} & T_{22} & T_{23} \\ T_{31} & T_{32} & T_{33} \end{bmatrix} = \begin{bmatrix} -p & 0 & 0 \\ 0 & -p & 0 \\ 0 & 0 & -p \end{bmatrix} + \begin{bmatrix} \tau_{11} & \tau_{12} & \tau_{13} \\ \tau_{21} & \tau_{22} & \tau_{23} \\ \tau_{31} & \tau_{32} & \tau_{33} \end{bmatrix}, \tag{1.141}
$$

and, in index notation,

$$
T_{ij} = -p\, \delta_{ij} + \tau_{ij}\,. \tag{1.142}
$$

The diagonal components, T_{ii}, of $\mathbf{T}$ are *normal stresses*, and the nondiagonal ones are *shear stresses*.

Equation (1.140) is the standard decomposition of the stress tensor, inasmuch as the measurable quantities are, in general, the total stress components T_{ij} and not p or τ_{ij}. For educational purposes, the following decomposition appears to be more illustrative:

$$
\mathbf{T} \;=\; -p_H\,\mathbf{I} \qquad - p_E\,\mathbf{I} \qquad + \boldsymbol{\tau}^N \qquad + \boldsymbol{\tau}^{SH} \tag{1.143}
$$

$$
\begin{bmatrix} Total \\ Stress \end{bmatrix} = \begin{bmatrix} Hydrostatic \\ Pressure \\ Stress \end{bmatrix} + \begin{bmatrix} Extra \\ Pressure \\ Stress \end{bmatrix} + \begin{bmatrix} Viscous \\ Normal \\ Stress \end{bmatrix} + \begin{bmatrix} Viscous \\ Shear \\ Stress \end{bmatrix}
$$

or, in matrix form,

$$
\begin{bmatrix} T_{11} & T_{12} & T_{13} \\ T_{21} & T_{22} & T_{23} \\ T_{31} & T_{32} & T_{33} \end{bmatrix} = \begin{bmatrix} -p_H & 0 & 0 \\ 0 & -p_H & 0 \\ 0 & 0 & -p_H \end{bmatrix} + \begin{bmatrix} -p_E & 0 & 0 \\ 0 & -p_E & 0 \\ 0 & 0 & -p_E \end{bmatrix}
$$

$$
+ \begin{bmatrix} \tau_{11} & 0 & 0 \\ 0 & \tau_{22} & 0 \\ 0 & 0 & \tau_{33} \end{bmatrix} + \begin{bmatrix} 0 & \tau_{12} & \tau_{13} \\ \tau_{21} & 0 & \tau_{23} \\ \tau_{31} & \tau_{32} & 0 \end{bmatrix} \tag{1.144}
$$

The *hydrostatic pressure stress*, $-p_H\mathbf{I}$, is the only nonzero stress component in static equilibrium; it is due to the weight of the fluid and is a function of the position or elevation z, i.e.,

$$
p_H(z) \;=\; p_0 - \rho g\,(z - z_0)\,, \tag{1.145}
$$

where p_0 is the reference pressure at $z = z_0$, ρ is the density of the fluid, and g is the gravitational acceleration.

The *extra pressure stress*, $-p_E\mathbf{I}$, arises in flowing media due to the perpendicular motion of the particles towards a material surface, and is proportional to the convective momentum carried by the moving molecules. In *inviscid motions*, where either the viscosity of the medium is vanishingly small or the velocity gradients are negligible, the hydrostatic and extra pressure stresses are the only nonzero stress components.

The *viscous normal stress*, τ^N, is due to accelerating or decelerating perpendicular motions towards material surfaces and is proportional to the viscosity of the medium and the velocity gradient along the streamlines.

Finally, the *viscous shear stress*, τ^{SH}, is due to shearing motions of adjacent material layers next to material surfaces. It is proportional to the viscosity of the medium and to the velocity gradient in directions perpendicular to the streamlines. In *stretching* or *extensional flows*, where there are no velocity gradients in the directions perpendicular to the streamlines, the viscous shear stress is zero and thus τ^N is the only nonzero viscous stress component. In *shear flows*, such as flows in rectilinear channels and pipes, τ^N vanishes.

In summary, the stress (or force per unit area) is the result of the momentum carried by N molecules across the surface according to Newton's law of motion:

$$\mathbf{n} \cdot \mathbf{T} = \mathbf{f} = \frac{\mathbf{F}}{\Delta S} = \frac{1}{\Delta S} \sum_{i=1}^{N} \frac{d}{dt}(m_i \mathbf{u}_i) \ . \tag{1.146}$$

Any flow is a superposition of the above mentioned motions and, therefore, the appropriate stress expression is that of Eqs. (1.140) and (1.143). Each of the stress components is expressed in terms of physical characteristics of the medium (i.e., viscosity, density, and elasticity, which are functions of temperature in nonisothermal situations) and the velocity field by means of the *constitutive equation* which is highlighted in Chapter 5.

The *strain tensor*, $\mathbf{C}$, represents the state of strain in a medium and is commonly called the *Cauchy strain tensor*. Its inverse, $\mathbf{B} = \mathbf{C}^{-1}$, is known as the *Finger strain tensor*. Both tensors are of primary use in non-Newtonian fluid mechanics. Dotted with the unit normal to a surface, the Cauchy strain tensor (or the Finger strain tensor) yields the strain of the surface due to shearing and stretching. The components of the two tensors are the spatial derivatives of the coordinates with respect to the coordinates at an earlier (Cauchy) or later (Finger) time of the moving fluid particle [9].

The *velocity gradient tensor*, $\nabla \mathbf{u}$, measures the rate of change of the separation vector, $\mathbf{r}_{AB}$, between neighboring fluid particles at A and B, according to

$$\nabla \mathbf{u} = \nabla \frac{d\mathbf{r}_{AB}}{dt} \ , \tag{1.147}$$

Shear flow: channel flow

Extensional flow: opposing jets

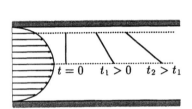
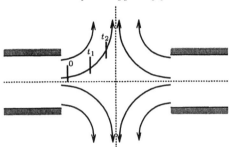

Figure 1.18. *Rotational (weak) and irrotational (strong) deformation of material lines in shear and extensional flows, respectively.*

and represents the rate of change of the magnitude (stretching or compression) and the orientation (rotation) of the material vector $\mathbf{r}_{AB}$. $\nabla\mathbf{u}$ is the dyadic tensor of the generalized derivative vector ∇ and the velocity vector $\mathbf{u}$, as explained in Section 1.4. Like any tensor, $\nabla\mathbf{u}$ can be decomposed into a symmetric, $\mathbf{D}$, and an antisymmetric part, $\mathbf{S}$:[7]

$$\nabla\mathbf{u} = \mathbf{D} + \mathbf{\Omega}. \tag{1.148}$$

The symmetric tensor

$$\mathbf{D} = \frac{1}{2}\left[\nabla\mathbf{u} + (\nabla\mathbf{u})^T\right] \tag{1.149}$$

is the *rate of strain* (or *rate of deformation*) *tensor*, and represents the state of the intensity or rate of strain. The antisymmetric tensor

$$\mathbf{S} = \frac{1}{2}\left[\nabla\mathbf{u} - (\nabla\mathbf{u})^T\right] \tag{1.150}$$

is the *vorticity tensor*.[8] If $\mathbf{n}$ is the unit normal to a surface, then the dot product $\mathbf{n} \cdot \mathbf{D}$ yields the rate of change of the distances in three mutually perpendicular

---

[7]Some authors define the rate-of-strain and the vorticity tensors as

$$\mathbf{D} = \nabla\mathbf{u} + (\nabla\mathbf{u})^T \quad \text{and} \quad \mathbf{S} = \nabla\mathbf{u} - (\nabla\mathbf{u})^T,$$

so that

$$2\,\nabla\mathbf{u} = \mathbf{D} + \mathbf{\Omega}.$$

[8]Other symbols used for the rate-of-strain and the vorticity tensors are $\mathbf{d}$, $\dot{\gamma}$, and $\mathbf{E}$ for $\mathbf{D}$, and $\mathbf{\Omega}$, ω, and $\mathbf{\Xi}$ for $\mathbf{S}$.

Tensor	Orientation	Operation	Result or Vector − Flux
Stress, **T**	unit normal, **n**	**n** · **T**	Traction
Rate of strain, **D**	unit normal, **n**	**n** · **D**	Rate of stretching
	unit tangent, **t**	**t** · **D**	Rate of rotation
Viscous Stress, τ	velocity gradient, ∇**u**	$\tau : \nabla$**u**	Scalar viscous dissipation

Table 1.3. *Vector-tensor operations producing measurable result or flux.*

directions. The dot product **n** · **S** gives the rate of change of orientation along these directions.

In *purely shear flows* the only strain is rotational. The distance between two particles on the same streamline does not change, whereas the distance between particles on different streamlines changes linearly with traveling time. Thus there is both stretching (or compression) and rotation of material lines (or material vectors), and the flow is characterized as *rotational or weak flow*. In *extensional flows*, the separation vectors among particles on the same streamline change their length exponentially, whereas the separation vectors among particles on different streamlines do not change their orientation. These flows are *irrotational or strong flows*. Figure 1.18 illustrates the deformation of material lines, defined as one-dimensional collections of fluid particles that can be shortened, elongated, and rotated, in rotational shear flows and in irrotational extensional flows.

The rate of strain tensor represents the strain state and is zero in *rigid-body motion* (*translation* and *rotation*), since this induces no strain (deformation). The vorticity tensor represents the state of rotation, and is zero in strong irrotational flows. Based on these remarks, we can say that strong flows are those in which the vorticity tensor is zero; the directions of maximum strain do not rotate to directions of less strain and, therefore, the maximum (strong) strain does not have the opportunity to relax. Weak flows are those of nonzero vorticity; in this case, the directions of maximum strain rotate, and the strain relaxes. Table 1.3 lists some examples of tensor action arising in Mechanics.

Example 1.3.5. Strong and weak flows

In steady channel flow (see Fig. 1.18), the velocity components are given by

$$u_x = a\left(1 - y^2\right), \quad u_y = 0 \quad \text{and} \quad u_z = 0 .$$

Let (x_0, y_0, z_0) and (x, y, z) be the positions of a particle at times $t=0$ and t, respectively. By integrating the velocity components with respect to time, one gets:

$$u_x = \frac{dx}{dt} = a\left(1 - y^2\right) \quad\Longrightarrow\quad x = x_0 + a\left(1 - y^2\right)t\,;$$
$$u_y = 0 \quad\Longrightarrow\quad y = y_0\,;$$
$$u_z = 0 \quad\Longrightarrow\quad z = z_0\,.$$

The fluid particle at (x, y_0, z_0) is separated linearly with time from that at (x_0, y_0, z_0), and, thus, the resulting strain is small. The matrix form of the velocity gradient tensor in Cartesian coordinates is

$$\nabla \mathbf{u} = \begin{bmatrix} \dfrac{\partial u_x}{\partial x} & \dfrac{\partial u_y}{\partial x} & \dfrac{\partial u_z}{\partial x} \\[2mm] \dfrac{\partial u_x}{\partial y} & \dfrac{\partial u_y}{\partial y} & \dfrac{\partial u_z}{\partial y} \\[2mm] \dfrac{\partial u_x}{\partial z} & \dfrac{\partial u_y}{\partial z} & \dfrac{\partial u_z}{\partial z} \end{bmatrix}\,, \tag{1.151}$$

and, therefore,

$$\nabla \mathbf{u} = \begin{bmatrix} 0 & 0 & 0 \\ -2ay & 0 & 0 \\ 0 & 0 & 0 \end{bmatrix}\,;\quad \mathbf{D} = \begin{bmatrix} 0 & -ay & 0 \\ -ay & 0 & 0 \\ 0 & 0 & 0 \end{bmatrix}\,;\quad \mathbf{S} = \begin{bmatrix} 0 & ay & 0 \\ -ay & 0 & 0 \\ 0 & 0 & 0 \end{bmatrix}\,.$$

The vorticity tensor is nonzero and thus the flow is weak.

Let us now consider the extensional flow of Fig. 1.18. The velocity components are given by

$$u_x = \varepsilon x\,,\quad u_y = -\varepsilon y \quad \text{and} \quad u_z = 0\,;$$

therefore,

$$u_x = \frac{dx}{dt} = \varepsilon x \quad\Longrightarrow\quad x = x_0\, e^{\varepsilon t}\,;$$
$$u_y = \frac{dy}{dt} = -\varepsilon y \quad\Longrightarrow\quad y = y_0\, e^{-\varepsilon t}\,;$$
$$u_z = 0 \quad\Longrightarrow\quad z = z_0\,.$$

Since the fluid particle at (x, y, z_0) is separated exponentially with time from that at (x_0, y_0, z_0), the resulting strain (stretching) is large. The velocity-gradient, rate of strain, and vorticity tensors are:

$$\nabla \mathbf{u} = \begin{bmatrix} \varepsilon & 0 & 0 \\ 0 & -\varepsilon & 0 \\ 0 & 0 & 0 \end{bmatrix}\,;\quad \mathbf{D} = \begin{bmatrix} \varepsilon & 0 & 0 \\ 0 & -\varepsilon & 0 \\ 0 & 0 & 0 \end{bmatrix}\,;\quad \mathbf{S} = \begin{bmatrix} 0 & 0 & 0 \\ 0 & 0 & 0 \\ 0 & 0 & 0 \end{bmatrix}\,.$$

Since the vorticity tensor is zero, the flow is strong. $\qquad\square$

1.4   Differential Operators

The nabla operator ∇, already encountered in previous sections, is a differential operator. In a Cartesian system of coordinates (x_1, x_2, x_3), defined by the orthonormal basis $(\mathbf{e}_1, \mathbf{e}_2, \mathbf{e}_3)$,

$$\nabla \equiv \mathbf{e}_1 \frac{\partial}{\partial x_1} + \mathbf{e}_2 \frac{\partial}{\partial x_2} + \mathbf{e}_3 \frac{\partial}{\partial x_3} = \sum_{i=1}^{3} \mathbf{e}_i \frac{\partial}{\partial x_i}, \tag{1.152}$$

or, in index notation,

$$\nabla \equiv \frac{\partial}{\partial x_i}. \tag{1.153}$$

The nabla operator is a vector operator which acts on scalar, vector, or tensor fields. The result of its action is another field, the order of which depends on the type of the operation. In the following, we will first define the various operations of ∇ in Cartesian coordinates, and then discuss their forms in curvilinear coordinates.

The *gradient* of a differentiable scalar field f, denoted by ∇f or *gradf*, is a vector field:

$$\nabla f = \left(\sum_{i=1}^{3} \mathbf{e}_i \frac{\partial}{\partial x_i} \right) f = \sum_{i=1}^{3} \mathbf{e}_i \frac{\partial f}{\partial x_i} = \mathbf{e}_1 \frac{\partial f}{\partial x_1} + \mathbf{e}_2 \frac{\partial f}{\partial x_2} + \mathbf{e}_3 \frac{\partial f}{\partial x_3}. \tag{1.154}$$

The gradient ∇f can be viewed as a generalized derivative in three dimensions; it measures the spatial change of f occurring within a distance $d\mathbf{r}(dx_1, dx_2, dx_3)$.

The *gradient* of a differentiable vector field $\mathbf{u}$ is a dyadic tensor field:

$$\nabla \mathbf{u} = \left(\sum_{i=1}^{3} \mathbf{e}_i \frac{\partial}{\partial x_i} \right) \left(\sum_{j=1}^{3} u_j \mathbf{e}_j \right) = \sum_{i=1}^{3} \sum_{j=1}^{3} \frac{\partial u_j}{\partial x_i} \mathbf{e}_i \mathbf{e}_j. \tag{1.155}$$

As explained in Section 1.3.3, if $\mathbf{u}$ is the velocity, then $\nabla \mathbf{u}$ is called the *velocity-gradient tensor*.

The *divergence* of a differentiable vector field $\mathbf{u}$, denoted by $\nabla \cdot \mathbf{u}$ or *divu*, is a scalar field

$$\nabla \cdot \mathbf{u} = \left(\sum_{i=1}^{3} \mathbf{e}_i \frac{\partial}{\partial x_i} \right) \cdot \left(\sum_{j=1}^{3} u_j \mathbf{e}_j \right) = \sum_{i=1}^{3} \frac{\partial u_i}{\partial x_i} \delta_{ij} = \frac{\partial u_1}{\partial x_1} + \frac{\partial u_2}{\partial x_2} + \frac{\partial u_3}{\partial x_3}. \tag{1.156}$$

$\nabla \cdot \mathbf{u}$ measures changes in magnitude or flux through a point. If $\mathbf{u}$ is the velocity, then $\nabla \cdot \mathbf{u}$ measures the rate of volume expansion per unit volume; hence, it is zero for incompressible fluids. The following identity is easy to prove:

$$\nabla \cdot (f\mathbf{u}) = \nabla f \cdot \mathbf{u} + f \nabla \cdot \mathbf{u}. \tag{1.157}$$

The *curl* or *rotation* of a differentiable vector field **u**, denoted by $\nabla \times \mathbf{u}$ or *curl***u** or *rot***u**, is a vector field:

$$\nabla \times \mathbf{u} = \left(\sum_{i=1}^{3} \mathbf{e}_i \frac{\partial}{\partial x_i} \right) \times \left(\sum_{j=1}^{3} u_j \mathbf{e}_j \right) = \begin{vmatrix} \mathbf{e}_1 & \mathbf{e}_2 & \mathbf{e}_3 \\ \frac{\partial}{\partial x_1} & \frac{\partial}{\partial x_2} & \frac{\partial}{\partial x_3} \\ u_1 & u_2 & u_3 \end{vmatrix} \qquad (1.158)$$

or

$$\nabla \times \mathbf{u} = \left(\frac{\partial u_3}{\partial x_2} - \frac{\partial u_2}{\partial x_3} \right) \mathbf{e}_1 + \left(\frac{\partial u_1}{\partial x_3} - \frac{\partial u_3}{\partial x_1} \right) \mathbf{e}_2 + \left(\frac{\partial u_2}{\partial x_1} - \frac{\partial u_1}{\partial x_2} \right) \mathbf{e}_3 . \qquad (1.159)$$

The field $\nabla \times \mathbf{u}$ is often called the *vorticity* (or *chirality*) of **u**.

The *divergence* of a differentiable tensor field τ is a vector field:[9]

$$\nabla \cdot \tau = \left(\sum_{k=1}^{3} \mathbf{e}_k \frac{\partial}{\partial x_k} \right) \cdot \left(\sum_{i=1}^{3} \sum_{j=1}^{3} \tau_{ij} \mathbf{e}_i \mathbf{e}_j \right) = \sum_{i=1}^{3} \sum_{j=1}^{3} \frac{\partial \tau_{ij}}{\partial x_i} \mathbf{e}_j . \qquad (1.160)$$

Example 1.4.1. The divergence and the curl of the position vector

Consider the position vector in Cartesian coordinates,

$$\mathbf{r} = x\,\mathbf{i} + y\,\mathbf{j} + z\,\mathbf{k} . \qquad (1.161)$$

For its divergence and curl, we obtain:

$$\nabla \cdot \mathbf{r} = \frac{\partial x}{\partial x} + \frac{\partial y}{\partial y} + \frac{\partial z}{\partial z} \quad \Longrightarrow$$

$$\nabla \cdot \mathbf{r} = 3 , \qquad (1.162)$$

and

$$\nabla \times \mathbf{r} = \begin{vmatrix} \mathbf{i} & \mathbf{j} & \mathbf{k} \\ \frac{\partial}{\partial x} & \frac{\partial}{\partial y} & \frac{\partial}{\partial z} \\ x & y & z \end{vmatrix} \quad \Longrightarrow$$

$$\nabla \times \mathbf{r} = \mathbf{0} \qquad (1.163)$$

Equations (1.162) and (1.163) hold in all coordinate systems. $\qquad\qquad$ □

---

[9]The divergence of a tensor τ is sometimes denoted by *div*τ .

Other useful operators involving the nabla operator are the *Laplace operator* ∇^2 and the operator $\mathbf{u} \cdot \nabla$, where $\mathbf{u}$ is a vector field. The *Laplacian* of a scalar f with continuous second partial derivatives is defined as the divergence of the gradient:

$$\nabla^2 f \equiv \nabla \cdot (\nabla f) = \frac{\partial^2 f}{\partial x_1^2} + \frac{\partial^2 f}{\partial x_2^2} + \frac{\partial^2 f}{\partial x_3^2}, \tag{1.164}$$

i.e.,

$$\nabla^2 \equiv \nabla \cdot \nabla = \frac{\partial^2}{\partial x_1^2} + \frac{\partial^2}{\partial x_2^2} + \frac{\partial^2}{\partial x_3^2}. \tag{1.165}$$

A function whose Laplacian is identically zero is called *harmonic*.

If $\mathbf{u} = u_1\mathbf{e}_1 + u_2\mathbf{e}_2 + u_3\mathbf{e}_3$ is a vector field, then

$$\nabla^2 \mathbf{u} = \nabla^2 u_1 \mathbf{e}_1 + \nabla^2 u_2 \mathbf{e}_2 + \nabla^2 u_3 \mathbf{e}_3. \tag{1.166}$$

For the operator $\mathbf{u} \cdot \nabla$, we obtain:

$$\mathbf{u} \cdot \nabla = (u_1\mathbf{e}_1 + u_2\mathbf{e}_2 + u_3\mathbf{e}_3) \cdot \left(\mathbf{e}_1 \frac{\partial}{\partial x_1} + \mathbf{e}_2 \frac{\partial}{\partial x_2} + \mathbf{e}_3 \frac{\partial}{\partial x_3} \right) \implies$$

$$\mathbf{u} \cdot \nabla = u_1 \frac{\partial}{\partial x_1} + u_2 \frac{\partial}{\partial x_2} + u_3 \frac{\partial}{\partial x_3}. \tag{1.167}$$

The above expressions are valid only for Cartesian coordinate systems. In curvilinear coordinate systems, the basis vectors are not constant and the forms of ∇ are quite different, as explained in Example 1.4.3. Notice that gradient always raises the order by one (the gradient of a scalar is a vector, the gradient of a vector is a tensor, and so on), while divergence reduces the order of a quantity by one. A summary of useful operations in Cartesian coordinates (x, y, z) is given in Table 1.4.

For any scalar function f with continuous second partial derivatives, the curl of the gradient is zero,

$$\nabla \times (\nabla f) = \mathbf{0}. \tag{1.168}$$

For any vector function $\mathbf{u}$ with continuous second partial derivatives, the divergence of the curl is zero,

$$\nabla \cdot (\nabla \times \mathbf{u}) = 0. \tag{1.169}$$

Equations (1.168) and (1.169) are valid independent of the coordinate system. Their proofs are left as exercises for the reader (Problem 1.11). Other identities involving the nabla operator are given in Table 1.5.

In fluid mechanics, the *vorticity* $\boldsymbol{\omega}$ of the velocity vector $\mathbf{u}$ is defined as the curl of $\mathbf{u}$,

$$\boldsymbol{\omega} \equiv \nabla \times \mathbf{u}. \tag{1.170}$$

$$\nabla = \mathbf{i}\,\frac{\partial}{\partial x} + \mathbf{j}\,\frac{\partial}{\partial y} + \mathbf{k}\,\frac{\partial}{\partial z}$$

$$\nabla^2 = \frac{\partial^2}{\partial x^2} + \frac{\partial^2}{\partial y^2} + \frac{\partial^2}{\partial z^2}$$

$$\mathbf{u}\cdot\nabla = u_x\frac{\partial}{\partial x} + u_y\frac{\partial}{\partial y} + u_z\frac{\partial}{\partial z}$$

$$\nabla p = \frac{\partial p}{\partial x}\,\mathbf{i} + \frac{\partial p}{\partial y}\,\mathbf{j} + \frac{\partial p}{\partial z}\,\mathbf{k}$$

$$\nabla\cdot\mathbf{u} = \frac{\partial u_x}{\partial x} + \frac{\partial u_y}{\partial y} + \frac{\partial u_z}{\partial z}$$

$$\nabla\times\mathbf{u} = \left(\frac{\partial u_z}{\partial y} - \frac{\partial u_y}{\partial z}\right)\mathbf{i} + \left(\frac{\partial u_x}{\partial z} - \frac{\partial u_z}{\partial x}\right)\mathbf{j} + \left(\frac{\partial u_y}{\partial x} - \frac{\partial u_x}{\partial y}\right)\mathbf{k}$$

$$\nabla\mathbf{u} = \frac{\partial u_x}{\partial x}\,\mathbf{ii} + \frac{\partial u_y}{\partial x}\,\mathbf{ij} + \frac{\partial u_z}{\partial x}\,\mathbf{ik} + \frac{\partial u_x}{\partial y}\,\mathbf{ji}$$

$$+ \frac{\partial u_y}{\partial y}\,\mathbf{jj} + \frac{\partial u_z}{\partial y}\,\mathbf{jk} + \frac{\partial u_x}{\partial z}\,\mathbf{ki} + \frac{\partial u_y}{\partial z}\,\mathbf{kj} + \frac{\partial u_z}{\partial z}\,\mathbf{kk}$$

$$\mathbf{u}\cdot\nabla\mathbf{u} = \left(u_x\frac{\partial u_x}{\partial x} + u_y\frac{\partial u_x}{\partial y} + u_z\frac{\partial u_x}{\partial z}\right)\mathbf{i} + \left(u_x\frac{\partial u_y}{\partial x} + u_y\frac{\partial u_y}{\partial y} + u_z\frac{\partial u_y}{\partial z}\right)\mathbf{j}$$

$$+ \left(u_x\frac{\partial u_z}{\partial x} + u_y\frac{\partial u_z}{\partial y} + u_z\frac{\partial u_z}{\partial z}\right)\mathbf{k}$$

$$\nabla\cdot\boldsymbol{\tau} = \left(\frac{\partial \tau_{xx}}{\partial x} + \frac{\partial \tau_{yx}}{\partial y} + \frac{\partial \tau_{zx}}{\partial z}\right)\mathbf{i} + \left(\frac{\partial \tau_{xy}}{\partial x} + \frac{\partial \tau_{yy}}{\partial y} + \frac{\partial \tau_{zy}}{\partial z}\right)\mathbf{j}$$

$$+ \left(\frac{\partial \tau_{xz}}{\partial x} + \frac{\partial \tau_{yz}}{\partial y} + \frac{\partial \tau_{zz}}{\partial z}\right)\mathbf{k}$$

Table 1.4. *Summary of differential operators in Cartesian coordinates* (x, y, z); *p,* **u,** *and* $\boldsymbol{\tau}$ *are scalar, vector, and tensor fields, respectively.*

$$\nabla(\mathbf{u} \cdot \mathbf{v}) = (\mathbf{u} \cdot \nabla)\,\mathbf{v} + (\mathbf{v} \cdot \nabla)\,\mathbf{u} + \mathbf{u} \times (\nabla \times \mathbf{v}) + \mathbf{v} \times (\nabla \times \mathbf{u})$$

$$\nabla \cdot (f\mathbf{u}) = f\,\nabla \cdot \mathbf{u} + \mathbf{u} \cdot \nabla f$$

$$\nabla \cdot (\mathbf{u} \times \mathbf{v}) = \mathbf{v} \cdot (\nabla \times \mathbf{u}) - \mathbf{u} \cdot (\nabla \times \mathbf{v})$$

$$\nabla \cdot (\nabla \times \mathbf{u}) = 0$$

$$\nabla \times (f\mathbf{u}) = f\,\nabla \times \mathbf{u} + \nabla f \times \mathbf{u}$$

$$\nabla \times (\mathbf{u} \times \mathbf{v}) = \mathbf{u}\,\nabla \cdot \mathbf{v} - \mathbf{v}\,\nabla \cdot \mathbf{u} + (\mathbf{v} \cdot \nabla)\,\mathbf{u} - (\mathbf{u} \cdot \nabla)\,\mathbf{v}$$

$$\nabla \times (\nabla \times \mathbf{u}) = \nabla(\nabla \cdot \mathbf{u}) - \nabla^2 \mathbf{u}$$

$$\nabla \times (\nabla f) = \mathbf{0}$$

$$\nabla(\mathbf{u} \cdot \mathbf{u}) = 2\,(\mathbf{u} \cdot \nabla)\,\mathbf{u} + 2\mathbf{u} \times (\nabla \times \mathbf{u})$$

$$\nabla^2(fg) = f\,\nabla^2 g + g\,\nabla^2 f + 2\,\nabla f \cdot \nabla g$$

$$\nabla \cdot (\nabla f \times \nabla g) = 0$$

$$\nabla \cdot (f\,\nabla g - g\,\nabla f) = f\,\nabla^2 g - g\,\nabla^2 f$$

Table 1.5. *Useful identities involving the nabla operator; f and g are scalar fields, and $\mathbf{u}$ and $\mathbf{v}$ are vector fields. It is assumed that all the partial derivatives involved are continuous.*

$$\nabla = \mathbf{e}_r \frac{\partial}{\partial r} + \mathbf{e}_\theta \frac{1}{r}\frac{\partial}{\partial \theta} + \mathbf{e}_z \frac{\partial}{\partial z}$$

$$\nabla^2 = \frac{1}{r}\frac{\partial}{\partial r}\left(r\frac{\partial}{\partial r}\right) + \frac{1}{r^2}\frac{\partial^2}{\partial \theta^2} + \frac{\partial^2}{\partial z^2}$$

$$\mathbf{u}\cdot\nabla = u_r\frac{\partial}{\partial r} + \frac{u_\theta}{r}\frac{\partial}{\partial \theta} + u_z\frac{\partial}{\partial z}$$

$$\nabla p = \frac{\partial p}{\partial r}\mathbf{e}_r + \frac{1}{r}\frac{\partial p}{\partial \theta}\mathbf{e}_\theta + \frac{\partial p}{\partial z}\mathbf{e}_z$$

$$\nabla\cdot\mathbf{u} = \frac{1}{r}\frac{\partial}{\partial r}(ru_r) + \frac{1}{r}\frac{\partial u_\theta}{\partial \theta} + \frac{\partial u_z}{\partial z}$$

$$\nabla\times\mathbf{u} = \left(\frac{1}{r}\frac{\partial u_z}{\partial \theta} - \frac{\partial u_\theta}{\partial z}\right)\mathbf{e}_r + \left(\frac{\partial u_r}{\partial z} - \frac{\partial u_z}{\partial r}\right)\mathbf{e}_\theta + \left[\frac{1}{r}\frac{\partial}{\partial r}(ru_\theta) - \frac{1}{r}\frac{\partial u_r}{\partial \theta}\right]\mathbf{e}_z$$

$$\nabla\mathbf{u} = \frac{\partial u_r}{\partial r}\mathbf{e}_r\mathbf{e}_r + \frac{\partial u_\theta}{\partial r}\mathbf{e}_r\mathbf{e}_\theta + \frac{\partial u_z}{\partial r}\mathbf{e}_r\mathbf{e}_z + \left(\frac{1}{r}\frac{\partial u_r}{\partial \theta} - \frac{u_\theta}{r}\right)\mathbf{e}_\theta\mathbf{e}_r$$

$$+ \left(\frac{1}{r}\frac{\partial u_\theta}{\partial \theta} + \frac{u_r}{r}\right)\mathbf{e}_\theta\mathbf{e}_\theta + \frac{1}{r}\frac{\partial u_z}{\partial \theta}\mathbf{e}_\theta\mathbf{e}_z + \frac{\partial u_r}{\partial z}\mathbf{e}_z\mathbf{e}_r + \frac{\partial u_\theta}{\partial z}\mathbf{e}_z\mathbf{e}_\theta + \frac{\partial u_z}{\partial z}\mathbf{e}_z\mathbf{e}_z$$

$$\mathbf{u}\cdot\nabla\mathbf{u} = \left[u_r\frac{\partial u_r}{\partial r} + u_\theta\left(\frac{1}{r}\frac{\partial u_r}{\partial \theta} - \frac{u_\theta}{r}\right) + u_z\frac{\partial u_r}{\partial z}\right]\mathbf{e}_r$$

$$+ \left[u_r\frac{\partial u_\theta}{\partial r} + u_\theta\left(\frac{1}{r}\frac{\partial u_\theta}{\partial \theta} + \frac{u_r}{r}\right) + u_z\frac{\partial u_\theta}{\partial z}\right]\mathbf{e}_\theta$$

$$+ \left[u_r\frac{\partial u_z}{\partial r} + u_\theta\frac{1}{r}\frac{\partial u_z}{\partial \theta} + u_z\frac{\partial u_z}{\partial z}\right]\mathbf{e}_z$$

$$\nabla\cdot\boldsymbol{\tau} = \left[\frac{1}{r}\frac{\partial}{\partial r}(r\tau_{rr}) + \frac{1}{r}\frac{\partial \tau_{\theta r}}{\partial \theta} + \frac{\partial \tau_{zr}}{\partial z} - \frac{\tau_{\theta\theta}}{r}\right]\mathbf{e}_r$$

$$+ \left[\frac{1}{r^2}\frac{\partial}{\partial r}(r^2\tau_{r\theta}) + \frac{1}{r}\frac{\partial \tau_{\theta\theta}}{\partial \theta} + \frac{\partial \tau_{z\theta}}{\partial z} - \frac{\tau_{\theta r} - \tau_{r\theta}}{r}\right]\mathbf{e}_\theta$$

$$+ \left[\frac{1}{r}\frac{\partial}{\partial r}(r\tau_{rz}) + \frac{1}{r}\frac{\partial \tau_{\theta z}}{\partial \theta} + \frac{\partial \tau_{zz}}{\partial z}\right]\mathbf{e}_z$$

Table 1.6. *Summary of differential operators in cylindrical polar coordinates (r, θ, z); p, $\mathbf{u}$, and $\boldsymbol{\tau}$ are scalar, vector, and tensor fields, respectively.*

$$\nabla = \mathbf{e}_r \frac{\partial}{\partial r} + \mathbf{e}_\theta \frac{1}{r}\frac{\partial}{\partial \theta} + \mathbf{e}_\phi \frac{1}{r\sin\theta}\frac{\partial}{\partial \phi}$$

$$\nabla^2 = \frac{1}{r^2}\frac{\partial}{\partial r}\left(r^2\frac{\partial}{\partial r}\right) + \frac{1}{r^2\sin\theta}\frac{\partial}{\partial \theta}\left(\sin\theta\frac{\partial}{\partial \theta}\right) + \frac{1}{r^2\sin^2\theta}\frac{\partial^2}{\partial \phi^2}$$

$$\mathbf{u}\cdot\nabla = u_r\frac{\partial}{\partial r} + \frac{u_\theta}{r}\frac{\partial}{\partial \theta} + \frac{u_\phi}{r\sin\theta}\frac{\partial}{\partial \phi}$$

$$\nabla p = \frac{\partial p}{\partial r}\mathbf{e}_r + \frac{1}{r}\frac{\partial p}{\partial \theta}\mathbf{e}_\theta + \frac{1}{r\sin\theta}\frac{\partial p}{\partial \phi}\mathbf{e}_\phi$$

$$\nabla\cdot\mathbf{u} = \frac{1}{r^2}\frac{\partial}{\partial r}(r^2 u_r) + \frac{1}{r\sin\theta}\frac{\partial}{\partial \theta}(u_\theta\sin\theta) + \frac{1}{r\sin\theta}\frac{\partial u_\phi}{\partial \phi}$$

$$\nabla\times\mathbf{u} = [\frac{1}{r\sin\theta}\frac{\partial}{\partial \theta}(u_\phi\sin\theta) - \frac{1}{r\sin\theta}\frac{\partial u_\theta}{\partial \phi}]\mathbf{e}_r + [\frac{1}{r\sin\theta}\frac{\partial u_r}{\partial \phi} - \frac{1}{r}\frac{\partial}{\partial r}(ru_\phi)]\mathbf{e}_\theta$$
$$+[\frac{1}{r}\frac{\partial}{\partial r}(ru_\theta) - \frac{1}{r}\frac{\partial u_r}{\partial \theta}]\mathbf{e}_\phi$$

$$\nabla\mathbf{u} = \frac{\partial u_r}{\partial r}\mathbf{e}_r\mathbf{e}_r + \frac{\partial u_\theta}{\partial r}\mathbf{e}_r\mathbf{e}_\theta + \frac{\partial u_\phi}{\partial r}\mathbf{e}_r\mathbf{e}_\phi + \left(\frac{1}{r}\frac{\partial u_r}{\partial \theta} - \frac{u_\theta}{r}\right)\mathbf{e}_\theta\mathbf{e}_r$$
$$+\left(\frac{1}{r}\frac{\partial u_\theta}{\partial \theta} + \frac{u_r}{r}\right)\mathbf{e}_\theta\mathbf{e}_\theta + \frac{1}{r}\frac{\partial u_\phi}{\partial \theta}\mathbf{e}_\theta\mathbf{e}_\phi + \left(\frac{1}{r\sin\theta}\frac{\partial u_r}{\partial \phi} - \frac{u_\phi}{r}\right)\mathbf{e}_\phi\mathbf{e}_r$$
$$+\left(\frac{1}{r\sin\theta}\frac{\partial u_\theta}{\partial \phi} - \frac{u_\phi}{r}\cot\theta\right)\mathbf{e}_\phi\mathbf{e}_\theta + \left(\frac{1}{r\sin\theta}\frac{\partial u_\phi}{\partial \phi} + \frac{u_r}{r} + \frac{u_\theta}{r}\cot\theta\right)\mathbf{e}_\phi\mathbf{e}_\phi$$

$$\mathbf{u}\cdot\nabla\mathbf{u} = [u_r\frac{\partial u_r}{\partial r} + u_\theta\left(\frac{1}{r}\frac{\partial u_r}{\partial \theta} - \frac{u_\theta}{r}\right) + u_\phi\left(\frac{1}{r\sin\theta}\frac{\partial u_r}{\partial \phi} - \frac{u_\phi}{r}\right)]\mathbf{e}_r$$
$$+[u_r\frac{\partial u_\theta}{\partial r} + u_\theta\left(\frac{1}{r}\frac{\partial u_\theta}{\partial \theta} + \frac{u_r}{r}\right) + u_\phi\left(\frac{1}{r\sin\theta}\frac{\partial u_\theta}{\partial \phi} - \frac{u_\phi}{r}\cot\theta\right)]\mathbf{e}_\theta$$
$$+[u_r\frac{\partial u_\phi}{\partial r} + u_\theta\frac{1}{r}\frac{\partial u_\phi}{\partial \theta} + u_\phi\left(\frac{1}{r\sin\theta}\frac{\partial u_\phi}{\partial \phi} + \frac{u_r}{r} + \frac{u_\theta}{r}\cot\theta\right)]\mathbf{e}_\phi$$

$$\nabla\cdot\boldsymbol{\tau} = [\frac{1}{r^2}\frac{\partial}{\partial r}(r^2\tau_{rr}) + \frac{1}{r\sin\theta}\frac{\partial}{\partial \theta}(\tau_{\theta r}\sin\theta) + \frac{1}{r\sin\theta}\frac{\partial \tau_{\phi r}}{\partial \phi} - \frac{\tau_{\theta\theta} + \tau_{\phi\phi}}{r}]\mathbf{e}_r$$
$$+[\frac{1}{r^3}\frac{\partial}{\partial r}(r^3\tau_{r\theta}) + \frac{1}{r\sin\theta}\frac{\partial}{\partial \theta}(\tau_{\theta\theta}\sin\theta) + \frac{1}{r\sin\theta}\frac{\partial \tau_{\phi\theta}}{\partial \phi} + \frac{\tau_{\theta r} - \tau_{r\theta} - \tau_{\phi\phi}\cot\theta}{r}]\mathbf{e}_\theta$$
$$+[\frac{1}{r^3}\frac{\partial}{\partial r}(r^3\tau_{r\phi}) + \frac{1}{r\sin\theta}\frac{\partial}{\partial \theta}(\tau_{\theta\phi}\sin\theta) + \frac{1}{r\sin\theta}\frac{\partial \tau_{\phi\phi}}{\partial \phi} + \frac{\tau_{\phi r} - \tau_{r\phi} - \tau_{\phi\theta}\cot\theta}{r}]\mathbf{e}_\phi$$

Table 1.7. *Summary of differential operators in spherical polar coordinates* (r, θ, ϕ); *p,* **u**, *and* $\boldsymbol{\tau}$ *are scalar, vector, and tensor fields, respectively.*

Other symbols used for the vorticity, in the fluid mechanics literature, are ζ, ξ and Ω. If, in a flow, the vorticity vector is zero everywhere, then the flow is said to be *irrotational*. Otherwise, i.e., if the vorticity is not zero, at least in some regions of the flow, then the flow is said to be *rotational*. For example, if the velocity field can be expressed as the gradient of a scalar function, i.e., if $\mathbf{u}=\nabla f$, then according to Eq. (1.168),

$$\boldsymbol{\omega} \equiv \nabla \times \mathbf{u} = \nabla \times (\nabla f) = \mathbf{0},$$

and, thus, the flow is irrotational.

A vector field $\mathbf{u}$ is said to be *solenoidal* if its divergence is everywhere zero, i.e., if

$$\nabla \cdot \mathbf{u} = 0. \tag{1.171}$$

From Eq. (1.169), we deduce that the vorticity vector is solenoidal, since

$$\nabla \cdot \boldsymbol{\omega} = \nabla \cdot (\nabla \times \mathbf{u}) = 0.$$

Example 1.4.2. Physical significance of differential operators

Consider an infinitesimal volume ΔV bounded by a surface ΔS. The gradient of a scalar field f can be defined as

$$\nabla f \equiv \lim_{\Delta V \to 0} \frac{\int_{\Delta S} \mathbf{n} f \, dS}{\Delta V}, \tag{1.172}$$

where $\mathbf{n}$ is the unit vector normal to the surface ΔS. The gradient here represents the *net vector flux* of the scalar quantity f at a point where the volume ΔV of surface ΔS collapses in the limit. At that point, the above equation reduces to Eq. (1.154).

The divergence of the velocity vector $\mathbf{u}$ can be defined as

$$\nabla \cdot \mathbf{u} \equiv \lim_{\Delta V \to 0} \frac{\int_{\Delta S} (\mathbf{n} \cdot \mathbf{u}) \, dS}{\Delta V}, \tag{1.173}$$

and represents the *scalar flux* of the vector $\mathbf{u}$ at a point, which is equivalent to the *local rate of expansion* (see Example 1.5.3).

Finally, the vorticity of $\mathbf{u}$ may be defined as

$$\nabla \times \mathbf{u} \equiv \lim_{\Delta V \to 0} \frac{\int_{\Delta S} (\mathbf{n} \times \mathbf{u}) dS}{\Delta V}, \tag{1.174}$$

and represents *the vector net flux* of the scalar angular component at a point, which tends to rotate the fluid particle at the point where ΔV collapses.

$\square$

Example 1.4.3. The nabla operator in cylindrical polar coordinates

(a) Express the nabla operator

$$\nabla = \mathbf{i}\, \frac{\partial}{\partial x} + \mathbf{j}\, \frac{\partial}{\partial y} + \mathbf{k}\, \frac{\partial}{\partial z} \qquad (1.175)$$

in cylindrical polar coordinates.

(b) Determine ∇c and $\nabla \cdot \mathbf{u}$, where c is a scalar and $\mathbf{u}$ is a vector.

(c) Derive the operator $\mathbf{u} \cdot \nabla$ and the dyadic product $\nabla \mathbf{u}$ in cylindrical polar coordinates.

Solution:

(a) From Table 1.1, we have:

$$
\begin{aligned}
\mathbf{i} &= \cos\theta\, \mathbf{e}_r - \sin\theta\, \mathbf{e}_\theta \\
\mathbf{j} &= \sin\theta\, \mathbf{e}_r + \cos\theta\, \mathbf{e}_\theta \\
\mathbf{k} &= \mathbf{e}_z
\end{aligned}
$$

Therefore, we just need to convert the derivatives with respect to x, y, and z into derivatives with respect to r, θ, and z. Starting with the expressions of Table 1.1 and using the chain rule, we get:

$$
\begin{aligned}
\frac{\partial}{\partial x} &= \frac{\partial r}{\partial x}\frac{\partial}{\partial r} + \frac{\partial \theta}{\partial x}\frac{\partial}{\partial \theta} = \cos\theta\, \frac{\partial}{\partial r} - \frac{\sin\theta}{r}\, \frac{\partial}{\partial \theta} \\
\frac{\partial}{\partial y} &= \sin\theta\, \frac{\partial}{\partial r} + \frac{\cos\theta}{r}\, \frac{\partial}{\partial \theta} \\
\frac{\partial}{\partial z} &= \frac{\partial}{\partial z}
\end{aligned}
$$

Substituting now into Eq. (1.175) gives

$$
\begin{aligned}
\nabla ={}& (\cos\theta\, \mathbf{e}_r - \sin\theta\, \mathbf{e}_\theta) \left(\cos\theta\, \frac{\partial}{\partial r} - \frac{\sin\theta}{r}\, \frac{\partial}{\partial \theta} \right) \\
&+ (\sin\theta\, \mathbf{e}_r + \cos\theta\, \mathbf{e}_\theta) \left(\sin\theta\, \frac{\partial}{\partial r} + \frac{\cos\theta}{r}\, \frac{\partial}{\partial \theta} \right) + \mathbf{e}_z\, \frac{\partial}{\partial z}\, .
\end{aligned}
$$

After some simplifications and using the trigonometric identity $\sin^2\theta + \sin^2\theta = 1$, we get

$$\nabla = \mathbf{e}_r\, \frac{\partial}{\partial r} + \mathbf{e}_\theta\, \frac{1}{r}\frac{\partial}{\partial \theta} + \mathbf{e}_z\, \frac{\partial}{\partial z} \qquad (1.176)$$

(b) The gradient of the scalar c is given by

$$\nabla c = \mathbf{e}_r \frac{\partial c}{\partial r} + \mathbf{e}_\theta \frac{1}{r} \frac{\partial c}{\partial \theta} + \mathbf{e}_z \frac{\partial c}{\partial z} . \tag{1.177}$$

For the divergence of the vector $\mathbf{u}$, we have

$$\nabla \cdot \mathbf{u} = \left(\mathbf{e}_r \frac{\partial}{\partial r} + \mathbf{e}_\theta \frac{1}{r} \frac{\partial}{\partial \theta} + \mathbf{e}_z \frac{\partial}{\partial z} \right) \cdot (u_r \mathbf{e}_r + u_\theta \mathbf{e}_\theta + u_z \mathbf{e}_z) .$$

Noting that the only nonzero spatial derivatives of the unit vectors are

$$\frac{\partial \mathbf{e}_r}{\partial \theta} = \mathbf{e}_\theta \quad \text{and} \quad \frac{\partial \mathbf{e}_\theta}{\partial \theta} = -\mathbf{e}_r$$

(see Eq. 1.17), we obtain

$$\begin{aligned}
\nabla \cdot \mathbf{u} &= \frac{\partial u_r}{\partial r} + \mathbf{e}_\theta \cdot \frac{1}{r} \left(u_r \frac{\partial \mathbf{e}_r}{\partial \theta} + \frac{\partial u_\theta}{\partial \theta} \mathbf{e}_\theta + u_\theta \frac{\partial \mathbf{e}_\theta}{\partial \theta} \right) + \frac{\partial u_z}{\partial z} \\
&= \frac{\partial u_r}{\partial r} + \frac{1}{r} \frac{\partial u_\theta}{\partial \theta} + \mathbf{e}_\theta \cdot \frac{1}{r} (u_r \mathbf{e}_\theta - u_\theta \mathbf{e}_r) + \frac{\partial u_z}{\partial z} \\
&= \frac{\partial u_r}{\partial r} + \frac{1}{r} \frac{\partial u_\theta}{\partial \theta} + \frac{u_r}{r} + \frac{\partial u_z}{\partial z} \qquad \Longrightarrow
\end{aligned}$$

$$\nabla \cdot \mathbf{u} = \frac{1}{r} \frac{\partial}{\partial r} (r u_r) + \frac{1}{r} \frac{\partial u_\theta}{\partial \theta} + \frac{\partial u_z}{\partial z} . \tag{1.178}$$

(c)

$$\mathbf{u} \cdot \nabla = (u_r \mathbf{e}_r + u_\theta \mathbf{e}_\theta + u_z \mathbf{e}_z) \left(\mathbf{e}_r \frac{\partial}{\partial r} + \mathbf{e}_\theta \frac{1}{r} \frac{\partial}{\partial \theta} + \mathbf{e}_z \frac{\partial}{\partial z} \right) \qquad \Longrightarrow$$

$$\mathbf{u} \cdot \nabla = u_r \frac{\partial}{\partial r} + \frac{u_\theta}{r} \frac{\partial}{\partial \theta} + u_z \frac{\partial}{\partial z} . \tag{1.179}$$

Finally, for the dyadic product $\nabla \mathbf{u}$ we have

$$\begin{aligned}
\nabla \mathbf{u} &= \left(\mathbf{e}_r \frac{\partial}{\partial r} + \mathbf{e}_\theta \frac{1}{r} \frac{\partial}{\partial \theta} + \mathbf{e}_z \frac{\partial}{\partial z} \right) (u_r \mathbf{e}_r + u_\theta \mathbf{e}_\theta + u_z \mathbf{e}_z) \\
&= \mathbf{e}_r \mathbf{e}_r \frac{\partial u_r}{\partial r} + \mathbf{e}_r \mathbf{e}_\theta \frac{\partial u_\theta}{\partial r} + \mathbf{e}_r \mathbf{e}_z \frac{\partial u_z}{\partial r} \\
&\quad + \mathbf{e}_\theta \mathbf{e}_r \frac{1}{r} \frac{\partial u_r}{\partial \theta} + \mathbf{e}_\theta \frac{1}{r} u_r \frac{\partial \mathbf{e}_r}{\partial \theta} + \mathbf{e}_\theta \mathbf{e}_\theta \frac{1}{r} \frac{\partial \mathbf{e}_\theta}{\partial \theta} + \mathbf{e}_\theta \frac{1}{r} u_\theta \frac{\partial \mathbf{e}_\theta}{\partial \theta} + \mathbf{e}_\theta \mathbf{e}_z \frac{1}{r} \frac{\partial u_z}{\partial \theta} \\
&\quad + \mathbf{e}_z \mathbf{e}_r \frac{\partial u_r}{\partial z} + \mathbf{e}_z \mathbf{e}_\theta \frac{\partial u_\theta}{\partial z} + \mathbf{e}_z \mathbf{e}_z \frac{\partial u_z}{\partial z} \qquad \Longrightarrow
\end{aligned}$$

$$
\begin{aligned}
\nabla \mathbf{u} \;=\; & \mathbf{e}_r \mathbf{e}_r \frac{\partial u_r}{\partial r} + \mathbf{e}_r \mathbf{e}_\theta \frac{\partial u_\theta}{\partial r} + \mathbf{e}_r \mathbf{e}_z \frac{\partial u_z}{\partial r} \\
& + \mathbf{e}_\theta \mathbf{e}_r \frac{1}{r}\left(\frac{\partial u_r}{\partial \theta} - u_\theta\right) + \mathbf{e}_\theta \mathbf{e}_\theta \frac{1}{r}\left(\frac{\partial u_\theta}{\partial \theta} + u_r\right) + \mathbf{e}_\theta \mathbf{e}_z \frac{1}{r}\frac{\partial u_z}{\partial \theta} \\
& + \mathbf{e}_z \mathbf{e}_r \frac{\partial u_r}{\partial z} + \mathbf{e}_z \mathbf{e}_\theta \frac{\partial u_\theta}{\partial z} + \mathbf{e}_z \mathbf{e}_z \frac{\partial u_z}{\partial z} .
\end{aligned}
\tag{1.180}
$$

□

Any other differential operation in curvilinear coordinates is evaluated following the procedures of Example 1.4.3. In Tables 1.6 and 1.7, we provide the most important differential operations in cylindrical and spherical coordinates, respectively.

1.4.1   The Substantial Derivative

The time derivative represents the rate of change of a physical quantity experienced by an observer who can be either stationary or moving. In the case of fluid flow, a nonstationary observer may be moving exactly as a fluid particle or not. Hence, at least three different time derivatives can be defined in fluid mechanics and in transport phenomena. The classical example of fish concentration in a lake, provided in [4], is illustrative of the similarities and differences between these time derivatives. Let $c(x, y, t)$ be the fish concentration in a lake. For a stationary observer, say standing on a bridge and looking just at a spot of the lake beneath him, the time derivative is determined by the number of fish arriving and leaving the spot of observation, i.e., the total change in concentration. Thus the total time derivative is identical to the *partial derivative*,

$$
\frac{dc}{dt} = \left(\frac{\partial c}{\partial t}\right)_{x,y} ,
\tag{1.181}
$$

and is only a function of the local change of concentration. Imagine now the observer riding a boat which can move with relative velocity $\mathbf{u}^{Rel}$ *with respect to that of the water.* Hence, if $\mathbf{u}^{Boat}$ and $\mathbf{u}^{Water}$ are the velocities of the boat and the water, respectively, then

$$
\mathbf{u}^{Rel} = \mathbf{u}^{Boat} + \mathbf{u}^{Water} .
\tag{1.182}
$$

The concentration now is a function not only of the time t, but also of the position of the boat $\mathbf{r}(x, y)$. The position of the boat is a function of time and, in fact,

$$
\frac{d\mathbf{r}}{dt} = \mathbf{u}^{Rel}
\tag{1.183}
$$

and so

$$\frac{dx}{dt} = u_x^{Rel} \quad \text{and} \quad \frac{dy}{dt} = u_y^{Rel}. \tag{1.184}$$

Thus, in this case, the *total time derivative* or the change experienced by the moving observer is,

$$\frac{d}{dt}[c(t,x,y)] \equiv \left(\frac{\partial c}{\partial t}\right)_{x,y} + \left(\frac{\partial c}{\partial x}\right)_{t,y}\frac{dx}{dt} + \left(\frac{\partial c}{\partial y}\right)_{t,y}\frac{dy}{dt} =$$
$$= \left(\frac{\partial c}{\partial t}\right)_{x,y} + u_x^{Rel}\left(\frac{\partial c}{\partial x}\right)_{t,y} + u_y^{Rel}\left(\frac{\partial c}{\partial y}\right)_{t,x}. \tag{1.185}$$

Imagine now the observer turning off the engine of the boat so that $\mathbf{u}^{Boat} = 0$ and $\mathbf{u}^{Rel} = \mathbf{u}^{Water}$. Then,

$$\frac{d}{dt}[c(t,x,y)] = \left(\frac{\partial c}{\partial t}\right)_{x,y} + \left(\frac{\partial c}{\partial x}\right)_{t,y}\frac{dx}{dt} + \left(\frac{\partial c}{\partial y}\right)_{t,x}\frac{dy}{dt}$$
$$= \left(\frac{\partial c}{\partial t}\right)_{x,y} + u_x^{Water}\left(\frac{\partial c}{\partial x}\right)_{t,y} + u_y^{Water}\left(\frac{\partial c}{\partial y}\right)_{t,x}$$
$$= \frac{\partial c}{\partial t} + \mathbf{u} \cdot \nabla c.$$

This derivative is called the *substantial derivative* and is denoted by D/Dt:

$$\frac{Dc}{Dt} \equiv \frac{\partial c}{\partial t} + \mathbf{u} \cdot \nabla c. \tag{1.186}$$

(The terms *substantive, material,* or *convective* are sometimes used for the *substantial derivative.*) The substantial derivative expresses the total time change of a quantity experienced by an observer following the motion of the liquid. It consists of a local change, $\partial c/\partial t$, which vanishes under steady conditions (i.e., same number of fish arrive and leave the spot of observation), and of a traveling change, $\mathbf{u} \cdot \nabla c$, which of course is zero for a stagnant liquid or uniform concentration. Thus, for a steady-state process,

$$\frac{Dc}{Dt} = \mathbf{u} \cdot \nabla c = u_1\frac{\partial c}{\partial x_1} + u_2\frac{\partial c}{\partial x_2} + u_3\frac{\partial c}{\partial x_3}. \tag{1.187}$$

For stagnant liquid or uniform concentration,

$$\frac{Dc}{Dt} = \left(\frac{\partial c}{\partial t}\right)_{x,y,z} = \frac{dc}{dt}. \tag{1.188}$$

Example 1.4.4. Substantial derivative[10]

Let $T(x, y)$ be the surface temperature of a stationary lake. Assume that you attach a thermometer to a boat and take a path through the lake defined by $x = a(t)$ and $y = b(t)$. Find an expression for the rate of change of the thermometer temperature in terms of the lake temperature.

Solution:

$$\frac{dT(x, y)}{dt} = \left(\frac{\partial T}{\partial t}\right)_{x,y} + \left(\frac{\partial T}{\partial x}\right)_{t,y}\frac{dx}{dt} + \left(\frac{\partial T}{\partial y}\right)_{t,x}\frac{dy}{dt}$$

$$= 0 + \left(\frac{\partial T}{\partial x}\right)_y\frac{da}{dt} + \left(\frac{\partial T}{\partial y}\right)_x\frac{db}{dt} .$$

Limiting cases:

$$\text{If} \quad T(x, y) = c, \quad \text{then} \quad \frac{dT}{dt} = 0 .$$

$$\text{If} \quad T(x, y) = f(x), \quad \text{then} \quad \frac{dT}{dt} = \frac{dT}{dx}\frac{da}{dt} = \frac{df}{dx}\frac{da}{dt} .$$

$$\text{If} \quad T(x, y) = g(y), \quad \text{then} \quad \frac{dT}{dt} = \frac{dT}{dy}\frac{db}{dt} = \frac{dg}{dy}\frac{db}{dt} .$$

Notice that the local time derivative is zero because $T(x, y)$ is not a function of time.

□

The forms of the substantial derivative operator in the three coordinate systems of interest are tabulated in Table 1.8.

1.5 Integral Theorems

The Gauss or divergence theorem

The Gauss theorem is one of the most important integral theorems of vector calculus. It can be viewed as a generalization of the *fundamental theorem of calculus* which states that

$$\int_a^b \frac{d\phi}{dx}dx = \phi(b) - \phi(a) , \tag{1.189}$$

where $\phi(x)$ is a scalar, one-dimensional function which obviously must be differentiable. Equation (1.189) can also be written as follows:

$$\mathbf{i} \int_a^b \left(\frac{d\phi}{dx}\right) dx = \mathbf{i}\left[\phi(b) - \phi(a)\right] = \left[\mathbf{n}\phi(x)\right]_a^b , \tag{1.190}$$

---

[10]Taken from Ref. [6].

Coordinate system	$\dfrac{D}{Dt} \equiv \dfrac{\partial}{\partial t} + \mathbf{u} \cdot \nabla$
(x, y, z)	$\dfrac{\partial}{\partial t} + u_x \dfrac{\partial}{\partial x} + u_y \dfrac{\partial}{\partial y} + u_z \dfrac{\partial}{\partial z}$
(r, θ, z)	$\dfrac{\partial}{\partial t} + u_r \dfrac{\partial}{\partial r} + \dfrac{u_\theta}{r} \dfrac{\partial}{\partial \theta} + u_z \dfrac{\partial}{\partial z}$
(r, θ, ϕ)	$\dfrac{\partial}{\partial t} + u_r \dfrac{\partial}{\partial r} + \dfrac{u_\theta}{r} \dfrac{\partial}{\partial \theta} + \dfrac{u_\phi}{r \sin \theta} \dfrac{\partial}{\partial \phi}$

Table 1.8. *The substantial derivative operator in various coordinate systems.*

$$\int_a^b \mathbf{i} \left(\frac{d\phi}{dx} \right) dx = [\mathbf{n}\phi(x)]_a^b$$

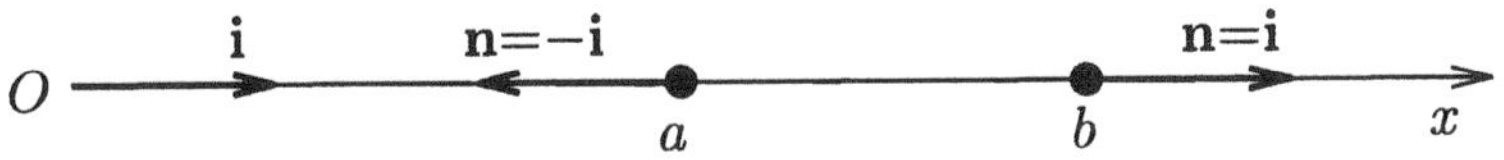

Figure 1.19. *The fundamental theorem of calculus.*

where $\mathbf{n}$ is the unit vector pointing outwards from the one-dimensional interval of integration, $a \leq x \leq b$, as shown in Fig. 1.19.

Equation (1.190) can be extended to two dimensions as follows. Consider the square S defined by $a \leq x \leq b$ and $c \leq y \leq d$, and a function $\phi(x, y)$ with continuous first partial derivatives. Then

$$
\begin{aligned}
\int_S \nabla \phi \, dS &= \int_c^d \int_a^b \left(\mathbf{i} \frac{\partial \phi}{\partial x} + \mathbf{j} \frac{\partial \phi}{\partial y} \right) dx dy = \mathbf{i} \int_c^d \int_a^b \frac{\partial \phi}{\partial x} dx dy + \mathbf{j} \int_c^d \int_a^b \frac{\partial \phi}{\partial y} dx dy \\
&= \mathbf{i} \int_c^d [\phi(b, y) - \phi(a, y)] dy + \mathbf{j} \int_a^b [\phi(x, d) - \phi(x, c)] dx \\
&= \int_c^d [\mathbf{n}\phi(x, y)]_a^b dy + \int_a^b [\mathbf{n}\phi(x, y)]_c^d dx \quad \Longrightarrow
\end{aligned}
$$

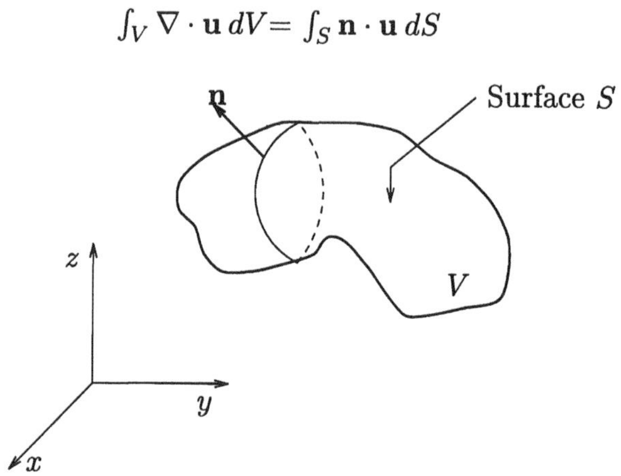

$$\int_V \nabla \cdot \mathbf{u} \, dV = \int_S \mathbf{n} \cdot \mathbf{u} \, dS$$

Figure 1.20. *The Gauss or divergence theorem.*

$$\int_S \nabla \phi \, dS = \int_C \mathbf{n} \, \phi \, d\ell \,, \qquad (1.191)$$

where $\mathbf{n}$ is the outward unit normal to the boundary C of S, and ℓ is the arc length around C. Note that Eq. (1.191) is valid for any surface S on the plane bounded by a curve C. Similarly, if V is an arbitrary closed region bounded by a surface S, and $\phi(x, y, z)$ is a scalar function with continuous first partial derivatives, one gets:

$$\int_V \nabla \phi \, dV = \int_S \mathbf{n} \, \phi \, dS \,, \qquad (1.192)$$

where $\mathbf{n}$ is the unit normal pointing outward from the surface S, as depicted in Fig. 1.20. Equation (1.192) is known as the *Gauss* or *divergence theorem*. The Gauss theorem holds not only for tensor fields of zeroth order (i.e., scalar fields), but also for tensors of higher order (i.e., vector and second-order tensor fields). If $\mathbf{u}$ and $\boldsymbol{\tau}$ are vector and tensor fields, respectively, with continuous first partial derivatives, the Gauss theorem takes the following forms:

$$\int_V \nabla \cdot \mathbf{u} \, dV = \int_S \mathbf{n} \cdot \mathbf{u} \, dS \,, \qquad (1.193)$$

and

$$\int_V \nabla \cdot \boldsymbol{\tau} \, dV = \int_S \mathbf{n} \cdot \boldsymbol{\tau} \, dS \,. \qquad (1.194)$$

In words, the Gauss theorem states that the volume integral of the divergence of a vector or tensor field over an arbitrary control volume V is equal to the flow rate of

$$\int_S \mathbf{n} \cdot (\nabla \times \mathbf{u})\, dS = \oint_C \mathbf{t} \cdot \mathbf{u}\, d\ell$$

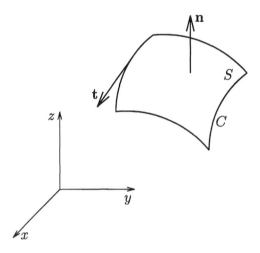

Figure 1.21. *The Stokes theorem.*

the field across the surface S bounding the domain V. If a vector field $\mathbf{u}$ happens to be solenoidal, $\nabla \cdot \mathbf{u} = 0$ and, hence, the flow rate of $\mathbf{u}$ across S is zero:

$$\int_S \mathbf{n} \cdot \mathbf{u}\, dS = 0 \,.$$

The Stokes theorem

Consider a surface S bounded by a closed curve C, and designate one of its sides as the outside. At any point of the outside, we define the unit normal $\mathbf{n}$ to point outwards; thus, $\mathbf{n}$ does not cross the surface S. Let us also assume that the unit tangent $\mathbf{t}$ to the boundary C is directed in such a way that the surface S is always on the left (Fig. 1.21). In this case, the surface S is said to be *oriented* according to the right-handed convention. The *Stokes theorem* states that the flow rate of the vorticity, $\nabla \times \mathbf{u}$, of a differentiable vector field $\mathbf{u}$ through S is equal to the circulation of $\mathbf{u}$ along the boundary C of S:

$$\int_S \mathbf{n} \cdot (\nabla \times \mathbf{u})\, dS = \oint_C \mathbf{t} \cdot \mathbf{u}\, d\ell \,. \tag{1.195}$$

Another form of the Stokes theorem is

$$\int_S (\nabla \times \mathbf{u}) \cdot d\mathbf{S} = \oint_C \mathbf{u} \cdot d\mathbf{r} \,, \tag{1.196}$$

where $d\mathbf{S}=\mathbf{n}dS$, $d\mathbf{r}=\mathbf{t}d\ell$, and $\mathbf{r}$ is the position vector.

One notices that the Gauss theorem expresses the volume integral of a differentiated quantity in terms of a surface integral which does not involve differentiation. Similarly, the Stokes theorem transforms a surface integral to a line integral eliminating the differential operator. The analogy with the fundamental theorem of calculus in Eq. (1.189) is obvious.

In the special case $\nabla \times \mathbf{u}=\mathbf{0}$, Eq. (1.196) indicates that the circulation of $\mathbf{u}$ is zero:

$$\oint_C \mathbf{u} \cdot d\mathbf{r} = 0 \,. \tag{1.197}$$

If $\mathbf{u}$ represents a force field which acts on one object, Eq. (1.197) implies that the work done in moving the object from one point to another is independent of the path joining the two points. Such a force field is called *conservative*. The necessary and sufficient condition for a force field to be conservative is $\nabla \times \mathbf{u}=\mathbf{0}$.

Example 1.5.1. Green's identities

Consider the vector field $\phi\nabla\psi$, where ϕ and ψ are scalar functions with continuous second partial derivatives. Applying the Gauss theorem, we get

$$\int_V \nabla \cdot (\phi\nabla\psi) \, dV = \int_S (\phi\nabla\psi) \cdot \mathbf{n} \, dS \,.$$

Using the identity

$$\nabla \cdot (\phi\nabla\psi) = \phi\nabla^2\psi + \nabla\phi \cdot \nabla\psi \,,$$

we derive *Green's first identity*:

$$\int_V \left(\phi\nabla^2\psi + \nabla\phi \cdot \nabla\psi \right) dV = \int_S (\phi\nabla\psi) \cdot \mathbf{n} \, dS \,. \tag{1.198}$$

Interchanging ϕ with ψ and subtracting the resulting new relation from the above equation yield *Green's second identity*:

$$\int_V \left(\phi\nabla^2\psi - \psi\nabla^2\phi \right) dV = \int_S (\phi\nabla\psi - \psi\nabla\phi) \cdot \mathbf{n} \, dS \,. \tag{1.199}$$

□

The Reynolds transport theorem

Consider a function $f(x,t)$ involving a parameter t. The derivative of the definite integral of $f(x,t)$ from $x=a(t)$ to $x=b(t)$ with respect to t is given by *Leibnitz's formula*:

$$\frac{d}{dt} \int_{x=a(t)}^{x=b(t)} f(x,t) \, dx = \int_{a(t)}^{b(t)} \frac{\partial f}{\partial t} \, dx + f(b,t)\frac{db}{dt} - f(a,t)\frac{da}{dt} \,. \tag{1.200}$$

In many cases, the parameter t can be viewed as the time. In such a case, the limits of integration a and b are functions of time moving with velocities $\frac{da}{dt}$ and $\frac{db}{dt}$, respectively. Therefore, another way to write Eq. (1.200) is

$$\mathbf{i}\, \frac{d}{dt} \int_{x=a(t)}^{x=b(t)} f(x,t)\, dx \;=\; \mathbf{i} \int_{a(t)}^{b(t)} \frac{\partial f}{\partial t}\, dx \;+\; [\mathbf{n} \cdot (f\mathbf{u})]_{a(t)}^{b(t)}\,, \qquad (1.201)$$

where $\mathbf{n}$ is the unit vector pointing outwards from the one-dimensional interval of integration, and $\mathbf{u}$ denotes the velocity of the endpoints.

The generalization of Eq. (1.201) in the three-dimensional space is provided by the *Reynolds Transport Theorem*. If $V(t)$ is a closed three-dimensional region bounded by a surface $S(t)$ moving with velocity $\mathbf{u}$, $\mathbf{r}$ is the position vector, and $f(\mathbf{r}, t)$ is a scalar function, then

$$\frac{d}{dt} \int_{V(t)} f(\mathbf{r}, t)\, dV \;=\; \int_{V(t)} \frac{\partial f}{\partial t}\, dV \;+\; \int_{S(t)} \mathbf{n} \cdot (f\mathbf{u})\, dS\,. \qquad (1.202)$$

The theorem is valid for vectorial and tensorial fields as well. If the boundary is fixed, $\mathbf{u}=0$, and the surface integral of Eq. (1.202) is zero. In this case, the theorem simply says that one can interchange the order of differentiation and integration.

Example 1.5.2. Conservation of mass

Assume that a balloon containing a certain amount of a gas moves in the air and is deformed as it moves. The mass m of the gas is then given by

$$m \;=\; \int_{V(t)} \rho\, dV\,,$$

where $V(t)$ is the region occupied by the balloon at time t, and ρ is the density of the gas. Since the mass of the gas contained in the balloon is constant,

$$\frac{dm}{dt} \;=\; \frac{d}{dt} \int_{V(t)} \rho\, dV \;=\; 0.$$

From the Reynolds transport theorem, we get:

$$\int_{V(t)} \frac{\partial \rho}{\partial t} dV \;+\; \int_{S(t)} \mathbf{n} \cdot (\rho\mathbf{u})\, dS \;=\; 0\,,$$

where $\mathbf{u}$ is the velocity of the gas, and $S(t)$ is the surface of the balloon. The surface integral is transformed to a volume one by means of the Gauss theorem to give:

$$\int_{V(t)} \frac{\partial \rho}{\partial t}\, dV \;+\; \int_{V(t)} \nabla \cdot (\rho\mathbf{u})\, dV \;=\; 0 \qquad \Longrightarrow$$

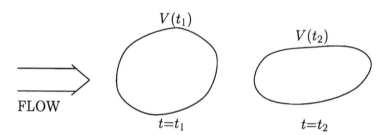

Figure 1.22. *A control volume $V(t)$ moving with the fluid.*

$$\int_{V(t)} \left[\frac{\partial \rho}{\partial t} + \nabla \cdot (\rho \mathbf{u}) \right] dV = 0 .$$

Since the above result is true for any arbitrary volume $V(t)$,

$$\frac{\partial \rho}{\partial t} + \nabla \cdot (\rho \mathbf{u}) = 0 . \tag{1.203}$$

This is the well known *continuity equation* resulting from the conservation of mass of the gas. This equation is valid for both *compressible* and *incompressible* fluids. If the fluid is incompressible, then ρ=const., and Eq. (1.203) is reduced to

$$\nabla \cdot \mathbf{u} = 0 . \tag{1.204}$$

□

Example 1.5.3. Local rate of expansion

Consider an imaginary three-dimensional region $V(t)$ containing a certain amount of fluid and moving together with the fluid, as illustrated in Fig. 1.22. Such a region is called a moving *control volume* (see Chapter 2). As the balloon in the previous example, the size and the shape of the control volume may change depending on the flow. We shall show that the local rate of expansion (or contraction) of the fluid per unit volume is equal to the divergence of the velocity field.

Applying the Reynolds transport theorem with f=1, we find

$$\frac{d}{dt} \int_{V(t)} dV = 0 + \int_{(t)} \mathbf{n} \cdot \mathbf{u} \, dS \quad \Longrightarrow$$

$$\frac{dV(t)}{dt} = \int_{S(t)} \mathbf{n} \cdot \mathbf{u} \, dS . \tag{1.205}$$

By means of the Gauss theorem, Eq. (1.205) becomes

$$\frac{dV(t)}{dt} = \int_{V(t)} \nabla \cdot \mathbf{u} \, dV \, . \tag{1.206}$$

Using now the mean-value theorem for integrals, we obtain

$$\frac{1}{V(t)} \frac{dV(t)}{dt} = \frac{1}{V(t)} \nabla \cdot \mathbf{u}\big|_{\mathbf{r}^*} \, , \tag{1.207}$$

where $\mathbf{r}^*$ is a point within $V(t)$. Taking the limit as $V(t) \to 0$, i.e., allowing $V(t)$ to shrink to a specific point, we find that

$$\lim_{V(t) \to 0} \frac{1}{V(t)} \frac{dV(t)}{dt} = \nabla \cdot \mathbf{u} \, , \tag{1.208}$$

where $\nabla \cdot \mathbf{u}$ is evaluated at the point in question. This result provides a physical interpretation for the divergence of the velocity vector as the *local rate of expansion* or *rate of dilatation* of the fluid. This rate is, of course, zero for incompressible fluids.   □

1.6   Problems

1.1. The vector $\mathbf{v}$ has the representation $\mathbf{v} = (x^2 + y^2)\,\mathbf{i} + xy\,\mathbf{j} + \mathbf{k}$ in Cartesian coordinates. Find the representation of $\mathbf{v}$ in cylindrical coordinates that share the same origin.

1.2. Sketch the vector $\mathbf{u} = 3\,\mathbf{i} + 6\,\mathbf{j}$ with respect to the Cartesian system. Find the dot products of $\mathbf{u}$ with the two basis vectors $\mathbf{i}$ and $\mathbf{j}$ and compare them with its components. Then, show the operation which projects a two-dimensional vector on a basis vector and the one projecting a three-dimensional vector on each of the mutually perpendicular planes of the Cartesian system.

1.3. Prove the following identity for the vector triple product

$$\mathbf{a} \times (\mathbf{b} \times \mathbf{c}) = \mathbf{b}(\mathbf{a} \cdot \mathbf{c}) - \mathbf{c}(\mathbf{a} \cdot \mathbf{b}) \, , \tag{1.209}$$

spelled mnemonically "*abc* equals *back* minus *cab*".

1.4. Find the representation of $\mathbf{u} = u_x\,\mathbf{i} + u_y\,\mathbf{j}$ with respect to a new Cartesian system that shares the same origin but at angle θ with respect to the original one. This *rotation* can be represented by

$$\mathbf{u}' = \mathbf{A} \cdot \mathbf{u} \, , \tag{1.210}$$

where $\mathbf{u}'$ is the new vector representation. What is the form of the matrix $\mathbf{A}$? Repeat for a new Cartesian system *translated* at a distance L from the original system. What is the matrix $\mathbf{A}$ in this case?

Show that the motions of rigid-body rotation and translation described above do not change the magnitude of a vector. Does vector orientation change with these motions?

1.5. Convert the following velocity profiles from Cartesian to cylindrical coordinates sharing the same origin, or vice versa, accordingly:
  (a)   Flow in a channel of half-width H: $\mathbf{u} = c(y^2 - H^2)\,\mathbf{i}$;
  (b)   *Stagnation flow:* $\mathbf{u} = cx\,\mathbf{i} - cx\,\mathbf{j}$;
  (c)   *Plug flow:* $\mathbf{u} = c\,\mathbf{i}$;
  (d)   Flow in a pipe of radius R: $\mathbf{u} = c(r^2 - R^2)\,\mathbf{e}_z$;
  (e)   *Sink flow:* $\mathbf{u} = \frac{c}{r}\,\mathbf{e}_r$;
  (f)   *Swirling flow:* $\mathbf{u} = cr\,\mathbf{e}_\theta$;
  (g)   *Spiral flow:* $\mathbf{u} = f(z)\,\mathbf{e}_z + wr\,\mathbf{e}_\theta$.
Note that c and ω are constants.
Hint: first, sketch the geometry of the flow and set the common origin of the two coordinate systems.

1.6. A small test membrane in a moving fluid is oriented in three directions in succession, and the tractions are measured and tabulated as follows (η is a constant):

Direction in which the test surface faces	Measured traction on the test surface (force/area)
$\mathbf{e}_1 = (\mathbf{i} + \mathbf{j})/\sqrt{2}$	$2(\eta - 1)(\mathbf{i} + \mathbf{j})$
$\mathbf{e}_2 = (\mathbf{i} - \mathbf{j})/\sqrt{2}$	$2(-\eta + 1)(\mathbf{i} - \mathbf{j})$
$\mathbf{e}_3 = \mathbf{k}$	$-\sqrt{2}\,\mathbf{k}$

(a) Establish whether the three orientations of the test surface are mutually perpendicular.

(b) Could this fluid be in a state of mechanical equilibrium? State the reason for your answer.

(c) What is the state of fluid stress at the point of measurement?

(d) Are there any *shear* stresses at the point of measurement? Indicate your reasoning.

(e) What is the stress tensor with respect to the basis $\{\mathbf{e}_1, \mathbf{e}_2, \mathbf{e}_3\}$?

1.7. Measurements of force per unit area were made on three mutually perpendicular test surfaces at point P with the following results:

Direction in which the test surface faces	Measured traction on the test surface (force/area)
i	i
j	$3\mathbf{j} - \mathbf{k}$
k	$-\mathbf{j} + 3\mathbf{k}$

(a) What is the state of stress at P?
(b) What is the traction acting on the surface with normal $\mathbf{n} = \mathbf{i} + \mathbf{j}$?
(c) What is the normal stress acting on this surface?

1.8. If $\tau = \mathbf{ii} + 3\mathbf{jj} - \mathbf{jk} - \mathbf{kj} + 3\mathbf{kk}$, or, in matrix notation,

$$\tau = \begin{bmatrix} 1 & 0 & 0 \\ 0 & 3 & -1 \\ 0 & -1 & 3 \end{bmatrix},$$

determine the invariants, and the magnitudes and directions of the principal stresses of τ. Check the values of the invariants using the principal stress magnitudes.

1.9. In an *extensional* (*stretching* or *compressing*) flow, the state of stress is fully determined by the diagonal tensor

$$\mathbf{T} = a\,\mathbf{e}_1\mathbf{e}_1 + a\,\mathbf{e}_2\mathbf{e}_2 - 2a\,\mathbf{e}_3\mathbf{e}_3 ,$$

where a is a constant.
(a) Show that there are three mutually perpendicular directions along which the resulting stresses are normal.
(b) What are the values of these stresses?
(c) How do these directions and corresponding stress values relate to the principal ones?

Consider now a *shear flow*, in which the stress tensor is given by $\mathbf{T} = -p\mathbf{I} + \tau$, where p is the pressure, and τ is an off-diagonal tensor:

$$\tau = \mathbf{e}_1\mathbf{e}_2 + 2\mathbf{e}_1\mathbf{e}_3 + 3\mathbf{e}_2\mathbf{e}_3 + \mathbf{e}_2\mathbf{e}_1 + 2\mathbf{e}_3\mathbf{e}_1 + 3\mathbf{e}_3\mathbf{e}_2 .$$

(d) What are the resulting stresses on the surfaces of orientations $\mathbf{e}_1, \mathbf{e}_2$, and $\mathbf{e}_3$?
(e) Are these orientations principal directions? If not, which are the principal directions?
(f) What are the principal values?

1.10. Consider a point at which the state of stress is given by the dyadic $\mathbf{ab} + \mathbf{ba}$, where the vectors $\mathbf{a}$ and $\mathbf{b}$ are not collinear. Let $\mathbf{i}$ be in the direction of $\mathbf{a}$ and $\mathbf{j}$ be perpendicular to $\mathbf{i}$ in the plane of $\mathbf{a}$ and $\mathbf{b}$. Let also $\mathbf{e}_\omega \equiv \mathbf{i}\cos\omega + \mathbf{j}\sin\omega$ stand for an arbitrary direction in the plane of $\mathbf{a}$ and $\mathbf{b}$.[11]

(a) Show that $\mathbf{t}(\omega) \equiv \mathbf{i}\sin\omega - \mathbf{j}\cos\omega$ is perpendicular to $\mathbf{e}_\omega$.

(b) Find expressions for the normal and shear stresses on an area element facing in the $+\mathbf{e}_\omega$ direction, in terms of ω and the x- and y-components of $\mathbf{a}$ and $\mathbf{b}$.

(c) By differentiation with respect to ω, find the directions and magnitudes of maximum and minimum normal stress. Show that these directions are perpendicular.

(d) Show that the results in (c) are the same as the eigenvectors and eigenvalues of the dyadic $\mathbf{ab} + \mathbf{ba}$ in two dimensions.

(e) Find the directions and magnitudes of maximum and minimum shear stresses. Show that the two directions are perpendicular.

1.11. If f is a scalar field and $\mathbf{u}$ is a vector field, both with continuous second partial derivatives, prove the following identities in Cartesian coordinates:

   (a) $\nabla \times \nabla f = \mathbf{0}$ (the curl of the gradient of f is zero);

   (b) $\nabla \cdot (\nabla \times \mathbf{u}) = 0$ (the divergence of the curl of $\mathbf{u}$ is zero).

1.12. Calculate the following quantities in Cartesian coordinates:

(a) The divergence $\nabla \cdot \mathbf{I}$ of the unit tensor $\mathbf{I}$.

(b) The *Newtonian stress tensor*

$$\boldsymbol{\tau} \equiv \eta\left[(\nabla\mathbf{u}) + (\nabla\mathbf{u})^T\right], \tag{1.211}$$

where η is the *viscosity*, and $\mathbf{u}$ is the velocity vector.

(c) The divergence $\nabla \cdot \boldsymbol{\tau}$ of the Newtonian stress tensor.

1.13. Prove the following identity in Cartesian coordinates:

$$\nabla \times \nabla \times \mathbf{u} = \nabla(\nabla \cdot \mathbf{u}) - \nabla^2\mathbf{u}. \tag{1.212}$$

1.14. If p is a scalar and $\mathbf{u}$ is a vector field,

(a) find the form of $\nabla \times \mathbf{u}$ in cylindrical coordinates;

(b) find ∇p and $\nabla \cdot \mathbf{u}$ in spherical coordinates.

1.15 Calculate the velocity-gradient and the vorticity tensors for the following two-dimensional flows and comment on their forms:

   (a)   *Shear flow:* $u_x = 1 - y$, $u_y = u_z = 0$;

   (b)   *Extensional flow:* $u_x = ax$, $u_y = -ay$, $u_z = 0$.

Also find the principal directions and values of both tensors. Are these related?

1.16. Derive the appropriate expression for the rate of change in fish concentration recorded by a marine biologist on a submarine traveling with velocity $\mathbf{u}^{SUB}$ *with*

---

[11]Taken from Ref. [2]

respect to the water. What is the corresponding expression when the submarine travels consistently at $z=h$ below sea level?

1.17. The concentration c of fish away from a feeding point in a lake is given by $c(x, y) = 1/(x^2 + y^2)$. Find the total change of fish concentration detected by an observer riding a boat traveling at speed $u=10$ m/sec straight away from the feeding point. What is the corresponding change detected by a stationary observer?

1.18. Calculate the velocity and the acceleration for the one-dimensional, linear motion of the position vector described by

$$\mathbf{r}(t) = \mathbf{i}\, x(t) = \mathbf{i}\, x_0 e^{at} \,,$$

with respect to an observer who
(a) is stationary at $x=x_0$;
(b) is moving with the velocity of the motion;
(c) is moving with velocity V in the same direction;
(d) is moving with velocity V in the opposite direction.
Hint: you may use the kinematic relation, $dx=u(t)dt$, to simplify things.

1.19. A parachutist falls initially with a speed of 300 km/h; once his parachute opens, his speed is reduced to 20 km/h. Determine the temperature change experienced by the parachutist in these two stages if the atmospheric temperature decreases with elevation z according to

$$T(z) = T_o - az \,,$$

where T_0 is the sea-level temperature, and $a=0.01^\circ \text{C/m}$.

1.20. The flow of an *incompressible* Newtonian fluid is governed by the *continuity* and the *momentum* equations,

$$\nabla \cdot \mathbf{u} = 0 \,, \tag{1.213}$$

and

$$\rho \left(\frac{\partial \mathbf{u}}{\partial t} + \mathbf{u} \cdot \nabla \mathbf{u} \right) \equiv \rho \frac{D\mathbf{u}}{Dt} = -\nabla p + \eta \nabla^2 \mathbf{u} + \rho \mathbf{g} \,, \tag{1.214}$$

where ρ is the density, and $\mathbf{g}$ is the gravitational acceleration. Simplify the momentum equation for irrotational flows ($\nabla \times \mathbf{u}=\mathbf{0}$). You may need to invoke both the continuity equation and vector identities to simplify the terms $\mathbf{u} \cdot \nabla \mathbf{u}$ and $\nabla^2 \mathbf{u} = \nabla \cdot (\nabla \mathbf{u})$.

1.21. By means of the Stokes theorem, examine the existence of vorticity in the following flows:

(a)   *Plug flow:* $\mathbf{u} = c\,\mathbf{i}$;

(b)   *Radial flow:* $\mathbf{u} = \frac{c}{r}\,\mathbf{e}_r$;

(c)   *Torsional flow:* $\mathbf{u} = cr\,\mathbf{e}_\theta$;

(d)   *Shear flow:* $\mathbf{u} = f(y)\,\mathbf{i}$;

(e)   *Extensional flow:* $\mathbf{u} = f(x)\,(\mathbf{i} - \mathbf{j})$.

Hint: you may use any convenient closed curve in the flow field.

1.22. Use the divergence theorem to show that

$$V = \frac{1}{3}\int_S \mathbf{n} \cdot \mathbf{r}\, dS , \qquad (1.215)$$

where S is the surface enclosing the region V, $\mathbf{n}$ is the unit normal pointing outward from S, and $\mathbf{r}$ is the position vector. Then, use Eq. (1.215) to find the volume of
(i) a rectangular parallelepiped with sides a, b, and c;
(ii) a right circular cone with height H and base radius R;
(iii) a sphere of radius R.
Use Eq. (1.215) to derive *Archimedes' principle* of buoyancy from the hydrostatic pressure on a submerged body.

1.23. Show by direct calculation that the divergence theorem does not hold for the vector field $\mathbf{u}(r,\theta,z) = \mathbf{e}_r/r$ in a cylinder of radius R and height H. Why does the theorem fail? Show that the theorem does hold for any annulus of radii R_0 and R where $0 < R_0 < R$. What restrictions must be placed on a surface so that the divergence theorem applies to a vector-valued function $\mathbf{v}(r,\theta,z)$.

1.24. Show that the Stokes theorem does not hold for $\mathbf{u} = (y\,\mathbf{i} - x\,\mathbf{j})/(x^2 + y^2)$ on a circle of radius R centered at the origin of the xy-plane. Why does the theorem fail? Show that the theorem does hold for the circular ring of radii R_0 and R, where $0 < R_0 < R$. In general, what restrictions must be placed on a closed curve so that Stokes' theorem will hold for any differentiable vector-valued function $\mathbf{v}(x,y)$?

1.25. Let C be a closed curve lying in the xy-plane and enclosing an area A, and $\mathbf{t}$ be the unit tangent to C. What condition must the differentiable vector field $\mathbf{u}$ satisfy such that

$$\oint_C \mathbf{u} \cdot \mathbf{t}\, d\ell = A ? \qquad (1.216)$$

Give some examples of vector fields having this property. Then use line integrals to find formulas for the area of rectangles, right triangles and circles. Show that the area enclosed by the plane curve C is

$$A = \frac{1}{2}\oint_C (\mathbf{r} \times \mathbf{t}) \cdot \mathbf{k}\, d\ell \qquad (1.217)$$

where $\mathbf{r}$ is the position vector and $\mathbf{k}$ is the unit vector in the z-direction.

1.7 References

1. M.R. Spiegel, *Vector Analysis and an Introduction to Tensor Analysis*, Schaum's Outline Series in Mathematics, McGraw-Hill, New York, 1959.

2. L.E. Scriven, *Fluid Mechanics Lecture Notes*, University of Minnesota, 1980.

3. G. Strang, *Linear Algebra and Its Applications*, Academic Press, Inc., Orlando, 1980.

4. R.B. Bird, W.E. Stewart, and E.N. Lightfoot, *Transport Phenomena*, John Wiley & Sons, New York, 1960.

5. H.M. Schey, *Div, Grad, Curl, and All That*, Norton and Company, New York, 1973.

6. R.L. Panton, *Incompressible Flow*, John Wiley & Sons, New York, 1996.

7. M.M. Lipschutz, *Differential Geometry*, Schaum's Outline Series in Mathematics, McGraw-Hill, New York, 1969.

8. G.E. Mase, *Theory and Problems of Continuum Mechanics*, Schaum's Outline Series in Engineering, McGraw-Hill, New York, 1970.

9. R.B. Bird, R.C. Armstrong, and O. Hassager, *Dynamics of Polymeric Liquids*, John Wiley & Sons, New York, 1987.

Chapter 2

INTRODUCTION TO THE CONTINUUM FLUID

2.1 Properties of the Continuum Fluid

A flow can be of statistical (i.e., molecular) or of continuum nature, depending on the involved length and time scales. Fluid mechanics is normally concerned with the macroscopic behavior of fluids on length scales significantly larger than the mean distance between molecules, and on time scales significantly larger than those associated with molecular vibrations. In such a case, a fluid can be approximated as a *continuum*, i.e., as a hypothetical infinitely divisible substance, and can be treated strictly by macroscopic methods. As a consequence of the *continuum hypothesis*, a fluid property is assumed to have a definite value at every point in space. This unique value is defined as the average over a very large number of molecules surrounding a given point within a small distance, which is still large compared with the mean intermolecular distance. Such a collection of molecules occupying a very small volume is called *fluid particle*. Hence, the velocity of a particle is considered equal to the mean velocity of the molecules it contains. The velocity so defined can also be considered to be the velocity of the fluid at the center of mass of the fluid particle. The continuum assumption implies that the values of the various fluid properties are continuous functions of position and of time. This assumption breaks down in rarefied gas flow where the mean free path of the molecules may be of the same order of magnitude as the physical dimensions of the flow. In this case, a microscopic or statistical approach must be used.

Properties are macroscopic, observable quantities that characterize a state. They are called *extensive* if they depend on the amount of fluid; otherwise, they are called *intensive*. Therefore, mass, weight, volume, and internal energy are extensive properties, whereas temperature, pressure, and density are intensive properties. The *temperature*, T, is a measure of thermal energy and may vary with position and time. The *pressure*, p, is also a function of position and time, defined as the limit of the ratio of the normal force, ΔF_n, acting on a surface to the area ΔA of the surface, as $\Delta A \to 0$,

$$p \equiv \lim_{\Delta A \to 0} \frac{\Delta F_n}{\Delta A} \, . \tag{2.1}$$

Hence, the pressure is a kind of *normal stress*. Similarly, the *shear stress* is defined as the limit of the tangential component of the force, ΔF_t, divided by ΔA, as $\Delta A \to 0$. Shear and normal stresses are considered in detail in Chapter 5.

Under equilibrium conditions, i.e., in a static situation, pressure results from random molecular collisions with the surface and is called *equilibrium* or *thermodynamic pressure*. Under flow conditions, i.e., in a dynamic situation, the pressure resulting from the directed molecular collisions with the surface is different from the thermodynamic pressure and is called *mechanical pressure*. The thermodynamic pressure can be determined from *equations of state* such as the *ideal gas law* for gases and the *van der Waals equation* for liquids. The mechanical pressure can be determined only by means of energy-like *conservation equations* that take into account not just the potential and the thermal energy associated with equilibrium, but also the kinetic energy associated with flow and deformation. The general relationship between thermodynamic and mechanical pressures is considered in Chapter 5.

The density

A fundamental property of continuum is the *mass density*. The density of a fluid at a point is defined as

$$\rho \equiv \lim_{\Delta V \to L^3} \left(\frac{\Delta m}{\Delta V} \right) , \tag{2.2}$$

where Δm is the mass of a very small volume ΔV surrounding the point, and L is a very small characteristic length which, however, is significantly larger than the mean distance between molecules. Density can be inverted to give the *specific volume*

$$\hat{V} \equiv \frac{1}{\rho} \, , \tag{2.3}$$

or the *molecular volume*

$$V_M \equiv \frac{\hat{V}}{M} \, , \tag{2.4}$$

where M is the *molecular weight*.

The density of a homogeneous fluid is a function of temperature T, pressure p, and molecular weight:

$$\rho = \rho(T, p, M) \, . \tag{2.5}$$

Equation (2.5) is an *equation of state* at equilibrium. An example of such an equation is the *ideal gas law*,

$$\rho = \frac{pM}{RT}, \tag{2.6}$$

where R is the *ideal gas constant* which is equal to 8314 Nm/(Kg mole K).

The density of an *incompressible fluid* is independent of the pressure. The density of a *compressible fluid* depends on the pressure and may vary in time and space, even under isothermal conditions. A measure of the change in volume and, therefore, in density of a certain mass of fluid subjected to pressure or normal forces, under constant temperature, is provided by the *isothermal compressibility* of the fluid, defined by

$$\beta \equiv -\frac{1}{V}\left(\frac{\partial V}{\partial p}\right)_T = -\left(\frac{\partial \ln V}{\partial p}\right)_T. \tag{2.7}$$

The compressibility of steel is around 5×10^{-12} m^2/N, that of water is 5×10^{-10} m^2/N, and that of air is identical to the inverse of its pressure (around 10^{-3} m^2/N at atmospheric pressure). Under isothermal conditions, solids, liquids, and gases are virtually incompressible at low pressures. Gases are compressible at moderate pressures, and their density is a strong function of pressure. Under nonisothermal conditions, all materials behave like compressible ones unless their *coefficient of thermal expansion*,

$$\alpha \equiv \left(\frac{\partial V}{\partial T}\right)_p, \tag{2.8}$$

is negligible.

Example 2.1.1. Air-density variations

The basic pressure-elevation relation of fluid statics is given by

$$\frac{dp}{dz} = -\rho g, \tag{2.9}$$

where g is the gravitational acceleration and z is the elevation. Assuming that air is an ideal gas, we can calculate the air density distribution as follows. Substituting Eq. (2.6) into Eq. (2.9), we get

$$\frac{dp}{dz} = -\frac{pMg}{RT} \implies \frac{dp}{p} = -\frac{Mg}{RT}dz.$$

If p_0 and ρ_0 denote the pressure and the density, respectively, at $z=0$, then

$$\int_{p_0}^{p(z)} \frac{dp}{p} = -\frac{Mg}{RT}\int_0^z dz \implies p = p_0 \exp\left(-\frac{Mgz}{RT}\right),$$

and

$$\rho = \rho_0 \exp\left(-\frac{Mgz}{RT}\right).$$

In reality, the temperature changes with elevation according to

$$T(z) = T_0 - az$$

where a is called the *atmospheric lapse rate* [1]. If the temperature variation is taken into account,

$$\int_{p_0}^{p(z)} \frac{dp}{p} = -\frac{Mg}{R} \int_0^z \frac{dz}{T_0 - az}$$

which yields

$$\frac{p(z)}{p_0} = \left(\frac{T_0 - az}{T_0}\right)^{\frac{Mg}{aR}}$$

and, therefore,

$$\frac{\rho(z)}{\rho_0} = \frac{p(z) T_0}{p_0 T(z)} = \left(\frac{T_0 - az}{T_0}\right)^{\frac{Mg}{aR} - 1}.$$

Thus, the density changes with elevation according to

$$\frac{1}{\rho_0}\frac{d\rho}{dz} = \left(-\frac{a}{T_0}\right)\left(\frac{Mg}{aR} - 1\right)\left(\frac{T_0 - az}{T_0}\right)^{\frac{Mg}{aR} - 1}.$$

□

The viscosity

A fluid in *static equilibrium* is under normal stress, which is the hydrostatic or thermodynamic pressure given by Eq. (2.1). As explained in Chapter 1, the total stress tensor, $\mathbf{T}$, consists of an isotropic pressure stress component, $-p\mathbf{I}$, and of an anisotropic viscous stress component, τ,

$$\mathbf{T} = -p\,\mathbf{I} + \tau. \tag{2.10}$$

The stress tensor τ comes from the relative motion of fluid particles and is zero in static equilibrium. When there is relative motion of fluid particles, the *velocity-gradient tensor*, $\nabla \mathbf{u}$, and the *rate-of-strain tensor*,

$$\mathbf{D} \equiv \frac{1}{2}[\nabla \mathbf{u} + (\nabla \mathbf{u})^T], \tag{2.11}$$

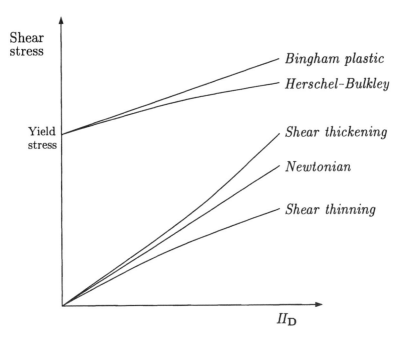

Figure 2.1. *Behavior of various non-Newtonian fluids.*

are not zero. Incompressible *Newtonian* fluids follow *Newton's law of viscosity* (discussed in detail in Chapter 5) which states that the viscous stress tensor τ is proportional to the rate-of-strain tensor,

$$\tau = 2\eta\,\mathbf{D} = \eta\left[\nabla\mathbf{u} + (\nabla\mathbf{u})^T\right] \tag{2.12}$$

or, equivalently,

$$\left[\nabla\mathbf{u} + (\nabla\mathbf{u})^T\right] = \frac{\tau}{\eta}\,. \tag{2.13}$$

The proportionality constant, η, which is a coefficient of *momentum transfer* in Eq. (2.12) and *resistance* in Eq. (2.13), is called *dynamic viscosity* or, simply, *viscosity*. The dynamic viscosity divided by density is called *kinematic viscosity* and is usually denoted by ν:

$$\nu \equiv \frac{\eta}{\rho}\,. \tag{2.14}$$

A fluid is called *ideal* or *inviscid* if its viscosity is zero; fluids of nonzero viscosity are called *viscous*. Viscous fluids not obeying Newton's law are generally called *non-Newtonian fluids*. These are classified into *generalized Newtonian* and *viscoelastic fluids.* Note that the same qualifiers are used to describe the corresponding flow, e.g., *ideal flow, Newtonian flow, viscoelastic flow* etc.

Generalized Newtonian fluids are viscous *inelastic* fluids that still follow Eq. (2.12), but the viscosity itself is a function of the rate of strain tensor **D**; more precisely, the viscosity is a function of the second invariant of **D**, $\eta = \eta(II_\mathbf{D})$. A fluid is said to be *shear thinning* if its viscosity is a decreasing function of $II_\mathbf{D}$; when the opposite is true, the fluid is said to be *shear thickening*. *Bingham-plastic fluids* are generalized Newtonian fluids that exhibit *yield stress*. The material flows only when the applied shear stress exceeds the finite yield stress. A *Herschel–Bulkley fluid* is a generalization of the Bingham fluid where, upon deformation, the viscosity is either shear thinning or shear thickening. The dependence of the shear stress on $II_\mathbf{D}$ is illustrated in Fig. 2.1, for various non-Newtonian fluids.

Fluids that have both viscous and *elastic properties* are called *viscoelastic fluids*. Many fluids of industrial importance such as polymeric liquids, solutions, melts, or suspensions fall into this category. Fluids exhibiting elastic properties are often referred to as *memory fluids*.

The field of Fluid Mechanics that studies the relation between stress and deformation, called the *constitutive equation*, is called *Rheology* from the Greek words "rheo" (to flow) and "logos" (science or logic), and is the subject of many textbooks [2,3].

The surface tension

Surface tension, σ, is a thermodynamic property which measures the anisotropy of the interactions between molecules on the interface of two immiscible fluids A and B. At equilibrium, the *capillary pressure* (i.e., the effective pressure due to surface tension) on a curved interface is balanced by the difference between the pressures in the fluids across the interface. The jump in the fluid pressure is given by the celebrated *Young–Laplace equation* of capillarity [4],

$$\Delta p = p_B - p_A = \sigma \left(\frac{1}{R_1} + \frac{1}{R_2} \right), \tag{2.15}$$

where R_1 and R_2 are the principal radii of *curvature*, i.e., the radii of the two mutually perpendicular maximum circles which are tangent to the (two-dimensional) surface at the point of contact. In Chapter 4, these important principles are expanded to include liquids in relative motion.

Example 2.1.2. Capillary pressure

A spherical liquid droplet is in static equilibrium in stationary air at low pressure p_G. How does the pressure p inside the droplet change for droplets of different radii R, for infinite, finite and zero surface tension?

Solution:
In the case of spherical droplets, $R_1=R_2=R$, and the Young–Laplace equation is reduced to

$$p - p_G = \frac{2\sigma}{R}.$$

The above formula says that the pressure within the droplet is higher than the pressure of the air. The liquid pressure increases with the surface tension and decreases with the size of the droplet. As the surface tension increases, the pressure difference can be supported by bigger liquid droplets. As the pressure difference increases, smaller droplets are formed under constant surface tension.   □

Measurement of fluid properties

The density, the viscosity, and the surface tension of pure, incompressible Newtonian liquids are functions of temperature and, to a much lesser extent, functions of pressure. These properties, blended with processing conditions, define a set of *dimensionless numbers* which fully characterize the behavior of the fluid under flow and processing. Three of the most important dimensionless numbers of fluid mechanics are briefly discussed below.

The *Reynolds number* expresses the relative magnitude of inertia forces to viscous forces and is defined by

$$Re \equiv \frac{L\bar{u}\rho}{\eta}, \tag{2.16}$$

where L is a characteristic length of the flow geometry (i.e., the diameter of a tube), and $\bar{u}$ is a characteristic velocity of the flow (e.g., the mean velocity of the fluid).

The *Stokes number* represents the relative magnitude of gravity forces to viscous forces and is defined by

$$St \equiv \frac{\rho g L^2}{\eta \bar{u}}. \tag{2.17}$$

The *capillary number* expresses the relative magnitude of viscous forces to surface tension forces, and is defined by

$$Ca \equiv \frac{\eta \bar{u}}{\sigma}. \tag{2.18}$$

The first two dimensionless numbers, Re and St, arise naturally in the dimensionless conservation of momentum equation; the third, Ca, appears in the dimensionless stress condition on a free surface. The procedure of nondimensionalizing these equations is described in Chapter 7, along with the *asymptotic analysis* which is used to construct approximate solutions for limiting values of the dimensionless numbers [5]. The governing equations of motion under these limiting conditions are simplified

Property	Fluid	p=0.1 atm		p=1 atm			p=10 atm	
		4°C	20°C	4°C	20°C	40°C	20°C	40°C
Density (Kg/m^3)	Air	0.129	0.120	1.29	1.20	1.13	12	11.3
	Water	1000	998	1000	998	992	998	992
Viscosity (cP)	Air	0.0158	0.0175	0.0165	0.0181	0.0195	0.0184	0.0198
	Water	1.792	1.001	1.792	1.002	0.656	1.002	0.657
Surface tension	Air	-	-	-	-	-	-	-
with air (dyn/cm)	Water	75.6	73	75.6	73	69.6	73	69.6

Table 2.1. *Density, viscosity and surface tension of air and water at several process conditions.*

by eliminating terms that are multiplied or divided by the limiting dimensionless numbers, accordingly.

Flows of highly viscous liquids are characterized by a vanishingly small Reynolds number and are called *Stokes* or *creeping flows*. Most flows of polymers are creeping flows [6]. The Reynolds number also serves to distinguish between *laminar* and *turbulent* flow. Laminar flows are characterized by the parallel sliding motion of adjacent fluid layers without intermixing, and persist for Reynolds numbers below a critical value that depends on the flow. For example, for flow in a pipe, this critical value is 2,100. Beyond that value, eddies start to develop within the fluid layers that cause intermixing and chaotic, oscillatory fluid motion, which characterizes turbulent flow. Laminar flows at Reynolds numbers sufficiently high that viscous effects are negligible are called *potential* or *Euler* flows. The Stokes number is zero in strictly horizontal flows and high in vertical flows of heavy liquids. The capillary number appears in flows with free surfaces and interfaces [7]. The surface tension, and thus the capillary number, can be altered by the addition of surfactants to the flowing liquids.

The knowledge of the dimensionless numbers and the prediction of the flow behavior demand an *a priori* measurement of density, viscosity and surface tension of the liquid under consideration. Density is measured by means of *pycnometers*, the function of which is primarily based on the Archimedes' principle of buoyancy. Viscosity is measured by means of *viscometers* or *rheometers* in small-scale flows; the torque necessary to drive the flow and the resulting deformation are related according to Newton's law of viscosity. Surface tension is measured by *tensiometers*. These are sensitive devices that record the force which is necessary to overcome the surface tension force in order to form droplets and bubbles or to break thin films. More sophisticated methods, usually based on optical techniques, are employed when accuracy is vital [8]. The principles of operation of pycnometers, viscometers, and tensiometers are highlighted in several chapters starting with Chapter 4. Densities, viscosities, and surface tension of air and water at several process conditions are tabulated in Table 2.1.

2.2   Macroscopic and Microscopic Balances

The *control volume* is an arbitrary *synthetic cut* in space which can be either fixed or moving. It is appropriately chosen within or around the system under consideration in order to apply the laws that describe its behavior. In flow systems, these laws are the equations of conservation (or change) of mass, momentum, and energy. To obtain information on *average* or *boundary quantities* (e.g., of the velocity and the temperature fields *inside* the flow system) without a detailed analysis of the flow, the control volume is usually taken to contain or to coincide with the real flow system. The application of the principles of conservation to this finite system produces the *macroscopic conservation equations*.

However, in order to derive the equations that yield detailed distributions of fields of interest, the control volume must be of infinitesimal dimensions that can shrink to zero, yielding a point volume. This approach reduces the quantities to point variables. The application of the conservation principles to this infinitesimal system produces the *microscopic* or *differential conservation equations*. In this case, there is generally no contact between the imaginary boundaries of the control volume and the real boundaries of the system. It is always convenient to choose the shape of the infinitesimal control volume to be similar to that of the geometry of the actual system - a cube for a rectangular geometry, an annulus for a cylindrical geometry, and a spherical shell for a spherical geometry.

Conservation of mass

Consider an arbitrary, fixed control volume V bounded by a surface S, as shown in Fig. 2.2. According to the law of conservation of mass, the rate of increase of the mass of the fluid within the control volume V is equal to the net influx of fluid across the surface S:

$$\left[\begin{array}{c} \textit{Rate of change} \\ \textit{of mass within } V \end{array} \right] = \left[\begin{array}{c} \textit{Rate of addition} \\ \textit{of mass across } S \end{array} \right] . \qquad (2.19)$$

The mass m of the fluid contained in V is given by

$$m = \int_V \rho \, dV , \qquad (2.20)$$

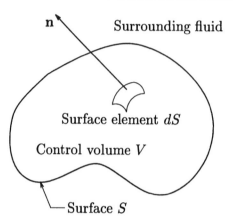

Figure 2.2. *Control volume in a flow field.*

and, hence, the rate of change in mass is

$$\frac{dm}{dt} = \frac{d}{dt} \int_V \rho \, dV \,. \tag{2.21}$$

Since the control volume V is fixed, the time derivative can be brought inside the integral:

$$\frac{dm}{dt} = \int_V \frac{\partial \rho}{\partial t} \, dV \,. \tag{2.22}$$

As for the mass rate across S, this is given by

$$-\int_S \mathbf{n} \cdot (\rho \mathbf{u}) \, dS \,,$$

where $\mathbf{n}$ is the outwardly directed unit vector normal to the surface S, and $\rho \mathbf{u}$ is the *mass flux* (i.e., mass per unit area per unit time). The minus sign accounts for the fact that the mass of the fluid contained in the control volume decreases when the flow is outward, i.e., when $\mathbf{n} \cdot (\rho \mathbf{u})$ is positive. By substituting the last expression and Eq. (2.22) in Eq. (2.19), we obtain the following form of the equation of mass conservation for a fixed control volume:

$$\frac{dm}{dt} = \int_V \frac{\partial \rho}{\partial t} \, dV = -\int_S \mathbf{n} \cdot (\rho \mathbf{u}) \, dS \,. \tag{2.23}$$

Example 2.2.1. Macroscopic balances

A reactant in water flows down the wall of a cylindrical tank in the form of thin

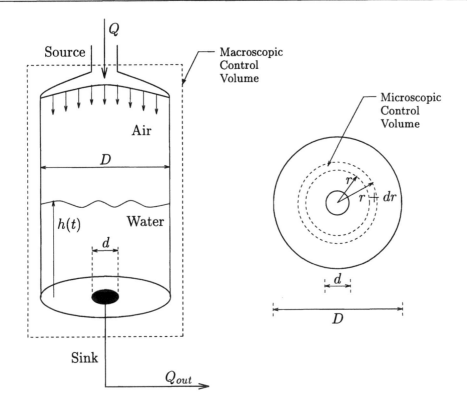

Figure 2.3. *Macroscopic and microscopic balances on a source-sink system.*

film at flow rate Q. The sink at the center of the bottom of diameter d discharges water at average velocity $\bar{u} = 2kh$, where k is a constant. Initially, the sink and the source are closed and the level of the water is h_0. What will be the level $h(t)$ after time t?

Solution:
We consider a control volume containing the flow system, as illustrated in Fig. 2.3. The rate of change in mass within the control volume is

$$\frac{dm}{dt} = \frac{d}{dt} \int_V \rho \, dV = \frac{d}{dt}(\rho V) = \rho \frac{d}{dt}\left(\frac{\pi D^2}{4}h\right) = \rho \frac{\pi D^2}{4}\frac{dh}{dt} .$$

We assume that water is incompressible. The net influx of mass across the surface S of the control volume is

$$-\int_S \mathbf{n} \cdot (\rho \mathbf{u}) \, dS = \rho \left(Q - Q_{out}\right) = \rho \left(Q - \frac{\pi d^2}{4}\bar{u}\right) = \rho \left(Q - \frac{\pi d^2}{2}kh\right) ,$$

where Q and Q_{out} are the *volumetric flow rates* at the inlet and the outlet, respectively, of the flow system (see Fig. 2.3). Therefore, the conservation of mass within the control volume gives:

$$\frac{\pi D^2}{4} \frac{dh}{dt} = Q - \frac{\pi d^2}{2} kh .$$ (2.24)

The solution to this equation, subjected to the initial condition

$$h(t = 0) = h_0 ,$$

is

$$h(t) = \frac{2Q}{\pi d^2 k} - \left(\frac{2Q}{\pi d^2 k} - h_0 \right) e^{-\left(\frac{2kd^2}{D^2} \right) t} .$$ (2.25)

The steady-state elevation is

$$h_{ss} = \lim_{t \to \infty} h(t) = \frac{2Q}{\pi d^2 k} .$$ (2.26)

Since, Eq. (2.24) is a macroscopic equation, its solution, given by Eq. (2.25), provides no information on the velocity from the wall to the sink, nor on the pressure distribution within the liquid. These questions are addressed in Example 2.2.2. □

Example 2.2.2. Microscopic balances

Assume now that the system of Example 2.2.1 is a kind of chemical reactor. Find an estimate of the *residence time* of a reactant particle (moving with the liquid) from the wall to the sink.

Solution:

The reactant flows down the vertical wall and enters the radial reacting flow at $r = D/2$ directed towards the cylindrical sink at $r = d/2$ (r is the distance from the center of the sink). If $u(r)$ is the pointwise radial velocity of the fluid, then

$$u(r) = \frac{dr}{dt} ,$$

and, therefore, the residence time of the fluid in the reaction field is given by

$$t = \int_{D/2}^{d/2} \frac{dr}{u(r)} .$$ (2.27)

Obviously, we need to calculate $u(r)$ as a function of r. The average velocity $\bar{u}$ found in Example 2.2.1 is of no use here. The velocity $u(r)$ can be found only by performing a microscopic balance. A convenient microscopic control volume is an annulus of radii r and $r + dr$ and of height dz, shown in Fig. 2.3. For this control volume, the conservation of mass states that

$$\frac{d}{dt}(\rho 2\pi r\, dr\, dz) = [2\pi r \rho u(r) dz]_{r+dr} - [2\pi r \rho u(r) dz]_r . \tag{2.28}$$

Assume, for the sake of simplicity, that the reactor operates at steady state, which means that $d/dt=0$ and $h=h_{ss}$. From Eq. (2.28), we get:

$$[ru(r)]_{r+dr} - [ru(r)]_r = 0 .$$

Dividing the above equation by dr, making the volume to shrink to zero by taking the limit as $dr \to 0$, and invoking the definition of the total derivative, we get a simple, ordinary differential equation:

$$\frac{d}{dr}[ru(r)] = \lim_{dr\to 0} \frac{[ru(r)]_{r+dr} - [ru(r)]_r}{dr} = 0 , \tag{2.29}$$

The solution of the above equation is

$$u(r) = \frac{c}{r} , \tag{2.30}$$

where c is a constant to be determined. The boundary condition at steady state demands that

$$Q = -2\pi \left. \frac{d}{2} h_{ss} u_r \right|_{r=d/2} = -\pi d h_{ss} \frac{2c}{d} \quad \Longrightarrow \quad c = -\frac{Q}{2\pi h_{ss}} .$$

The *velocity profile* is, therefore, given by

$$u(r) = -\frac{Q}{2\pi h_{ss}} \frac{1}{r} . \tag{2.31}$$

We can now substitute Eq. (2.31) in Eq. (2.27) and calculate the residence time:

$$t = -\int_{D/2}^{d/2} \frac{2\pi h_{ss}}{Q} r\, dr = \frac{\pi h_{ss}}{4Q} \left(D^2 - d^2 \right) . \tag{2.32}$$

The pressure distribution can be calculated using the *Bernoulli equation*, developed in Chapter 5. Along the radial streamline,

$$\frac{p(r)}{\rho} + \frac{u^2(r)}{2} = \left[\frac{p(r)}{\rho} + \frac{u^2(r)}{2} \right]_{r=\frac{D}{2}} . \tag{2.33}$$

For $d/D \ll 1$, it is reasonable to assume that at $r = D/2$, $u \approx 0$, and $p \approx 0$, and, therefore,

$$p(r) = -\frac{\rho}{2}u^2(r) = -\frac{\rho Q^2}{8\pi^2 h_{ss}^2}\frac{1}{r^2} < 0. \qquad (2.34)$$

Equation (2.34) predicts an increasingly negative pressure towards the sink. Under these conditions, *cavitation* and even *boiling* may occur when the pressure $p(r)$ is identical to the vapor pressure of the liquid. These phenomena, which are important in a diversity of engineering applications, cannot be predicted by macroscopic balances.   □

Conservation of linear momentum

An *isolated* solid body of mass m moving with velocity **u** possesses momentum, $\mathbf{J} \equiv m\mathbf{u}$. According to *Newton's law of motion*, the rate of change of momentum of the solid body is equal to the force **F** exerted on the mass m:

$$\frac{d\mathbf{J}}{dt} = \mathbf{F}, \quad \Longrightarrow \quad \frac{d}{dt}(m\mathbf{u}) = \mathbf{F}. \qquad (2.35)$$

The force **F** in Eq. (2.35) is a *body force*, i.e., an external force exerted on the mass m. The most common body force is the gravity force,

$$\mathbf{F}_G = m\,\mathbf{g}, \qquad (2.36)$$

which is directed to the center of the Earth (**g** is the acceleration of gravity). *Electromagnetic* forces are another kind of body force. Equation (2.35) describes the conservation of *linear* momentum of an isolated body or system:

$$\left[\begin{array}{c} \textit{Rate of change} \\ \textit{of momentum} \\ \textit{of an isolated system} \end{array} \right] = \left[\begin{array}{c} \textit{Body} \\ \textit{force} \end{array} \right]. \qquad (2.37)$$

In the case of a nonisolated flow system, i.e., a control volume V, momentum is convected across the bounding surface S due to (a) the flow of the fluid across S, and (b) the molecular motions and interactions at the boundary S. The law of conservation of momentum is then stated as follows:

$$\left[\begin{array}{c} \textit{Rate of} \\ \textit{increase of} \\ \textit{momentum} \\ \textit{within } V \end{array} \right] = \left[\begin{array}{c} \textit{Rate of} \\ \textit{inflow of} \\ \textit{momentum} \\ \textit{across } S \\ \textit{by bulk} \\ \textit{flow} \end{array} \right] + \left[\begin{array}{c} \textit{Rate of} \\ \textit{inflow of} \\ \textit{momentum} \\ \textit{across } S \\ \textit{by molecular} \\ \textit{processes} \end{array} \right] + \left[\begin{array}{c} \textit{Body} \\ \textit{force} \end{array} \right]. \qquad (2.38)$$

The momentum $\mathbf{J}$ of the fluid contained within a control volume V is given by

$$\mathbf{J} = \int_V \rho \mathbf{u} \, dV \,, \tag{2.39}$$

and, therefore,

$$\frac{d\mathbf{J}}{dt} = \frac{d}{dt} \int_V \rho \mathbf{u} \, dV \,. \tag{2.40}$$

The rate of addition of momentum due to the flow across S is

$$-\int_S \mathbf{n} \cdot (\rho \mathbf{u}) \mathbf{u} \, dS = -\int_S \mathbf{n} \cdot (\rho \mathbf{u}\mathbf{u}) \, dS \,,$$

where $\mathbf{n}$ is the unit normal pointing outwards from the surface S. The minus sign in the above expression accounts for the fact that the content of the control volume increases when the velocity vector $\mathbf{u}$ points inwards to the control volume. The dyadic tensor $\rho \mathbf{u}\mathbf{u}$ is the *momentum flux* (i.e., momentum per unit area per unit time). The momentum flux is obviously a symmetric tensor. Its component $\rho u_i u_j \mathbf{i}\mathbf{j}$ represents the j component of the momentum convected in the i direction per unit area per unit time.

The additional momentum flux due to molecular motions and interactions between the fluid and its surroundings is another symmetric tensor, the total stress tensor $\mathbf{T}$, defined in Eq. (2.10). Therefore, the rate of addition of momentum across S, due to molecular processes, is

$$\int_S \mathbf{n} \cdot \mathbf{T} \, dS = \int_S \mathbf{n} \cdot (-p\mathbf{I} + \boldsymbol{\tau}) \, dS \,. \tag{2.41}$$

As already mentioned, the anisotropic viscous stress tensor $\boldsymbol{\tau}$ accounts for the relative motion of fluid particles. In static equilibrium, the only nonzero stress contribution to the momentum flux comes from the hydrostatic pressure p. The vector $\mathbf{n} \cdot \mathbf{T}$ is the traction produced by $\mathbf{T}$ on a surface element of orientation $\mathbf{n}$. The term (2.41) is often interpreted physically as the resultant of the *surface* (or *contact*) *forces* exerted by the surrounding fluid on the fluid inside the control volume V. It is exactly the hydrodynamic force acting on the boundary S, as required by the principle of action-reaction (Newton's third law).

Assuming that the only body force acting on the fluid within the control volume V is due to gravity, i.e.,

$$\int_V \rho \mathbf{g} \, dV \,,$$

and substituting the above expressions into Eq. (2.38), we obtain the following form of the law of conservation of momentum:

$$\int_V \rho \mathbf{u} \, dV = -\int_S \mathbf{n} \cdot (\rho \mathbf{u}\mathbf{u}) \, dS + \int_S \mathbf{n} \cdot (-p\mathbf{I} + \boldsymbol{\tau}) \, dS + \int_V \rho \mathbf{g} \, dV \,. \tag{2.42}$$

The surface integrals of Eqs. (2.23) and (2.42) can be converted to volume integrals by means of the Gauss divergence theorem. As explained in Chapter 3, this step is necessary for obtaining the differential forms of the corresponding conservation equations.

2.3   Local Fluid Kinematics

Fluids cannot support any shear stress without deforming or flowing, and continue to flow as long as shear stresses persist. The effect of the externally applied shear stress is dissipated away from the boundary due to the viscosity. This gives rise to a relative motion between different fluid particles. The relative motion forces fluid material lines that join two different fluid particles to stretch (or compress) and to rotate as the two fluid particles move with different velocities. In general, the induced deformation gives rise to normal and shear stresses, similar to internal stresses developed in a stretched or twisted rubber cylinder. The difference between the two cases is that, when the externally applied forces are removed, the rubber cylinder returns to its original undeformed and unstressed state, whereas the fluid remains in its deformed state. In the field of *rheology*, it is said that rubber exhibits perfect *memory* of its rest or undeformed state, whereas viscous inelastic liquids, which include the Newtonian liquids, exhibit no memory at all. Viscoelastic materials exhibit *fading memory* and their behavior is between that of ideal elastic rubber and that of viscous inelastic liquids. These distinct behaviors are determined by the *constitutive equation*, which relates deformation to stress.

Since the conservation equations and the constitutive equation are expressed in terms of relative *kinematics*, i.e., velocities, gradients of velocities, strains, and rates of strain, it is important to choose the most convenient way to quantify these variables. The interconnection between these variables requires the investigation and representation of the relative motion of a fluid particle with respect to its neighbors.

Flow kinematics, i.e., the relative motion of fluid particles, can be described by using either a *Lagrangian* or an *Eulerian* description. In the Lagrangian or *material description*, the motion of individual particles is tracked; the position $\mathbf{r}^*$ of a *marked* fluid particle is considered to be a function of time and of its label, such as its initial position $\mathbf{r}_0^*$, $\mathbf{r}^* = \mathbf{r}^*(\mathbf{r}_0^*, t)$. For a fixed $\mathbf{r}_0^*$, we have

$$\mathbf{r}^* = \mathbf{r}^*(t) , \qquad (2.43)$$

which is a parametric equation describing the locus of the marked particle, called a *path line*. The independent variables in Lagrangian formulations are the position of a marked fluid particle and time, t. This is analogous to an observer riding a fluid particle and marking his/her position while he/she records the traveling time and other quantities of interest. For example, the pressure p in Lagrangian variables is given by $p=p(\mathbf{r}_0^*, t)$.

In the Eulerian description, dependent variables, such as the velocity vector and pressure, are considered to be functions of *fixed* spatial coordinates and of time, e.g., $\mathbf{u}=\mathbf{u}(\mathbf{r}, t)$, $p=p(\mathbf{r}, t)$, etc. If all dependent variables are independent of time, the flow is said to be *steady*.

Since both Lagrangian and Eulerian variables describe the same flow, there must be a relation between the two. This relation is expressed by the *substantial derivative* which, in the Lagrangian description, is identical to the common total derivative. The Lagrangian acceleration, $\mathbf{a}^*$, is related to the Eulerian acceleration, $\mathbf{a}=\partial\mathbf{u}/\partial t$, as follows:

$$\mathbf{a}^* = \frac{D\mathbf{u}}{Dt} = \frac{\partial\mathbf{u}}{\partial t} + \mathbf{u}\cdot\nabla\mathbf{u}\,. \tag{2.44}$$

Note that the velocity $\mathbf{u}$ in the above equation is the Eulerian one. In steady flows, the Eulerian acceleration, $\mathbf{a}=\partial\mathbf{u}/\partial t$, is zero, whereas the Lagrangian one, $\mathbf{a}^*$, may not be so if finite spatial velocity gradients exist.

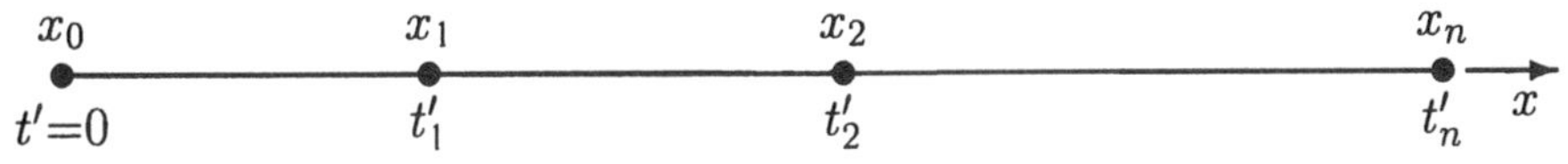

Figure 2.4. *Positions of a fluid particle in one-dimensional motion.*

We will illustrate the two flow descriptions using an idealized one-dimensional example. Consider steady motion of fluid particles along the x-axis, such that

$$x_i^* = x_{i-1}^* + c\,(t_i - t_{i-1})^2\,, \tag{2.45}$$

where x_i^* is the position of a fluid particle at time t_i (Fig. 2.4), and c is a positive constant. The Lagrangian description of motion gives the position of the particle in terms of its initial position, x_0^*, and the lapsed *traveling time*, t',

$$x^*(x_0^*, t') = x_0^* + c\,t'^2\,. \tag{2.46}$$

The velocity of the particle is

$$u^*(x_0^*, t') = \frac{dx^*}{dt'} = 2c\,t'\,, \tag{2.47}$$

which, in this case, is independent of x_0^*. The corresponding acceleration is

$$a^*(x_0^*, t') = \frac{du^*}{dt'} = 2c > 0\,. \tag{2.48}$$

The *separation distance* between two particles 1 and 2 (see Fig. 2.4),

$$\Delta x^* = x_2^* - x_1^* = c\left(t_2'^2 - t_1'^2\right),$$

(2.49)

changes with time according to

$$\frac{d\Delta x^*}{dt'} = 2c\left(t_2' - t_1'\right) = u_2^* - u_1^* > 0,$$

(2.50)

and is, therefore, continuously stretched, given that $u_2^* > u_1^*$. The velocity gradient is

$$\frac{du^*}{dx^*} = \frac{1}{u^*(t')}\frac{du^*}{dt'} = \frac{2c}{2ct'} = \frac{1}{t'}.$$

(2.51)

In the above expressions, the traveling time t' is related to the traveling distance by the simple kinematic argument,

$$dx^* = u(t')dt',$$

(2.52)

and is different from the time t which characterizes an unsteady flow under the Eulerian description.

In the Eulerian description, the primary variable is

$$u(x) = 2c^{1/2}(x - x_0)^{1/2}.$$

(2.53)

Note that time, t, does not appear due to the fact that the motion is steady. Equation (2.44) is easily verified in this steady, one-dimensional flow:

$$\frac{\partial u}{\partial t} + u\frac{\partial u}{\partial x} = 0 + 2c^{1/2}(x - x_0)^{1/2}\frac{1}{2}2c^{1/2}(x - x_0)^{-1/2} = 2c = a^*.$$

The Eulerian description may not be convenient to describe path lines but it is more appropriate than the Lagrangian description in calculating *streamlines*. These are lines to which the velocity vector is tangent at any instant. Hence, streamlines can be calculated by

$$\mathbf{u} \times d\mathbf{r} = \mathbf{0},$$

(2.54)

where $\mathbf{r}$ is the position vector describing the streamline. In Cartesian coordinates, Eq. (2.54) is reduced to

$$\frac{dx}{u_x} = \frac{dy}{u_y} = \frac{dz}{u_z}.$$

(2.55)

When the flow is steady, a path line coincides with the streamline that passes through $\mathbf{r}_0^*$. The surface formed instantaneously by all the streamlines that pass through a given closed curve in the fluid is called *streamtube*.

From Eq. (2.55), the equation of a streamline in the xy-plane is given by

$$\frac{dx}{u_x} = \frac{dy}{u_y} \quad \Longrightarrow \quad u_y\,dx - u_x\,dy = 0\,. \tag{2.56}$$

A useful concept related to streamlines, in two-dimensional, bidirectional flows, is the *stream function*. In the case of incompressible flow,[1] the stream function, $\psi(x,y)$, is defined by[2]

$$u_x = -\frac{\partial\psi}{\partial y} \quad \text{and} \quad u_y = \frac{\partial\psi}{\partial x}\,. \tag{2.57}$$

An important feature of the stream function is that it automatically satisfies the continuity equation,

$$\frac{\partial u_x}{\partial x} + \frac{\partial u_y}{\partial y} = 0\,, \tag{2.58}$$

as can easily be verified. The stream function is a useful tool in solving creeping, two-dimensional, bidirectional flows. Its definitions and use, for various classes of incompressible flow, are examined in detail in Chapter 10.

Substituting Eqs. (2.57) into Eq. (2.56), we get

$$d\psi = \frac{\partial\psi}{\partial x}dx + \frac{\partial\psi}{\partial y}dy = 0\,. \tag{2.59}$$

Therefore, the stream function, ψ, is constant along a streamline. Moreover, from the definition of a streamline, we realize that there is no flow across a streamline. The volume flow rate, Q, per unit distance in the z direction across a curve connecting two streamlines (see Fig. 2.5) is the integral of $d\psi$ along the curve. Since the

---

[1] For steady, compressible flow in the xy-plane, the stream function is defined by

$$\rho\,u_x = -\frac{\partial\psi}{\partial y} \quad \text{and} \quad \rho\,u_y = \frac{\partial\psi}{\partial x}\,.$$

In this case, the difference $\psi_2 - \psi_1$ is the mass flow rate (per unit depth) between the two streamlines.

[2] Note that many authors define the stream function with the opposite sign, i.e.,

$$u_x = \frac{\partial\psi}{\partial y} \quad \text{and} \quad u_y = -\frac{\partial\psi}{\partial x}\,.$$

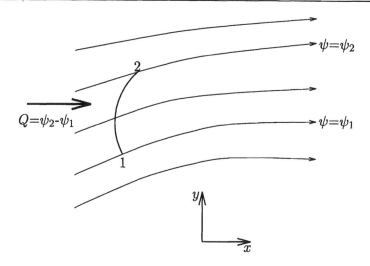

Figure 2.5. *Volume flow rate per unit depth across a curve connecting two stream-lines.*

differential of ψ is exact, this integral depends only on the end points of integration, i.e.,

$$Q = \int_1^2 d\psi = \psi_2 - \psi_1 . \tag{2.60}$$

Example 2.3.1. Stagnation flow

Consider the steady, two-dimensional *stagnation flow* against a solid wall, shown in Fig. 2.6. Outside a thin boundary layer near the wall, the position of a particle, located initially at $\mathbf{r}_0^*(x_0^*, y_0^*)$, obeys the following relations:

$$x^*(x_0^*, t') = x_0^* \, e^{\varepsilon t'} \quad \text{and} \quad y^*(y_0^*, t') = y_0^* \, e^{-\varepsilon t'} , \tag{2.61}$$

which is, of course, the Lagrangian description of the flow. The corresponding velocity components are

$$u_x^*(x_0^*, t') = \frac{dx^*}{dt'} = \varepsilon x_0^* \, e^{\varepsilon t'} \quad \text{and} \quad u_y^*(y_0^*, t') = \frac{dy^*}{dt'} = -\varepsilon y_0^* \, e^{-\varepsilon t'} . \tag{2.62}$$

Eliminating the traveling time t' from the above equations results in the equation of the path line,

$$x^* y^* = x_0^* y_0^* , \tag{2.63}$$

which is a hyperbola, in agreement with the physics of the flow.

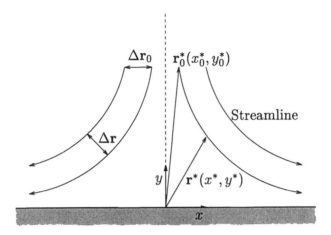

Figure 2.6. *Stagnation flow.*

In the Eulerian description, the velocity components are:

$$u_x = \varepsilon x \quad \text{and} \quad u_y = -\varepsilon y \,. \tag{2.64}$$

The streamlines of the flow are calculated by means of Eq. (2.55):

$$\frac{dx}{u_x} = \frac{dy}{u_y} \quad \Longrightarrow \quad \frac{dx}{\varepsilon x} = \frac{dy}{-\varepsilon y} \quad \Longrightarrow \quad xy = x_0 y_0 \,. \tag{2.65}$$

Equations (2.65) and (2.63) are identical: since the flow is steady, streamlines and path lines coincide.   □

The Lagrangian description is considered a more natural choice to represent the actual kinematics and stresses experienced by fluid particles. However, the use of this description in solving complex flow problems is limited due to the fact that it requires tracking of fluid particles along *a priori* unknown streamlines. The approach is particularly convenient in flows of viscoelastic liquids, i.e., of fluids with memory that require particle tracking and calculation of deformation and stresses along streamlines. The Eulerian formulation is, in general, more convenient to use because it deals only with local or present kinematics. In most cases, all variables of interest, such as strain (deformation), rate of strain, stress, vorticity, streamlines and others, can be calculated from the velocity field. An additional advantage of the Eulerian description is that it involves time as a variable only in unsteady flows, whereas the Lagrangian description uses traveling time even in steady-state flows. Finally, quantities following the motion of the liquid can be reproduced easily from the Eulerian variables by means of the substantial derivative.

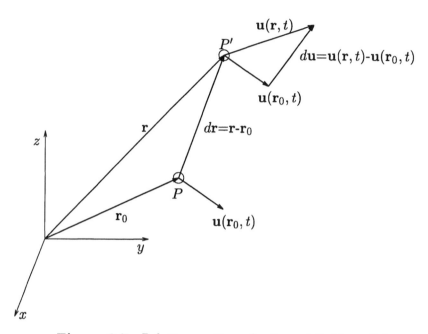

Figure 2.7. *Relative motion of adjacent fluid particles.*

2.4   Elementary Fluid Motions

The relative motion of fluid particles gives rise to velocity gradients that are directly responsible for strain (deformation). Strain, in turn, creates internal shear and extensional stresses that are quantified by the constitutive equation. Therefore, it is important to study how relative motion between fluid particles arises and how this relates to strain and stress.

Consider the adjacent fluid particles P and P' of Fig. 2.7 located at points $\mathbf{r_0}$ and $\mathbf{r}$, respectively, and assume that the distance $d\mathbf{r}=\mathbf{r}-\mathbf{r_0}$ is vanishingly small. The velocity $\mathbf{u}(\mathbf{r},t)$ of the particle P' can be *locally* decomposed into four elementary motions:

(a) *rigid-body translation*;
(b) *rigid-body rotation*;
(c) *isotropic expansion*; and
(d) *pure straining motion without change of volume.*

Actually, this decomposition is possible for any vector $\mathbf{u}$ in the three-dimensional space.

Expanding $\mathbf{u}(\mathbf{r}, t)$ in a Taylor series with respect to $\mathbf{r}$ about $\mathbf{r}_0$, we get

$$\mathbf{u}(\mathbf{r}, t) = \mathbf{u}(\mathbf{r}_0, t) + d\mathbf{r} \cdot \nabla \mathbf{u} + O[(d\mathbf{r})^2], \qquad (2.66)$$

where $\nabla \mathbf{u}$ is the velocity gradient tensor. Retaining only the linear term, we have

$$\mathbf{u}(\mathbf{r}, t) = \mathbf{u}(\mathbf{r}_0, t) + d\mathbf{u}, \qquad (2.67)$$

where the velocity $\mathbf{u}(\mathbf{r}_0, t)$ of P represents, of course, *rigid-body translation*, and

$$d\mathbf{u} = d\mathbf{r} \cdot \nabla \mathbf{u} \qquad (2.68)$$

represents the *relative velocity* of particle P' with respect to P. The rigid-body translation component, $\mathbf{u}(\mathbf{r}_0, t)$, does not give rise to any strain or stress, and can be omitted by placing the frame origin or the observer on a moving particle. All the information for the relative velocity $d\mathbf{u}$ is contained in the velocity gradient tensor. The relative velocity can be further decomposed into two components corresponding to rigid-body rotation and pure straining motion, respectively. Recall that $\nabla \mathbf{u}$ can be written as the sum of a symmetric and an antisymmetric tensor,

$$\nabla \mathbf{u} = \mathbf{D} + \mathbf{S}, \qquad (2.69)$$

where

$$\mathbf{D} \equiv \frac{1}{2} [\nabla \mathbf{u} + (\nabla \mathbf{u})^T] \qquad (2.70)$$

is the symmetric rate-of-strain tensor, and

$$\mathbf{S} \equiv \frac{1}{2} [\nabla \mathbf{u} - (\nabla \mathbf{u})^T] \qquad (2.71)$$

is the antisymmetric vorticity tensor. Substituting Eqs. (2.69) to (2.71) in Eq. (2.68), we get

$$d\mathbf{u} = d\mathbf{r} \cdot (\mathbf{D} + \mathbf{S}) = d\mathbf{r} \cdot \frac{1}{2} [\nabla \mathbf{u} + (\nabla \mathbf{u})^T] + d\mathbf{r} \cdot \frac{1}{2} [\nabla \mathbf{u} - (\nabla \mathbf{u})^T]. \qquad (2.72)$$

The first term,

$$\mathbf{u}^{(s)} = d\mathbf{r} \cdot \mathbf{D} = d\mathbf{r} \cdot \frac{1}{2} [\nabla \mathbf{u} + (\nabla \mathbf{u})^T] \qquad (2.73)$$

represents the *pure straining motion* of P' about P. The second term

$$\mathbf{u}^{(r)} = d\mathbf{r} \cdot \mathbf{S} = d\mathbf{r} \cdot \frac{1}{2} [\nabla \mathbf{u} - (\nabla \mathbf{u})^T] \qquad (2.74)$$

represents the *rigid-body rotation* of P' about P. A flow in which $\mathbf{D}$ is zero everywhere corresponds to rigid-body motion (including translation and rotation). Rigid-body motion does not alter the shape of fluid particles, resulting only in their displacement. On the other hand, straining motion results in deformation of fluid particles.

Note that the matrix forms of $\nabla \mathbf{u}$, $\mathbf{D}$, and $\mathbf{S}$ in Cartesian coordinates are given by

$$
\nabla \mathbf{u} = \begin{bmatrix} \dfrac{\partial u_x}{\partial x} & \dfrac{\partial u_y}{\partial x} & \dfrac{\partial u_z}{\partial x} \\ \dfrac{\partial u_x}{\partial y} & \dfrac{\partial u_y}{\partial y} & \dfrac{\partial u_z}{\partial y} \\ \dfrac{\partial u_x}{\partial z} & \dfrac{\partial u_y}{\partial z} & \dfrac{\partial u_z}{\partial z} \end{bmatrix},
\tag{2.75}
$$

$$
\mathbf{D} = \frac{1}{2} \begin{bmatrix} 2\dfrac{\partial u_x}{\partial x} & \left(\dfrac{\partial u_x}{\partial y}+\dfrac{\partial u_y}{\partial x}\right) & \left(\dfrac{\partial u_x}{\partial z}+\dfrac{\partial u_z}{\partial x}\right) \\ \left(\dfrac{\partial u_x}{\partial y}+\dfrac{\partial u_y}{\partial x}\right) & 2\dfrac{\partial u_y}{\partial y} & \left(\dfrac{\partial u_y}{\partial z}+\dfrac{\partial u_z}{\partial y}\right) \\ \left(\dfrac{\partial u_x}{\partial z}+\dfrac{\partial u_z}{\partial x}\right) & \left(\dfrac{\partial u_y}{\partial z}+\dfrac{\partial u_z}{\partial y}\right) & 2\dfrac{\partial u_z}{\partial z} \end{bmatrix},
\tag{2.76}
$$

and

$$
\mathbf{S} = \frac{1}{2} \begin{bmatrix} 0 & -\left(\dfrac{\partial u_y}{\partial x}-\dfrac{\partial u_x}{\partial y}\right) & \left(\dfrac{\partial u_x}{\partial z}-\dfrac{\partial u_z}{\partial x}\right) \\ \left(\dfrac{\partial u_y}{\partial x}-\dfrac{\partial u_x}{\partial y}\right) & 0 & -\left(\dfrac{\partial u_z}{\partial y}-\dfrac{\partial u_y}{\partial z}\right) \\ -\left(\dfrac{\partial u_x}{\partial z}-\dfrac{\partial u_z}{\partial x}\right) & \left(\dfrac{\partial u_z}{\partial y}-\dfrac{\partial u_y}{\partial z}\right) & 0 \end{bmatrix}.
\tag{2.77}
$$

Any antisymmetric tensor has only three independent components and may, therefore, be associated with a vector, referred to as the *dual vector* of the antisymmetric tensor. The dual vector of the vorticity tensor $\mathbf{S}$ is the vorticity vector,

$$
\boldsymbol{\omega} \equiv \nabla \times \mathbf{u}.
\tag{2.78}
$$

In Cartesian coordinates, it is easy to verify that, if

$$
\boldsymbol{\omega} = \omega_x \mathbf{i} + \omega_y \mathbf{j} + \omega_z \mathbf{k},
\tag{2.79}
$$

then

$$
\mathbf{S} = \frac{1}{2} \begin{bmatrix} 0 & -\omega_z & \omega_y \\ \omega_z & 0 & -\omega_x \\ -\omega_y & \omega_x & 0 \end{bmatrix}
\tag{2.80}
$$

and

$$d\mathbf{r} \cdot \mathbf{S} = \frac{1}{2}\,\boldsymbol{\omega} \times d\mathbf{r}\,. \tag{2.81}$$

The vorticity tensor in Eq. (2.74) can be replaced by its dual vorticity vector, according to

$$\mathbf{u}^{(r)} = d\mathbf{r} \cdot \frac{1}{2}[\nabla\mathbf{u} - (\nabla\mathbf{u})^T] = \frac{1}{2}\,(\nabla \times \mathbf{u}) \times d\mathbf{r} = \frac{1}{2}\,\boldsymbol{\omega} \times d\mathbf{r}\,. \tag{2.82}$$

In *irrotational* flows, the vorticity $\boldsymbol{\omega}$ is everywhere zero and, as a result, the rigid-body rotation component $\mathbf{u}^{(r)}$ is zero. If the vorticity is not everywhere zero, then the flow is called *rotational*. The rigid-body rotation component $\mathbf{u}^{(r)}$ also obeys the relation

$$\mathbf{u}^{(r)} \equiv \boldsymbol{\Omega} \times d\mathbf{r}\,, \tag{2.83}$$

where $\boldsymbol{\Omega}$ is the *angular velocity*. Therefore, the vorticity vector $\boldsymbol{\omega}$ is twice the angular velocity of the local rigid-body rotation. It should be emphasized that the vorticity acts as a measure of the *local* rotation of fluid particles, and it is not directly connected with the curvature of the streamlines, i.e., it is independent of any *global* rotation of the fluid.

It must be always kept in mind that the pure straining motion component $\mathbf{u}^{(s)}$ represents strain unaffected by rotation, i.e., strain experienced by an observer rotating with the local vorticity. The straining part of the velocity gradient tensor, which is the rate of strain tensor, can be broken into two parts: an extensional one representing isotropic expansion, and one representing pure straining motion without change of volume. In other words, the rate of strain tensor $\mathbf{D}$ can be written as the sum of a properly chosen diagonal tensor and a symmetric tensor of zero trace:

$$\mathbf{D} = \frac{1}{3}tr(\mathbf{D})\,\mathbf{I} + [\mathbf{D} - \frac{1}{3}tr(\mathbf{D})\,\mathbf{I}]\,. \tag{2.84}$$

The diagonal elements of the tensor $[\mathbf{D} - \frac{1}{3}tr(\mathbf{D})\mathbf{I}]$ represent normal or extensional strains on three mutually perpendicular surfaces. The off-diagonal elements represent shear strains in two directions on each of the three mutually perpendicular surfaces. Noting that

$$tr(\mathbf{D}) = \nabla \cdot \mathbf{u}\,, \tag{2.85}$$

Eq. (2.84) takes the form:

$$\mathbf{D} = \frac{1}{3}\nabla \cdot \mathbf{u}\,\mathbf{I} + \frac{1}{2}\,[\nabla\mathbf{u} + (\nabla\mathbf{u})^T - \frac{2}{3}(\nabla \cdot \mathbf{u})\,\mathbf{I}]\,. \tag{2.86}$$

Therefore, the strain velocity, $\mathbf{u}^{(s)}=d\mathbf{r}\cdot\mathbf{D}$, can be written as

$$\mathbf{u}^{(s)} = \mathbf{u}^{(e)} + \mathbf{u}^{(st)}, \qquad (2.87)$$

where

$$\mathbf{u}^{(e)} = d\mathbf{r}\cdot\frac{1}{3}\left(\nabla\cdot\mathbf{u}\right)\mathbf{I} \qquad (2.88)$$

represents isotropic expansion, and

$$\mathbf{u}^{(st)} = d\mathbf{r}\cdot\frac{1}{2}\left[\nabla\mathbf{u}+(\nabla\mathbf{u})^{T}-\frac{2}{3}\left(\nabla\cdot\mathbf{u}\right)\mathbf{I}\right] \qquad (2.89)$$

represents pure straining motion without change of volume.

In summary, the velocity of a fluid particle in the vicinity of the point $\mathbf{r}_0$ is decomposed as

$$\mathbf{u}(\mathbf{r},t) = \mathbf{u}(\mathbf{r}_0,t) + \mathbf{u}^{(r)} + \mathbf{u}^{(e)} + \mathbf{u}^{(st)}, \qquad (2.90)$$

or, in terms of the vorticity vector, the rate of strain tensor and the divergence of the velocity vector,

$$\mathbf{u}(\mathbf{r},t) = \mathbf{u}(\mathbf{r}_0,t) + \frac{1}{2}\boldsymbol{\omega}\times d\mathbf{r} + d\mathbf{r}\cdot\frac{1}{3}\nabla\cdot\mathbf{u}\,\mathbf{I} + d\mathbf{r}\cdot\frac{1}{2}\left[\nabla\mathbf{u}+(\nabla\mathbf{u})^{T}-\frac{2}{3}(\nabla\cdot\mathbf{u})\,\mathbf{I}\right]. \quad (2.91)$$

Alternative expressions for all the components of the velocity are given in Table 2.2.

The isotropic expansion component $\mathbf{u}^{(e)}$ accounts for any expansion or contraction due to compressibility. For incompressible fluids, $tr(\mathbf{D})=\nabla\cdot\mathbf{u}=0$, and, therefore, $\mathbf{u}^{(e)}$ is zero. In Example 1.5.3 we have shown that the *local rate of expansion* per unit volume is equal to the divergence of the velocity field,

$$\Delta = \lim_{V(t)\to 0}\frac{1}{V(t)}\frac{dV(t)}{dt} = \nabla\cdot\mathbf{u}. \qquad (2.92)$$

Since $\mathbf{D}$ is a symmetric tensor, it has three real eigenvalues, λ_1, λ_2, and λ_3, and three mutually orthogonal eigenvectors. Hence, in the system of the orthonormal basis $\{\mathbf{e}_1', \mathbf{e}_2', \mathbf{e}_3'\}$ of its eigenvectors, $\mathbf{D}$ takes the diagonal form:

$$\mathbf{D}' = \begin{bmatrix} \lambda_1 & 0 & 0 \\ 0 & \lambda_2 & 0 \\ 0 & 0 & \lambda_3 \end{bmatrix}. \qquad (2.93)$$

If $\mathbf{r}'=(r_1', r_2', r_3')$ is the position vector in the system $\{\mathbf{e}_1', \mathbf{e}_2', \mathbf{e}_3'\}$, then

$$\frac{\Delta d\mathbf{r}'}{\Delta t} = \mathbf{u}^{(s)} = d\mathbf{r}'\cdot\mathbf{D}'. \qquad (2.94)$$

Velocity in the vicinity of $\mathbf{r}_0$

$$\mathbf{u}(\mathbf{r},t) \;=\; \mathbf{u}(\mathbf{r}_0,t) \,+\, d\mathbf{u}$$

or

$$\mathbf{u}(\mathbf{r},t) \;=\; \mathbf{u}(\mathbf{r}_0,t) \,+\, \mathbf{u}^{(r)} \,+\, \mathbf{u}^{(s)}$$

or

$$\mathbf{u}(\mathbf{r},t) \;=\; \mathbf{u}(\mathbf{r}_0,t) \,+\, \mathbf{u}^{(r)} \,+\, \mathbf{u}^{(e)} \,+\, \mathbf{u}^{(st)}$$

Rigid − body translation

$$\mathbf{u}(\mathbf{r}_0,t)$$

Relative velocity

$$d\mathbf{u} \;=\; d\mathbf{r}\cdot\nabla\mathbf{u} \;=\; \mathbf{u}^{(r)} \,+\, \mathbf{u}^{(s)}$$

Rigid − body rotation

$$\mathbf{u}^{(r)} \;=\; d\mathbf{r}\cdot\mathbf{S} \;=\; d\mathbf{r}\cdot\tfrac{1}{2}\left[\nabla\mathbf{u} - (\nabla\mathbf{u})^T\right] \;=\; \tfrac{1}{2}\,\omega\times d\mathbf{r} \;=\; \Omega\times d\mathbf{r}$$

Pure straining motion

$$\mathbf{u}^{(s)} \;=\; d\mathbf{r}\cdot\mathbf{D} \;=\; d\mathbf{r}\cdot\tfrac{1}{2}\left[\nabla\mathbf{u} + (\nabla\mathbf{u})^T\right] \;=\; \mathbf{u}^{(e)} \,+\, \mathbf{u}^{(st)}$$

Isotropic expansion

$$\mathbf{u}^{(e)} \;=\; d\mathbf{r}\cdot\tfrac{1}{3}\,tr(\mathbf{D})\,\mathbf{I} \;=\; d\mathbf{r}\cdot\tfrac{1}{3}\left(\nabla\cdot\mathbf{u}\right)\mathbf{I}$$

Pure straining motion without change of volume

$$\mathbf{u}^{(st)} \;=\; d\mathbf{r}\cdot\left[\mathbf{D} - \tfrac{1}{3}\,tr(\mathbf{D})\,\mathbf{I}\right] \;=\; d\mathbf{r}\cdot\tfrac{1}{2}\left[\nabla\mathbf{u} + (\nabla\mathbf{u})^T - \tfrac{2}{3}\left(\nabla\cdot\mathbf{u}\right)\mathbf{I}\right]$$

Table 2.2. *Decomposition of the velocity* $\mathbf{u}(\mathbf{r},t)$ *of a fluid particle in the vicinity of the point* $\mathbf{r}_0$.

This vector equation is equivalent to three linear differential equations,

$$\frac{\Delta dr'_i}{\Delta t} = \lambda_i \, dr'_i \,, \quad i = 1, 2, 3 \,. \tag{2.95}$$

The rate of change of the unit length along the axis of $\mathbf{e}'_i$ at $t=0$ is, therefore, equal to λ_i. The vector field $d\mathbf{r}' \cdot \mathbf{D}'$ is merely expanding or contracting along each of the axes $\mathbf{e}'_i$. For the rate of change of the volume V of a rectangular parallelepiped whose sides dr'_1, dr'_2, and dr'_3 are parallel to the three eigenvectors of $\mathbf{D}$, we get

$$\frac{\Delta V}{\Delta t} = \frac{\Delta}{\Delta t}(dr'_1, dr'_2, dr'_3) = \frac{\Delta dr'_1}{\Delta t}dr'_2 dr'_3 + dr'_1\frac{\Delta dr'_2}{\Delta t}dr'_3 + dr'_1 dr'_2\frac{\Delta dr'_3}{\Delta t} \implies$$

$$\frac{\Delta V}{\Delta t} = (\lambda_1 + \lambda_2 + \lambda_3)\, V \,. \tag{2.96}$$

The trace of a tensor is invariant under orthogonal transformations. Hence,

$$\frac{1}{V}\frac{\Delta V}{\Delta t} = \lambda_1 + \lambda_2 + \lambda_3 = tr\mathbf{D}' = tr\mathbf{D} = \nabla \cdot \mathbf{u} \,. \tag{2.97}$$

This result is equivalent to Eq. (2.92).

Another way to see that $\mathbf{u}^{(e)}$ accounts for the local rate of expansion is to show that $\nabla \cdot \mathbf{u}^{(e)} = \Delta$. Recall that $\nabla \cdot \mathbf{u}$ is evaluated at $\mathbf{r}_0$, and $d\mathbf{r}$ is the position vector of particle P' with respect to a coordinate system centered at P. Hence,

$$\nabla \cdot \mathbf{u}^{(e)} = \nabla \cdot \left(d\mathbf{r} \cdot \frac{1}{3}(\nabla \cdot \mathbf{u})\,\mathbf{I}\right) = \frac{1}{3}(\nabla \cdot \mathbf{u})\,\nabla \cdot (d\mathbf{r} \cdot \mathbf{I}) = \frac{1}{3}(\nabla \cdot \mathbf{u})\,\nabla \cdot d\mathbf{r} \implies$$

$$\nabla \cdot \mathbf{u}^{(e)} = \nabla \cdot \mathbf{u} = \Delta \,. \tag{2.98}$$

Moreover, it is easily shown that the velocity $\mathbf{u}^{(e)}$ is irrotational, i.e., it produces no vorticity:

$$\nabla \times \mathbf{u}^{(e)} = \nabla \times \left(d\mathbf{r} \cdot \frac{1}{3}(\nabla \cdot \mathbf{u})\,\mathbf{I}\right) = \frac{1}{3}(\nabla \cdot \mathbf{u})\,\nabla \times (d\mathbf{r} \cdot \mathbf{I}) = \frac{1}{3}(\nabla \cdot \mathbf{u})\,\nabla \times d\mathbf{r} \implies$$

$$\nabla \times \mathbf{u}^{(e)} = \mathbf{0} \,. \tag{2.99}$$

In deriving Eqs. (2.98) and (2.99), the identities $\nabla \cdot d\mathbf{r} = 3$ and $\nabla \times d\mathbf{r} = 0$ were used (see Example 1.4.1).

Due to the conditions $\nabla \cdot \mathbf{u}^{(e)} = \Delta$ and $\nabla \times \mathbf{u}^{(e)} = \mathbf{0}$, the velocity $\mathbf{u}^e$ can be written as the gradient of a scalar field $\phi^{(e)}$,

$$\mathbf{u}_e = \nabla \phi^{(e)} \,, \tag{2.100}$$

which satisfies the Poisson equation:

$$\nabla^2 \phi^{(e)} = \Delta .$$ (2.101)

A solution to Eqs. (2.100) and (2.101) is given by

$$\phi^{(e)}(\mathbf{r}) = -\frac{1}{4\pi} \int_V \Delta(\mathbf{r}') \frac{1}{|\mathbf{r} - \mathbf{r}'|} dV(\mathbf{r}')$$ (2.102)

and

$$u^{(e)}(\mathbf{r}) = \frac{1}{4\pi} \int_V \Delta(\mathbf{r}') \frac{\mathbf{r} - \mathbf{r}'}{|\mathbf{r} - \mathbf{r}'|^3} dV(\mathbf{r}') ,$$ (2.103)

where V is the volume occupied by the fluid.

The curl of the rotational velocity $\mathbf{u}^{(r)}$ is, in fact, equal to the vorticity $\boldsymbol{\omega}$. Invoking the vector identity

$$\nabla \times (\mathbf{a} \times \mathbf{b}) = \mathbf{a} \nabla \cdot \mathbf{b} - \mathbf{b} \nabla \cdot \mathbf{a} + (\mathbf{b} \cdot \nabla) \mathbf{a} - (\mathbf{a} \cdot \nabla) \mathbf{b} ,$$

we get

$$\begin{aligned} \nabla \times \mathbf{u}^{(r)} &= \frac{1}{2} \nabla \times (\boldsymbol{\omega} \times d\mathbf{r}) \\ &= \frac{1}{2} [\boldsymbol{\omega} \nabla \cdot d\mathbf{r} - d\mathbf{r} \nabla \cdot \boldsymbol{\omega} + (d\mathbf{r} \cdot \nabla) \boldsymbol{\omega} - (\boldsymbol{\omega} \cdot \nabla) d\mathbf{r}] . \end{aligned}$$

Since $\nabla \cdot d\mathbf{r}=3$, $\nabla \cdot \boldsymbol{\omega}=0$ (the vorticity is solenoidal), $(d\mathbf{r} \cdot \nabla)\boldsymbol{\omega}=0$ (evaluated at $\mathbf{r}_0$), and $(\boldsymbol{\omega} \cdot \nabla)d\mathbf{r}=\boldsymbol{\omega}$, one gets

$$\nabla \times \mathbf{u}^{(r)} = \boldsymbol{\omega} .$$ (2.104)

Given that rigid motion is volume preserving, the divergence of the rotational velocity is zero,

$$\nabla \cdot \mathbf{u}^{(r)} = \frac{1}{2} \nabla \cdot (\boldsymbol{\omega} \times d\mathbf{r}) = \frac{1}{2} [d\mathbf{r} \cdot (\nabla \times \boldsymbol{\omega}) - \boldsymbol{\omega} \cdot \nabla \times d\mathbf{r}] = \frac{1}{2} (d\mathbf{r} \cdot \mathbf{0} - \boldsymbol{\omega} \cdot \mathbf{0}) \implies$$

$$\nabla \cdot \mathbf{u}^{(r)} = 0 ,$$ (2.105)

which can be verified by the fact that the vorticity tensor has zero trace. Equations (2.104) and (2.105) suggest a solution of the form,

$$\mathbf{u}^{(r)} = \nabla \times \mathbf{B}^{(r)} ,$$ (2.106)

where $\mathbf{B}^{(r)}$ is a vector potential for $\mathbf{u}^{(r)}$ that satisfies Eq. (2.105) identically. From Eq. (2.104), one gets

$$\nabla \times (\nabla \times \mathbf{B}^{(r)}) = \boldsymbol{\omega} \implies \nabla(\nabla \cdot \mathbf{B}^{(r)}) - \nabla^2 \mathbf{B}^{(r)} = \boldsymbol{\omega} .$$ (2.107)

If $\nabla \cdot \mathbf{B}^{(r)} = 0$,

$$\nabla^2 \mathbf{B}^{(r)} = -\boldsymbol{\omega} \,. \tag{2.108}$$

The solution to Eqs. (2.106) to (2.108) is given by

$$\mathbf{B}^{(r)}(\mathbf{r}) = \frac{1}{4\pi} \int_V \frac{\boldsymbol{\omega}}{|\mathbf{r} - \mathbf{r}'|} \, dV(\mathbf{r}') \tag{2.109}$$

and

$$\mathbf{u}^{(r)}(\mathbf{r}) = -\frac{1}{4\pi} \int_V \frac{(\mathbf{r} - \mathbf{r}') \times \boldsymbol{\omega}}{|\mathbf{r} - \mathbf{r}'|^3} \, dV(\mathbf{r}') \,, \tag{2.110}$$

which suggest that rotational velocity, at a point $\mathbf{r}$, is induced by the vorticity at neighboring points, $\mathbf{r}'$.

Due to the fact that the expansion, Δ, and the vorticity, $\boldsymbol{\omega}$, are accounted for by the expansion and rotational velocities, respectively, the straining velocity, $\mathbf{u}^{(st)}$, is both *solenoidal* and *irrotational*. Therefore,

$$\nabla \cdot \mathbf{u}^{(st)} = \frac{1}{2} \nabla \cdot \left\{ d\mathbf{r} \cdot [\nabla \mathbf{u} + (\nabla \mathbf{u})^T - \frac{2}{3} (\nabla \cdot \mathbf{u}) \, \mathbf{I}] \right\} = 0 \tag{2.111}$$

and

$$\nabla \times \mathbf{u}^{(st)} = \frac{1}{2} \nabla \times \left\{ d\mathbf{r} \cdot [\nabla \mathbf{u} + (\nabla \mathbf{u})^T - \frac{2}{3} (\nabla \cdot \mathbf{u}) \, \mathbf{I}] \right\} = \mathbf{0} \,. \tag{2.112}$$

A potential function $\phi^{(st)}$, such that

$$\mathbf{u}^{(st)} = \nabla \phi^{(st)} \,, \tag{2.113}$$

satisfies Eq. (2.112) and reduces Eq. (2.111) to the Laplace equation,

$$\nabla^2 \phi^{(st)} = 0 \,. \tag{2.114}$$

The Laplace equation has been studied extensively, and many solutions are known [9]. The key to the solution of potential flow problems is the selection of proper solutions that satisfy the boundary conditions. By means of the divergence and Stokes theorems, we get from Eqs. (2.111) and (2.112)

$$\int_V \nabla \cdot \mathbf{u}^{(st)} \, dV = \int_S \mathbf{n} \cdot \mathbf{u}^{(st)} \, dS = 0 \tag{2.115}$$

and

$$\int_S \mathbf{n} \cdot (\nabla \times \mathbf{u}^{(st)}) \, dS = \int_C \mathbf{t} \cdot \mathbf{u}^{(st)} \, d\ell = 0 \,. \tag{2.116}$$

It is clear that the solution $\mathbf{u}^{(st)}$ depends entirely on boundary data.

More details on the mechanisms, concepts, and closed form solutions of local and relative kinematics are given in numerous theoretical Fluid Mechanics [10-12], Rheology [13], and Continuum Mechanics [14] publications.

Example 2.4.1. Local kinematics of stagnation flow

Consider the two-dimensional flow of Fig. 2.6, with Eulerian velocities

$$u_x = \varepsilon x \quad \text{and} \quad u_y = -\varepsilon y .$$

For the velocity gradient tensor we get

$$\nabla \mathbf{u} = \begin{bmatrix} \varepsilon & 0 \\ 0 & -\varepsilon \end{bmatrix} = \varepsilon \, \mathbf{ii} - \varepsilon \, \mathbf{jj} .$$

Since $\nabla \mathbf{u}$ is symmetric,

$$\mathbf{D} = \nabla \mathbf{u} = \varepsilon \, \mathbf{ii} - \varepsilon \, \mathbf{jj} ,$$

and

$$\mathbf{S} = \mathbf{O} .$$

Therefore, the flow is irrotational. It is also incompressible, since

$$tr(\mathbf{D}) = \nabla \cdot \mathbf{u} = \varepsilon - \varepsilon = 0 .$$

For the velocities $\mathbf{u}^{(r)}$, $\mathbf{u}^{(e)}$, and $\mathbf{u}^{(st)}$, we find:

$$\mathbf{u}^{(r)} = dr \cdot \mathbf{S} = 0 ,$$
$$\mathbf{u}^{(e)} = dr \cdot \frac{1}{3} \left(\nabla \cdot \mathbf{u} \right) \mathbf{I} = 0 ,$$
$$\mathbf{u}^{(st)} = dr \cdot \frac{1}{2} [\nabla \mathbf{u} + (\nabla \mathbf{u})^T - \frac{2}{3} \left(\nabla \cdot \mathbf{u} \right) \mathbf{I}] = dr \cdot (\varepsilon \, \mathbf{ii} - \varepsilon \, \mathbf{jj}) .$$

Therefore, expansion and rotation are zero, and there is only extension of the material vector dr. If dr is of the form,

$$dr = adx \, \mathbf{i} + bdy \, \mathbf{j} ,$$

then

$$\mathbf{u}^{(st)} = a\varepsilon dx \, \mathbf{i} - b\varepsilon dy \, \mathbf{j} .$$

If, for instance, $dr = adx \, \mathbf{i}$, then $\mathbf{u}^{(st)} = a\varepsilon dx \, \mathbf{i}$ and extension is in the x direction. $\square$

Example 2.4.2. Local kinematics of rotational shear flow

We consider here shear flow in a channel of width $2H$. If the x-axis lies on the plane of symmetry and points in the direction of the flow, the Eulerian velocity profiles are

$$u_x = c\left(H^2 - y^2\right) \quad \text{and} \quad u_y = u_z = 0\,,$$

where c is a positive constant. The resulting velocity gradient tensor is

$$\nabla \mathbf{u} = \begin{bmatrix} 0 & 0 \\ -2c\,y & 0 \end{bmatrix} = -2c\,y\,\mathbf{ji}\,,$$

and thus

$$\mathbf{D} = \frac{1}{2}\left[\nabla\mathbf{u} + (\nabla\mathbf{u})^T\right] = \begin{bmatrix} 0 & -c\,y \\ -c\,y & 0 \end{bmatrix} = -c\,y\,(\mathbf{ij} + \mathbf{ji})\,,$$

and

$$\mathbf{S} = \frac{1}{2}\left[\nabla\mathbf{u} - (\nabla\mathbf{u})^T\right] = \begin{bmatrix} 0 & c\,y \\ -c\,y & 0 \end{bmatrix} = c\,y\,(\mathbf{ij} - \mathbf{ji})\,.$$

Since

$$tr(\mathbf{D}) = \nabla \cdot \mathbf{u} = 0\,,$$

the flow is incompressible.

If $d\mathbf{r}$ is of the form

$$d\mathbf{r} = adx\,\mathbf{i} + bdy\,\mathbf{j}\,,$$

then

$$\begin{aligned}
\mathbf{u}^{(r)} &= d\mathbf{r} \cdot \mathbf{S} = (adx\,\mathbf{i} + bdy\,\mathbf{j}) \cdot c\,y\,(\mathbf{ij} - \mathbf{ji}) = c\,y\,(-b\,dy\,\mathbf{i} + a\,dx\,\mathbf{j})\,, \\
\mathbf{u}^{(e)} &= d\mathbf{r} \cdot \frac{1}{3}\nabla \cdot \mathbf{u}\,\mathbf{I} = \mathbf{0}\,, \\
\mathbf{u}^{(st)} &= d\mathbf{r} \cdot \left[\mathbf{D} - \frac{1}{3}\nabla \cdot \mathbf{u}\,\mathbf{I}\right] = (adx\,\mathbf{i} + bdy\,\mathbf{j}) \cdot c\,y\,(-\mathbf{ij} - \mathbf{ji}) \\
&= -c\,y\,(b\,dy\,\mathbf{i} + a\,dx\,\mathbf{j})\,.
\end{aligned}$$

Despite the fact that the fluid is not rotating globally (the streamlines are straight lines), the flow is rotational,

$$\boldsymbol{\omega} = \nabla \times \mathbf{u} = -\frac{du_x}{dy}\,\mathbf{k} = 2c\,y\,\mathbf{k} \neq \mathbf{0}\,.$$

The vorticity is maximum along the wall ($y=H$), and zero along the centerline ($y=0$). The existence of vorticity gives rise to extensional strain. This is known as *vorticity induced extension*, to avoid confusion with the *strain induced extension*, represented by $du^{(e)}$. Unlike the latter, the vorticity induced extensional strain does not generate any normal stresses, but it does contribute to shear stresses.   □

The rate of strain tensor $\mathbf{D}$ results in extensional and shear strain. Consider again the relative velocity between the particles P and P' of Fig. 2.7,

$$d\mathbf{u} = d\mathbf{r} \cdot \nabla\mathbf{u} = (\nabla\mathbf{u})^T \cdot d\mathbf{r} . \tag{2.117}$$

By definition,

$$d\mathbf{u} \equiv \frac{D d\mathbf{r}}{Dt} \quad \Longrightarrow \quad \frac{D d\mathbf{r}}{Dt} = d\mathbf{r} \cdot \nabla\mathbf{u} = (\nabla\mathbf{u})^T \cdot d\mathbf{r} . \tag{2.118}$$

Let $\mathbf{a}$ be the unit vector in the direction of $d\mathbf{r}$ and $ds = |d\mathbf{r}|$, i.e., $d\mathbf{r} = \mathbf{a}ds$. Then, from Eq. (2.118) we get:

$$\frac{D\mathbf{a}ds}{Dt} = \mathbf{a}ds \cdot \nabla\mathbf{u} = (\nabla\mathbf{u})^T \cdot \mathbf{a}ds \quad \Longrightarrow \quad \mathbf{a}\frac{1}{ds}\frac{Dds}{Dt} = \mathbf{a}\cdot\nabla\mathbf{u} = (\nabla\mathbf{u})^T \cdot\mathbf{a} \quad \Longrightarrow$$

$$\frac{1}{ds}\frac{Dds}{Dt} = (\mathbf{a}\cdot\nabla\mathbf{u})\cdot\mathbf{a} = \mathbf{a}\cdot[(\nabla\mathbf{u})^T\cdot\mathbf{a}] \quad \Longrightarrow \quad \frac{1}{ds}\frac{Dds}{Dt} = \mathbf{a}\cdot\frac{1}{2}[\nabla\mathbf{u}+(\nabla\mathbf{u})^T]\cdot\mathbf{a} \quad \Longrightarrow$$

$$\frac{1}{ds}\frac{Dds}{Dt} = \mathbf{a}\cdot\mathbf{D}\cdot\mathbf{a} . \tag{2.119}$$

Equation (2.119) describes the extension of the material length ds with time. The term $\mathbf{a}\cdot\mathbf{D}\cdot\mathbf{a}$ is called *extensional strain rate*. The extensional strain rate of a material vector aligned with one Cartesian axis, $d\mathbf{r} = \mathbf{e}_i ds$, is equal to the corresponding diagonal element of $\mathbf{D}$:

$$\frac{1}{ds}\frac{Dds}{Dt}\bigg|_{\mathbf{e}_i ds} = \mathbf{e}_i \cdot \mathbf{D} \cdot \mathbf{e}_i = D_{ii} = \frac{\partial u_i}{\partial x_i} . \tag{2.120}$$

Similar expressions can be obtained for the shear (or angular) strain. The shearing of fluid particles depends on how the angle between material vectors evolves with time. If $\mathbf{a}$ and $\mathbf{b}$ are unit material vectors originally at right angles, i.e., $\mathbf{a} \cdot \mathbf{b} = 0$, then the angle θ between the two material vectors evolves according to

$$\frac{D\theta}{Dt}\bigg|_{\theta=\frac{\pi}{2}} = -2\,\mathbf{a}\cdot\mathbf{D}\cdot\mathbf{b} . \tag{2.121}$$

The right-hand side of the above equation is the *shear strain rate*. Since $\mathbf{D}$ is symmetric, the order of $\mathbf{a}$ and $\mathbf{b}$ in Eq. (2.121) is immaterial. The shear strain rate between material vectors along two axes x_i and x_j of the Cartesian coordinate system is opposite to the ij-component of the rate-of-strain tensor:

$$\left.\frac{D\theta}{Dt}\right|_{\mathbf{e}_i,\mathbf{e}_j} = -2\,\mathbf{e}_i \cdot \mathbf{D} \cdot \mathbf{e}_j = -2D_{ij} = -\left(\frac{\partial u_i}{\partial x_j} + \frac{\partial u_j}{\partial x_i}\right). \qquad (2.122)$$

Example 2.4.3. Deformation of material lines

We revisit here the two flows studied in Examples 2.4.1 and 2.4.2.

Irrotational extensional flow

For the material vector $d\mathbf{r}=\mathbf{a}ds$ with

$$\mathbf{a} = \frac{a_1\,\mathbf{i} + a_2\,\mathbf{j}}{\sqrt{a_1^2 + a_2^2}},$$

the extensional strain rate is

$$\frac{1}{ds}\frac{Dds}{Dt} = \mathbf{a} \cdot \mathbf{D} \cdot \mathbf{a} = \frac{a_1\,\mathbf{i} + a_2\,\mathbf{j}}{\sqrt{a_1^2 + a_2^2}} \cdot (\varepsilon\,\mathbf{i}\mathbf{i} - \varepsilon\,\mathbf{i}\mathbf{j}) \cdot \frac{a_1\,\mathbf{i} + a_2\,\mathbf{j}}{\sqrt{a_1^2 + a_2^2}} \implies$$

$$\frac{1}{ds}\frac{Dds}{Dt} = \frac{a_1^2 - a_2^2}{a_1^2 + a_2^2}\,\varepsilon.$$

We observe that if $a_1=\pm a_2$, the material length ds does not change with time. A material vector along the x direction ($d\mathbf{r}=\mathbf{i}ds$) changes its length according to

$$\frac{D(\ln ds)}{Dt} = \frac{1}{ds}\frac{Dds}{Dt} = \varepsilon \implies ds = (ds)_0\,e^{\varepsilon t}.$$

Similarly, for $d\mathbf{r}=\mathbf{j}ds$, we find that $ds=(ds)_0\,e^{-\varepsilon t}$.

The shear strain rate for $\mathbf{a}=\mathbf{i}$ and $\mathbf{b}=\mathbf{j}$ is

$$\left.\frac{D\theta}{Dt}\right|_{\mathbf{i},\mathbf{j}} = -2\,\mathbf{a} \cdot \mathbf{D} \cdot \mathbf{b} = -2\,\mathbf{i} \cdot (\varepsilon\,\mathbf{i}\mathbf{i} - \varepsilon\,\mathbf{j}\mathbf{j}) \cdot \mathbf{j} = 0,$$

in agreement with the fact that shearing is not present in irrotational extensional flows.

Rotational shear flow

We consider a material vector of arbitrary orientation,

$$d\mathbf{r} = \mathbf{a}ds = \frac{a_1\,\mathbf{i} + a_2\,\mathbf{j}}{\sqrt{a_1^2 + a_2^2}}\,ds,$$

for which

$$\frac{D(\ln ds)}{Dt} = \frac{1}{ds}\frac{Dds}{Dt} = \mathbf{a}\cdot\mathbf{D}\cdot\mathbf{a}$$

$$= \frac{a_1\,\mathbf{i}+a_2\,\mathbf{j}}{\sqrt{a_1^2+a_2^2}}\cdot[-c\,y\,(\mathbf{ij}+\mathbf{ji})]\cdot\frac{a_1\,\mathbf{i}+a_2\,\mathbf{j}}{\sqrt{a_1^2+a_2^2}} = -\frac{2a_1a_2}{a_1^2+a_2^2}\,c\,y\,,$$

or

$$\frac{D(\ln ds)}{Dt} = \frac{a_1a_2}{a_1^2+a_2^2}\frac{\partial u_x}{\partial y}\,.$$

We easily deduce that a material vector parallel to the x-axis does not change length. The shear strain rate for $\mathbf{a}=\mathbf{i}$ and $\mathbf{b}=\mathbf{j}$ is

$$\left.\frac{D\theta}{Dt}\right|_{\mathbf{i},\mathbf{j}} = -2\,\mathbf{a}\cdot\mathbf{D}\cdot\mathbf{b} = -2\,\mathbf{i}\cdot[-c\,y\,(\mathbf{ij}+\mathbf{ji})]\cdot\mathbf{j} = 2c\,y\,,$$

or

$$\left.\frac{D\theta}{Dt}\right|_{\mathbf{i},\mathbf{j}} = -\frac{\partial u_x}{\partial y}\,.$$

□

2.5 Problems

2.1. Repeat Example 2.1.2 for cylindrical droplets of radius R and length $L \gg R$. How does the inside pressure change with R, L, and σ?

2.2. The Eulerian description of a two-dimensional flow is given by

$$u_x = ay \quad \text{and} \quad u_y = 0\,,$$

where a is a positive constant.
(a) Calculate the Lagrangian kinematics and compare with the Eulerian ones.
(b) Calculate the velocity gradient, the rate of strain and the vorticity tensors.
(c) Find the deformation of material vectors parallel to the x- and y-axes.
(d) Find the deformation of material vectors diagonal to the two axes. Explain the physics behind your findings.

2.3. Write down the Young–Laplace equation for interfaces of the following configurations: spherical, cylindrical, planar, elliptical, parabolic, and hyperbolic.

2.4. The motion of a solid body on the xy-plane is described by

$$\mathbf{r}(t) = \mathbf{i}\,a\cos\omega t + \mathbf{j}\,b\sin\omega t\,,$$

where a, b, and ω are constants. How far is the body from the origin at any time t? Find the velocity and the acceleration vectors. Show that the body moves on an elliptical path.

2.5. Derive the equation that governs the pressure distribution in the atmosphere by means of momentum balance on an appropriate control volume. You must utilize the integral theorems of Chapter 1.

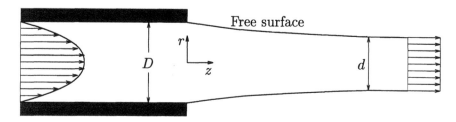

Figure 2.8. *Contraction of a round Newtonian jet at a high Reynolds number.*

2.6. Consider the *high Reynolds number* flow of a Newtonian jet issuing from a capillary of diameter D, as illustrated in Fig. 2.8. Upstream of the exit of the capillary, the flow is assumed to be *fully developed*, i.e., the axial velocity is parabolic,

$$u_z = \frac{32\,Q}{\eta\,D^4}\left(\frac{D^2}{4} - r^2\right),$$

where η is the viscosity of the liquid, ρ is its density, and Q denotes the volumetric flow rate. The liquid leaves the capillary as a free, round jet and, after some rearrangement, the flow downstream becomes *plug*, i.e.,

$$u_z = V.$$

Using appropriate conservation statements, calculate the velocity V and the final diameter d of the jet. Repeat the procedure for a plane jet issuing from a slit of thickness H and width W.

2.7. Use the substantial derivative,

$$\frac{D(ds)}{Dt} = \frac{\partial(ds)}{\partial t} + \mathbf{u}\cdot\nabla(ds) \tag{2.123}$$

to find how material lengths, ds, change along streamlines. Consider vectors tangent and perpendicular to streamlines. Apply your findings to the following flows:

(a) $u_x = \varepsilon x$ and $u_y = -\varepsilon y$;
(b) $u_x = ay$ and $u_y = 0$.

2.8. A material vector **a** enters perpendicularly a shear field given by $u_x = ay$ and $u_y = 0$. Describe its motion and deformation as it travels in the field. Repeat for the extensional field given by $u_x = \varepsilon x$ and $u_y = -\varepsilon y$.

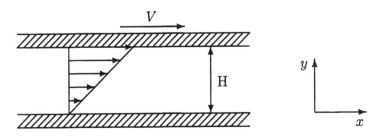

Figure 2.9. *Plane Couette flow.*

2.9. Calculate the configuration of a material square in the *plane Couette flow*, the geometry of which is depicted in Fig. 2.9. The lower wall is fixed, the upper wall is moving with speed V, and the x-component of the velocity is given by

$$u_x = \frac{y}{H} V . \tag{2.124}$$

Consider three entering locations: adjacent to each of the walls and at $y = H/2$. How would you use this flow to measure velocity, vorticity, and stress?

2.10. The velocity vector

$$\mathbf{u}(t) = \Omega(t)\, r\, \mathbf{e}_\theta + u_r(t)\, \mathbf{e}_r + u_z\, \mathbf{e}_z$$

describes a *spiral flow* in cylindrical coordinates.
(a) Calculate the acceleration vector $\mathbf{a}(t)$ and the position vector $\mathbf{r}(t)$.
(b) How do things change when $u_z = 0$, $\Omega(t) = \Omega_0$ and $u_r(t) = u_0$? Sketch a representative streamline.

2.6   References

1. S. Eskinazi, *Fluid Mechanics and Thermodynamics of our Environment*, Academic Press, New York, 1975.

2. R.B. Bird, R.C. Armstrong, and O. Hassager, *Dynamics of Polymeric Liquids: Fluid Mechanics*, John Wiley & Sons, New York, 1977.

3. R.I. Tanner, *Engineering Rheology*, Clarendon Press, Oxford, 1985.

4. V.G. Levich, *Physicochemical Hydrodynamics*, Prentice-Hall, Englewood Cliffs, 1962.

5. M. Van Dyke, *Perturbation Methods in Fluid Mechanics*, Academic Press, New York, 1964.

6. J.R.A. Pearson, *Mechanics of Polymer Processing*, Elsevier Publishers, London and New York, 1985.

7. B.V. Deryagin and S.M. Levi, *Film Coating Theory*, Focal Press, New York, 1964.

8. R.J. Goldstein, *Fluid Mechanics Measurements*, Hemisphere Publishing Corporation, New York, 1983.

9. P.R. Garabedian, *Partial Differential Equations*, Chelsea Publishing Company, New York, 1986.

10. G.K. Batchelor, *An Introduction to Fluid Dynamics*, Cambridge University Press, Cambridge, 1967.

11. C. Pozrikidis, *Introduction to Theoretical and Computational Fluid Dynamics*, Oxford University Press, New York, 1997.

12. R.L. Panton, *Incompressible Flow*, John Wiley & Sons, New York, 1996.

13. G. Astarita and G. Marrucci, *Principles of Non-Newtonian Fluid Mechanics*, McGraw-Hill, New York, 1974.

14. W. Prager, *Introduction to Mechanics of Continua*, Ginn, Boston, 1961.

Chapter 3

CONSERVATION LAWS

Initiation of relative fluid motion, and thus development of velocity gradients, occurs under the action of external force gradients, such as those due to pressure, elevation, shear stresses, density, electromagnetic forces, etc. For example, rain falls to earth due to elevation differences (i.e., gravity differential), and butter spreads thin on toast due to the shearing action of a knife. Additionally, industrial liquids are transferred by means of piping systems after being pushed by pumps or pulled by vacuum, both of which generate pressure differentials. Meteorological phenomena are primarily due to air circulation, as a result of density differences induced by nonisothermal conditions. Finally, conducting liquids flow in nonuniform magnetic fields.

3.1 Control Volume and Surroundings

Mass, momentum, and energy within a flowing medium may be transferred by *convection* and/or *diffusion*. Convection is due to bulk fluid motion, and diffusion is due to molecular motions which can take place independently of the presence of bulk motion. These transfer mechanisms are illustrated in Fig. 3.1, where, without loss of generality, we consider a stationary control volume interacting with its surroundings through the bounding surface, S. Due to the velocity $\mathbf{u}$, fluid entering or leaving the stationary control volume carries by means of convection:

(a) Net mass per unit time,

$$\dot{m}_C = \int_S \rho \left(\mathbf{n} \cdot \mathbf{u} \right) dS , \qquad (3.1)$$

where $\mathbf{n}$ is the local outward-pointing unit normal vector, and ρ is the fluid density (subscript C denotes flux by convection).

(b) Net momentum per unit time,

$$\dot{\mathbf{J}}_C = \int_S \rho \mathbf{u} \left(\mathbf{n} \cdot \mathbf{u} \right) dS , \qquad (3.2)$$

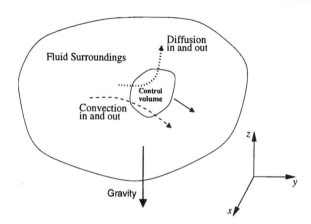

Figure 3.1. *Convection and diffusion between a control volume and its surroundings.*

where $\mathbf{J} = \rho\mathbf{u}$ is the momentum per unit volume.

(c) Net mechanical energy per unit time,

$$\dot{E}_C = \int_S \rho \left(\frac{\mathbf{u}^2}{2} + \frac{p}{\rho} + gz \right) \mathbf{n} \cdot \mathbf{u} \, dS \,, \qquad (3.3)$$

where the three scalar quantities in parentheses correspond to the kinetic energy, the flow work, and the potential energy per unit mass flow rate; p is the pressure, g is the gravitational acceleration, and z is the vertical distance.

(d) Net thermal energy per unit time,

$$\dot{H}_C = \int_S \rho U (\mathbf{n} \cdot \mathbf{u}) \, dS \,, \qquad (3.4)$$

where U is the *internal energy* per unit mass. This is defined as $dU \equiv C_v \, dT$, where C_v is the *specific heat* at constant volume, and T is the temperature.

(e) Total energy per unit time,

$$(\dot{E}_T)_C = \dot{E}_C + \dot{H}_C = \int_S \rho \left(\frac{\mathbf{u}^2}{2} + \frac{p}{\rho} + gz + U \right) (\mathbf{n} \cdot \mathbf{u}) \, dS \,. \qquad (3.5)$$

While *convection* occurs due to bulk motion, *diffusion* is independent of it, and it is entirely due to a gradient that drives to equilibrium. For instance, diffusion, commonly known as *conduction*, of heat occurs whenever there is a temperature gradient (i.e., potential), $\nabla T \neq \mathbf{0}$. Diffusion of mass occurs due to a concentration gradient, $\nabla c \neq \mathbf{0}$, and diffusion of momentum takes place due to velocity, or force gradients. Table 3.1 lists common examples of diffusion.

Quantity	Resistance	Result or Flux
Temperature, T	$1/k$	$-k\nabla T$
Solute, c	$1/D$	$-D\nabla c$
Potential, V	R	$-\frac{1}{R}\nabla V$
Velocity, $\mathbf{u}$	$1/\eta$	$\eta[\nabla\mathbf{u} + (\nabla\mathbf{u})^T]$

Table 3.1. *Common examples of diffusion.*

Common forms of diffusion in fluid mechanics are:

(a) Heat conduction, which according to *Fourier's law* is expressed as

$$\dot{H}_D = -\int_S k\,(\mathbf{n}\cdot\nabla T)\,dS\,, \tag{3.6}$$

where k is the *thermal conductivity* (subscript D denotes flux by diffusion).

(b) Momentum diffusion, which according to Newton's law of viscosity is expressed as

$$\mathbf{f} = \int_S \mathbf{n}\cdot\mathbf{T}\,dS\,, \tag{3.7}$$

where $\mathbf{f}$, $\mathbf{T}$, η and $\nabla\mathbf{u}$ are, respectively, the traction force per unit area, the local total stress tensor, the viscosity, and the velocity gradient tensor. Momentum diffusion also occurs under the action of body forces, according to Newton's law of gravity,

$$\mathbf{f} = \int_V \rho\,\mathbf{g}\,dV\,, \tag{3.8}$$

where $\mathbf{f}$ is the weight, and $\mathbf{g}$ is the gravitational acceleration vector.

Production, destruction, or *conversion* of fluid quantities may take place within a system or a control volume, such as mechanical energy conversion expressed by

$$\dot{E} = \int_V [\dot{W} - p(\nabla \cdot \mathbf{u}) - (\tau : \nabla \mathbf{u})] \, dV \neq 0 \,, \tag{3.9}$$

and thermal energy conversion given by

$$\dot{H} = \int_V (\tau : \nabla \mathbf{u} + p\nabla \cdot \mathbf{u}) \pm \dot{H}_r \, dV \neq 0 \,, \tag{3.10}$$

where $\dot{W}$ is the rate of production of work, and $\dot{H}_r$ is production or consumption of heat by *exothermic* and *endothermic* chemical reactions. While mechanical and thermal energy conversion within a control volume is finite, there is no total mass or momentum conversion.

According to the sign convention adopted here, mechanical energy is gained by work W done to (+) (e.g., by a pump) or by (-) the control volume (e.g., by a turbine). In addition, mechanical energy is lost to heat due to volume expansion $(\nabla \cdot \mathbf{u})$, and due to *viscous dissipation* ($\tau : \nabla \mathbf{u}$) as a result of friction between fluid layers moving at different velocities, and between the fluid and solid boundaries.

Overall change of fluid quantities within the control volume such as mass, momentum, and energy is expressed as

$$\frac{d}{dt} \int_V q \, dV \,, \tag{3.11}$$

where q is the considered property per unit volume, or the *density* of the property.

3.2   The General Equations of Conservation

The development of the conservation equations starts with the general *statement of conservation*

$$\left\{ \begin{array}{c} Rate\ of \\ change \end{array} \right\} = \left\{ \begin{array}{c} Net \\ convection \end{array} \right\} \pm \left\{ \begin{array}{c} Net \\ diffusion \end{array} \right\} \pm \left\{ \begin{array}{c} Production/ \\ Destruction \end{array} \right\}, \tag{3.12}$$

which, in mathematical terms, takes the form,

$$\frac{d}{dt} \int_V (\quad) \, dV = -\int_S (\quad) \mathbf{n} \cdot \mathbf{u} \, dS + \int_S k\nabla(\quad) \cdot \mathbf{n} \, dS + \int_V (\quad) \, dV \,. \tag{3.13}$$

Here, V and S are respectively the volume and the bounding surface of the control volume, $\mathbf{n}$ is the outward-pointing unit normal vector along S, $\mathbf{u}$ is the fluid velocity with respect to the control volume, k is a diffusion coefficient, and $\nabla(\ )$ is the driving gradient responsible for diffusion. By substituting the expressions of Section 3.1 in Eq. (3.13), the *integral forms* of the conservation equations are obtained as follows:

(a) Mass conservation

$$\frac{d}{dt} \int_V \rho \, dV = - \int_S \rho \, (\mathbf{n} \cdot \mathbf{u}) \, dS \,. \tag{3.14}$$

(b) Linear momentum conservation

$$\frac{d}{dt} \int_V \rho \mathbf{u} \, dV = - \int_S \rho \mathbf{u} \, (\mathbf{n} \cdot \mathbf{u}) \, dS + \int_S \mathbf{n} \cdot \mathbf{T} \, dS + \int_V \rho \mathbf{g} \, dV \,. \tag{3.15}$$

(c) Total energy conservation

$$\frac{d}{dt} \int_V \rho E_T \, dV = - \int_S \rho E_T \, (\mathbf{n} \cdot \mathbf{u}) \, dS + \int_S (\mathbf{n} \cdot \mathbf{T}) \cdot \mathbf{u} \, dS + \int_V \rho \, (\mathbf{u} \cdot \mathbf{g}) \, dV \,, \tag{3.16}$$

where the total energy is defined as the sum of the mechanical and internal energy, $E_T \equiv E + U$. The last two terms in Eq. (3.16) are the rate of work or power due to contact and body forces, respectively.

(d) Thermal energy change

$$\frac{d}{dt} \int_V \rho U \, dV = - \int_S \rho U (\mathbf{u} \cdot \mathbf{n}) \, dS + \int_V [(\boldsymbol{\tau} : \nabla \mathbf{u}) + p \, (\nabla \cdot \mathbf{u})] \, dV$$

$$\pm \int_V \dot{H}_r \, dV + \int_S k \nabla T \cdot \mathbf{n} \, dS \,, \tag{3.17}$$

where $\boldsymbol{\tau}$ is the viscous stress tensor related to the total stress tensor by $\mathbf{T} = -p\mathbf{I} + \boldsymbol{\tau}$.

(e) Mechanical energy change

$$\frac{d}{dt} \int_V \rho E \, dV = \frac{d}{dt} \int_V \rho (E_T - U) \, dV$$

$$= - \int_S \rho E(\mathbf{u} \cdot \mathbf{n}) \, dS + \int_S \mathbf{n} \cdot (\mathbf{u} \cdot T) \, dS - \int_V [\boldsymbol{\tau} : \nabla \mathbf{u} + p \nabla \cdot \mathbf{u}] \, dV$$

$$+ \int_V \rho(\mathbf{u} \cdot \mathbf{g}) \, dV \pm \int_V \dot{H}_r \, dV - \int_S k(\nabla T \cdot \mathbf{n}) \, dS \,. \tag{3.18}$$

The energy equations are typically expressed in terms of a measurable property such as temperature by means of $dU \equiv C_v dT$. For constant C_v, $U = U_0 + C_v(T - T_0)$, where T_0 is a reference temperature of known internal energy U_0.

The minus sign associated with the convection terms is a consequence of the sign convention adopted here: the unit normal vector is positive when pointing outwards. Therefore, a normal velocity *towards* the control volume results in a positive increase of a given quantity, i.e., $d/dt > 0$.

Example 3.2.1

Derive the conservation of mass equation by means of a control volume moving with the fluid velocity.

Solution:

The total change of mass within the control volume, given by

$$\frac{d}{dt} \int_V \rho \, dV = - \int_S \rho(\mathbf{n} \cdot \mathbf{u}^R) \, dS \,,$$

is zero because the relative velocity, $\mathbf{u}^R$, between the control volume and its surroundings, is zero. Furthermore, according to the Reynolds transport theorem,

$$\frac{d}{dt} \int_V \rho \, dV = \int_V \frac{\partial \rho}{\partial t} \, dV + \int_S \rho(\mathbf{n} \cdot \mathbf{u}) \, dS = 0 \,.$$

By invoking the divergence theorem, we get

$$\int_V \frac{\partial \rho}{\partial t} \, dV = - \int_S (\mathbf{n} \cdot \rho \mathbf{u}) \, dS = - \int_V \nabla \cdot (\rho \mathbf{u}) \, dV \qquad \Longrightarrow$$

$$\int_V \left[\frac{\partial \rho}{\partial t} + \nabla \cdot (\rho \mathbf{u}) \right] dV = 0 \,. \tag{3.19}$$

Since the control volume is arbitrary,

$$\frac{\partial \rho}{\partial t} + \nabla \cdot (\rho \mathbf{u}) = 0 \,, \tag{3.20}$$

which is the familiar form of the continuity equation.   □

Example 3.2.2. Flow in an inclined pipe

Apply the integral equations of the conservation of mass, momentum, and mechanical energy to study the steady incompressible flow in an inclined pipe (Fig. 3.2).

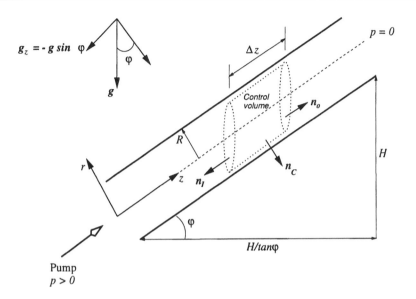

Figure 3.2. *Flow in an inclined pipe and stationary control volume.*

Solution:
For the selected control volume shown in Fig. 3.2, the rate of change of mass for incompressible or steady flow is

$$\frac{d}{dt} \int_V \rho \, dV = \int_V \frac{\partial \rho}{\partial t} \, dV = 0 \,.$$

Therefore, net convection of mass is zero, i.e.,

$$\int_S \rho(\mathbf{n} \cdot \mathbf{u}) \, dS = \int_{S_I} \rho(\mathbf{u}_I \cdot \mathbf{n}_I) \, dS_I + \int_{S_o} \rho(\mathbf{u}_o \cdot \mathbf{n}_o) \, dS_o + \int_{S_C} \rho(\mathbf{u}_C \cdot \mathbf{n}_C) \, dS_C = 0 \,,$$

where $\mathbf{n}_I, \mathbf{n}_o$, and $\mathbf{n}_C$ are, respectively, the unit normal vectors at the inlet, outlet, and cylindrical surfaces of the control volume. The velocities at the corresponding surfaces are denoted by $\mathbf{u}_I, \mathbf{u}_o$, and $\mathbf{u}_C$.

At the inlet, $\mathbf{n}_I \cdot \mathbf{u}_I = -u_n^I = -u_I(r)$; at the outlet $\mathbf{n}_o \cdot \mathbf{u}_o = u_n^o = u_o(r)$; $\mathbf{n}_C \cdot \mathbf{u}_C$ is the normal velocity to the cylindrical surface which is zero. Moreover,

$$dS_I = d(\pi r_I^2) = (2\pi r dr)_I \,, \quad dS_o = d(\pi r_o^2) = (2\pi r dr)_o \,, \quad dS_C = 2\pi R dz \,,$$

and

$$dV = d(\pi r^2)dz = 2\pi r dr dz \,.$$

The above expressions are substituted in the appropriate terms of the conservation of mass equation, Eq. (3.14), to yield

$$-2\pi \int_0^R [ru(r)]_I \, dr + 2\pi \int_0^R [ru(r)]_o \, dr + 0 = 0 \,,$$

and

$$\int_0^R \left([ru(r)]_I - [ru(r)]_o\right) \, dr = 0 \,.$$

Since the control volume is arbitrary, we must have

$$[ru(r)]_I = [ru(r)]_o \,,$$

which yields the well-known result for steady pipe flow, $u(r)_I = u(r)_o = u(r)$, i.e., the flow is characterized by a single velocity component which is parallel to the pipe wall and depends only on r.

For the same control volume, the rate of change of linear momentum for steady flow is

$$\frac{d}{dt} \int_V \rho \mathbf{u} \, dV = \int_V \rho \frac{\partial \mathbf{u}}{\partial t} \, dV = \mathbf{0} \,.$$

Convection of momentum in the flow direction (z-direction) is given by

$$
\begin{aligned}
\mathbf{e}_z \cdot \int_S \rho \mathbf{u}(\mathbf{n} \cdot \mathbf{u}) \, dS &= \rho \mathbf{e}_z \cdot \int_{S_I} \mathbf{u}_I(\mathbf{n}_I \cdot \mathbf{u}_I) \, dS_I + \rho \mathbf{e}_z \cdot \int_{S_o} \mathbf{u}_o(\mathbf{n}_o \cdot \mathbf{u}_o) \, dS_o \\
&\quad + \rho \mathbf{e}_z \cdot \int_{S_C} \mathbf{u}_C(\mathbf{n}_C \cdot \mathbf{u}_C) \, dS_C \\
&= -2\pi\rho \int_0^R ru_I^2(r) \, dr + 2\pi\rho \int_0^R ru_o^2(r) \, dr + 0 = 0 \,.
\end{aligned}
$$

The contact force (stress) contribution is

$$
\begin{aligned}
\mathbf{e}_z \cdot \int_S \mathbf{n} \cdot \mathbf{T} dS &= \mathbf{e}_z \cdot \int_{S_I} \mathbf{n}_I \cdot \mathbf{T}_I \, dS_I + \mathbf{e}_z \cdot \int_{S_o} \mathbf{n}_o \cdot \mathbf{T}_o \, dS_o + \mathbf{e}_z \cdot \int_{S_C} \mathbf{n}_C \cdot \mathbf{T}_C \, dS_C \\
&= \mathbf{e}_z \cdot \int_{S_I} \mathbf{n}_I \cdot (-p\mathbf{I} + \boldsymbol{\tau}) \, dS_I + \mathbf{e}_z \cdot \int_{S_o} \mathbf{n}_o \cdot (-p\mathbf{I} + \boldsymbol{\tau}) \, dS_o \\
&\quad + \mathbf{e}_z \cdot \int_{S_C} \mathbf{n}_C \cdot (-p\mathbf{I} + \boldsymbol{\tau}) \, dS_C \\
&= -2\pi \int_0^R (-p + \tau_{zz})_I \, rdr + 2\pi \int_0^R (-p + \tau_{zz})_o \, rdr + 2\pi(\Delta z)R\, \tau_{rz}^w \,,
\end{aligned}
$$

where τ_{rz}^w is the shear stress at the wall. By means of *macroscopic balances*, the various quantities are approximated by their average values. Therefore,

$$\mathbf{e}_z \cdot \int_S \mathbf{n} \cdot \mathbf{T}\, dS = -2\pi \frac{R^2}{2}[(-p + \tau_{zz})_I - (-p + \tau_{zz})_o] + 2\pi R\, \Delta z\, \tau_{rz}^w$$

$$= \pi R^2[-\Delta p + \Delta \tau_{zz}] + 2\pi R \tau_{rz}^w \Delta z \,,$$

where $\Delta p = p_o - p_I < 0$.

Finally, the body force contribution in the flow direction is

$$\mathbf{e}_z \cdot \int_V \rho \mathbf{g}\, dV = \mathbf{e}_z \cdot \left[\int_0^R \rho(g_r \mathbf{e}_r + g_z \mathbf{e}_z + g_\theta \mathbf{e}_\theta) 2\pi\, rdr \right] \Delta z$$

$$= -2\pi \Delta z \int_0^R \rho g \sin\phi\, rdr = -2\pi \frac{R^2}{2} \Delta z \rho g \sin\phi \,.$$

Therefore, the overall, *macroscopic momentum equation* is

$$-\frac{\Delta p}{\Delta z} + \frac{\Delta \tau_{zz}}{\Delta z} + \frac{2}{R}\tau_{rz}^w - \rho g \sin\phi = 0 \,.$$

$\square$

Example 3.2.3. Growing bubble

A spherical gas bubble of radius $R(t)$ grows within a liquid at a rate $\dot{R} = dR/dt$. The gas inside the bubble behaves as incompressible fluid. However, both the mass and volume change due to evaporation of liquid at the interface. By choosing appropriate control volumes show that:
(a) the gas velocity is zero;
(b) the mass flux at $r < R$ is $\rho_G \dot{R}$;
(c) the mass flux at $r > R$ is $-(\rho_L - \rho_G)\dot{R}(R^2/r^2)$.

Solution:
The problem is solved by applying the mass conservation equation,

$$\frac{d}{dt} \int_{V(t)} \rho\, dV = -\int_{S(t)} \mathbf{n} \cdot \rho(\mathbf{u} - \mathbf{u}_s)\, dS \,,$$

where V is the control volume bounded by the surface S, $\mathbf{u}$ is the velocity of the fluid under consideration, and $\mathbf{u}_s$ is the velocity of the surface bounding the control volume. In the following, the control volume is always a sphere. Therefore, the normal to the surface S is $\mathbf{n} = \mathbf{e}_r$.

(a) The control volume is fixed ($\mathbf{u}_s=0$) of radius r, and contains only gas, i.e., $r < R$. From the Reynolds transport theorem, we have

$$\frac{d}{dt} \int_V \rho_G \, dV = \int_V \frac{\partial \rho_G}{\partial t} \, dV = 0 \, .$$

Therefore, for the mass flux we get

$$-\int_S \mathbf{n} \cdot \rho_G(\mathbf{u} - \mathbf{u}_s) \, dS = \frac{d}{dt} \int_{V(t)} \rho \, dV = 0 \quad \Longrightarrow$$

$$\int_S \mathbf{n} \cdot \rho_G \mathbf{u} \, dS = 0 \quad \Longrightarrow \quad \mathbf{u} = \mathbf{0} \quad \text{for all } r < R \, .$$

(b) The control volume is moving with the bubble ($\mathbf{u}_s=\dot{R}\mathbf{e}_r$) and contains only gas ($r < R$). From the Reynolds transport theorem, we get

$$\frac{d}{dt} \int_{V(t)} \rho_G \, dV = \int_{V(t)} \frac{\partial \rho_G}{\partial t} \, dV + \int_{S(t)} \mathbf{n} \cdot (\rho_G \mathbf{u}_s) \, dS = 0 + \rho_G \dot{R} \, (4\pi r^2) = 4\pi \rho_G \dot{R} r^2 \, .$$

The mass flux is given by

$$\frac{d}{dt} \int_{V(t)} \rho_G \, dV = -\int_{S(t)} \mathbf{n} \cdot \rho_G(\mathbf{u} - \mathbf{u}_s) \, dS = q \, 4\pi r^2 \, ,$$

where q is the relative flux per unit area. Combining the above expressions, we get $q=\rho_G\dot{R}$.

(c) The control volume is fixed ($\mathbf{u}_s=0$) and contains the bubble ($r > R$). From the Reynolds transport theorem, we get

$$\begin{aligned}
\frac{d}{dt} \int_V \rho \, dV &= \frac{d}{dt} \int_{V_G(t)} \rho_G \, dV + \frac{d}{dt} \int_{V_L(t)} \rho_L \, dV \\
&= \int_{V_G(t)} \frac{\partial \rho_G}{\partial t} \, dV + \int_{S(R)} \mathbf{n} \cdot (\rho_G \mathbf{u}_s) \, dS \\
&\quad \int_{V_L(t)} \frac{\partial \rho_L}{\partial t} \, dV + \int_{S(r)} \mathbf{n} \cdot (\rho_L \mathbf{u}_s) \, dS + \int_{S(R)} \mathbf{n} \cdot (\rho_L \mathbf{u}_s) \, dS \\
&= 0 + \int_{S(R)} \mathbf{e}_r \cdot (\rho_G \mathbf{u}_s) \, dS + 0 + 0 - \int_{S(R)} \mathbf{e}_r \cdot (\rho_L \mathbf{u}_s) \, dS \\
&= \int_{S(R)} (\rho_G - \rho_L) \, u_s \, dS = -(\rho_L - \rho_G)\dot{R} \, (4\pi R^2) \, .
\end{aligned}$$

For the mass flux, we have

$$-\int_S \rho \mathbf{n} \cdot (\mathbf{u} - \mathbf{u}_s) \, dS = q \, 4\pi r^2 \,.$$

Combining the above two equations, we get

$$q = -(\rho_L - \rho_G)\dot{R}\frac{R^2}{r^2} \,.$$

□

3.3 The Differential Forms of the Conservation Equations

The integral forms of the conservation equations derived in Section 3.2, arise naturally from the conservation statement, Eq. (3.13). However, these equations are not convenient to use in complex flow problems. To address this issue, the conservation equations are expressed in differential form by invoking the integral theorems of Chapter 1.

The general form of the integral equation of change, with respect to a stationary control volume V bounded by a surface S, may be written as

$$\int_V \frac{\partial}{\partial t}(\quad)_1 \, dV = -\int_S \mathbf{n} \cdot (\quad)_1 \mathbf{u} \, dS + \int_S \mathbf{n} \cdot (\quad)_2 \, dS + \int_V (\quad)_3 \, dV \,. \quad (3.21)$$

Here $(\quad)_1$ is a scalar (e.g., energy or density) or a vector (e.g., momentum), $(\quad)_2$ is a vector (e.g., gradient of temperature) or a tensor (e.g., stress tensor), and $(\quad)_3$ is a vector (e.g., gravity) or a scalar (e.g., viscous dissipation or heat release by reaction).

By invoking the Gauss divergence theorem, the surface integrals of Eq. (3.21) are expressed as volume integrals:

$$\int_S \mathbf{n} \cdot (\quad)_1 \mathbf{u} \, dS = \int_V \nabla \cdot [(\quad)_1 \mathbf{u}] \, dV \,,$$

$$\int_S \mathbf{n} \cdot (\quad)_2 \, dS = \int_V \nabla \cdot (\quad)_2 \, dV \,.$$

Equation (3.21) then becomes

$$\int_V \left[\frac{\partial}{\partial t}(\quad)_1 + \nabla \cdot [(\quad)_1 \mathbf{u}] - \nabla \cdot (\quad)_2 - (\quad)_3 \right] dV = 0 \,. \quad (3.22)$$

Since the choice of the volume V is arbitrary, we deduce that

$$\frac{\partial}{\partial t}(\quad)_1 + \nabla \cdot [(\quad)_1 \mathbf{u}] - \nabla \cdot (\quad)_2 - (\quad)_3 = 0. \tag{3.23}$$

Equation (3.23) is the *differential analogue* of Eq. (3.21). It states that driving gradients $\nabla(\quad)_2$, or equivalent mechanisms, $(\quad)_{1,3}$, compete to generate change, $\partial(\quad)/\partial t$. The term $\nabla \cdot (\quad)_2$ contains the transfer or resistance coefficients according to Table 3.1. These coefficients are scalar quantities for isotropic media, vectors for media with two-directional anisotropies, and tensors for media with three-directional anisotropies. Typical transfer coefficients are the scalar viscosity of Newtonian liquids, the vector conductivity (and mass diffusivity) in long-fiber composite materials, and the tensor permeability of three-dimensional porous media. As shown below, particular conservation equations are obtained by filling the parentheses of Eq. (3.23) with the appropriate variables.

Mass conservation (continuity equation)

For any fluid, conservation of mass is expressed by the scalar equation

$$\frac{\partial}{\partial t}(\rho)_1 + \nabla \cdot [(\rho)_1 \mathbf{u}] \quad \Longrightarrow$$

$$\frac{\partial \rho}{\partial t} + \nabla \cdot (\rho \mathbf{u}) = 0. \tag{3.24}$$

Hence, a velocity profile represents an *admissible (real) flow*, if and only if it satisfies the continuity equation. For incompressible fluids, Eq. (3.24) reduces to

$$\nabla \cdot \mathbf{u} = 0. \tag{3.25}$$

Momentum equation

For any fluid, the momentum equation is

$$\frac{\partial}{\partial t}(\rho \mathbf{u})_1 + \nabla \cdot [(\rho \mathbf{u})_1 \mathbf{u}] - \nabla \cdot (\mathbf{T})_2 - (\rho \mathbf{g})_3 = 0. \tag{3.26}$$

Since $\mathbf{T} = -p\mathbf{I} + \boldsymbol{\tau}$, the momentum equation takes the form

$$\rho \left(\frac{\partial \mathbf{u}}{\partial t} + \mathbf{u} \cdot \nabla \mathbf{u} \right) = \nabla \cdot (-p\mathbf{I} + \boldsymbol{\tau}) + \rho \mathbf{g}. \tag{3.27}$$

Equation (3.27) is a vector equation and can be decomposed further into three scalar components by taking the scalar product with the basis vectors of an appropriate orthogonal coordinate system. By setting $\mathbf{g} = -g\nabla z$, where z is the distance from an arbitrary reference elevation in the direction of gravity, Eq. (3.27) can be also expressed as

$$\rho\frac{D\mathbf{u}}{Dt} = \rho\left(\frac{\partial\mathbf{u}}{\partial t} + \mathbf{u}\cdot\nabla\mathbf{u}\right) = \nabla\cdot(-p\mathbf{I} + \boldsymbol{\tau}) + \nabla(-\rho gz)\,, \qquad (3.28)$$

where D/Dt is the substantial derivative introduced in Chapter 1. The momentum equation then states that the acceleration of a particle following the motion is the result of a net force expressed by the gradient of pressure, viscous, and gravity forces.

Mechanical energy equation

This takes the form

$$\frac{\partial}{\partial t}\left(\rho\frac{\mathbf{u}^2}{2}\right) + \mathbf{u}\cdot\nabla\left(\rho\frac{\mathbf{u}^2}{2}\right) = p(\nabla\cdot\mathbf{u}) - \nabla\cdot(p\mathbf{u}) - \boldsymbol{\tau}:\nabla\mathbf{u}$$
$$+\nabla\cdot(\boldsymbol{\tau}\cdot\mathbf{u}) + \rho(\mathbf{u}\cdot\mathbf{g})\,. \qquad (3.29)$$

To derive the above equation, we used the identities

$$\mathbf{u}\cdot\nabla p = \nabla\cdot(p\mathbf{u}) - p\nabla\cdot\mathbf{u}\,, \qquad \mathbf{u}\cdot\nabla\cdot\boldsymbol{\tau} = \nabla\cdot(\boldsymbol{\tau}\cdot\mathbf{u}) - \boldsymbol{\tau}:\nabla\mathbf{u}$$

and the continuity equation, Eq. (3.24).

Thermal energy equation

Conservation of thermal energy is expressed by

$$\rho\left[\frac{\partial U}{\partial t} + \mathbf{u}\cdot\nabla U\right] = [\boldsymbol{\tau}:\nabla\mathbf{u} + p\nabla\cdot\mathbf{u}] + \nabla(\kappa\nabla T) \pm \dot{H}_r\,, \qquad (3.30)$$

where U is the internal energy per unit mass, and $\dot{H}_r$ is the heat of reaction.

Temperature equation

By invoking the definition of the internal energy, $dU \equiv C_v dT$, Eq. (3.30) becomes,

$$\rho C_v\left(\frac{\partial T}{\partial t} + \mathbf{u}\cdot\nabla T\right) = \boldsymbol{\tau}:\nabla\mathbf{u} + p\nabla\cdot\mathbf{u} + \nabla(k\nabla T) \pm \dot{H}_r\,. \qquad (3.31)$$

For heat conduction in solids, i.e., when $\mathbf{u} = 0$, $\nabla\mathbf{u} = 0$, and $C_v = C$, the resulting equation is

$$\rho C\frac{\partial T}{\partial t} = \nabla(k\nabla T) \pm \dot{H}_r\,. \qquad (3.32)$$

For *phase change*, the latent heat rate per unit volume must be added as a source term to the energy equation.

Total energy and enthalpy equations

By adding Eqs. (3.29) and (3.30) and rearranging terms, we get

$$\rho \left[\frac{\partial}{\partial t} \left(\frac{\mathbf{u}^2}{2} + U \right) + \mathbf{u} \cdot \nabla \left(\frac{\mathbf{u}^2}{2} + gz + U \right) \right] = -\nabla \cdot \mathbf{u} + \nabla \cdot (\boldsymbol{\tau} \cdot \mathbf{u}) + \nabla \cdot (k \nabla T). \quad (3.33)$$

By invoking the definition of *enthalpy*, $H \equiv U + p/\rho$, we get

$$\nabla H = \nabla U + \nabla (pV) = \nabla U + p \nabla \left(\frac{1}{\rho} \right) + \frac{1}{\rho} \nabla p . \quad (3.34)$$

Equation (3.33) then becomes

$$\rho \left[\frac{\partial}{\partial t} \left(\frac{\mathbf{u}^2}{2} + U \right) + \mathbf{u} \cdot \nabla \left(\frac{\mathbf{u}^2}{2} + gz + H \right) \right] = -p \nabla \cdot \mathbf{u} + \nabla \cdot (\boldsymbol{\tau} \cdot \mathbf{u}) + \nabla \cdot (k \nabla T) . \quad (3.35)$$

The term $(p \nabla \cdot \mathbf{u})$ represents work done by expansion or compression. This term is important for gases and compressible liquids, but vanishes for incompressible liquids. Notice also that the viscous dissipation term disappears from the total energy and enthalpy equations.

The equations of motion of any incompressible fluid are tabulated in Tables 3.2 to 3.4 for the usual orthogonal coordinate systems. The above equations are specialized for incompressible, laminar flow of Newtonian fluids by means of Newton's law of viscosity

$$\mathbf{T} = -p \mathbf{I} + \boldsymbol{\tau} = -p \mathbf{I} + \eta \left[\nabla \mathbf{u} + \nabla (\mathbf{u})^T \right] . \quad (3.36)$$

In the context of this book, we mostly deal with continuity, and the three components of the momentum equation. The first four equations under consideration are commonly known as *equations of motion*.

Example 3.3.1

Repeat Example 3.2.2 by using now the differential form of the equations of Table 3.3. First derive the appropriate differential equations by simplifying the conservation equations; then state appropriate assumptions based on the geometry, the symmetry of the problem, and your intuition.

Solution:

We employ a cylindrical coordinate system with the z-axis aligned with the axis of

Continuity equation

$$\frac{\partial u_x}{\partial x} + \frac{\partial u_y}{\partial y} + \frac{\partial u_z}{\partial z} = 0$$

Momentum equation

$x-$component :

$$\rho \left(\frac{\partial u_x}{\partial t} + u_x \frac{\partial u_x}{\partial x} + u_y \frac{\partial u_x}{\partial y} + u_z \frac{\partial u_x}{\partial z} \right) =$$

$$= -\frac{\partial p}{\partial x} + \left[\frac{\partial \tau_{xx}}{\partial x} + \frac{\partial \tau_{yx}}{\partial y} + \frac{\partial \tau_{zx}}{\partial z} \right] + \rho g_x$$

$y-$component :

$$\rho \left(\frac{\partial u_y}{\partial t} + u_x \frac{\partial u_y}{\partial x} + u_y \frac{\partial u_y}{\partial y} + u_z \frac{\partial u_y}{\partial} \right) =$$

$$= -\frac{\partial p}{\partial y} + \left[\frac{\partial \tau_{xy}}{\partial x} + \frac{\partial \tau_{yy}}{\partial y} + \frac{\partial \tau_{zy}}{\partial z} \right] + \rho g_y$$

$z-$component :

$$\rho \left(\frac{\partial u_z}{\partial t} + u_x \frac{\partial u_z}{\partial x} + u_y \frac{\partial u_z}{\partial y} + u_z \frac{\partial u_z}{\partial z} \right) =$$

$$= -\frac{\partial p}{\partial z} + \left[\frac{\partial \tau_{xz}}{\partial z} + \frac{\partial \tau_{yz}}{\partial y} + \frac{\partial \tau_{zz}}{\partial z} \right] + \rho g_z$$

Table 3.2. *The equations of motion for incompressible fluids in Cartesian coordinates.*

Continuity equation

$$\frac{1}{r}\frac{\partial}{\partial r} + (ru_r)\frac{1}{r}\frac{\partial u_\theta}{\partial \theta} + \frac{\partial u_z}{\partial z} = 0$$

Momentum equation

$r-$component :

$$\rho\left(\frac{\partial u_r}{\partial t} + u_r\frac{\partial u_r}{\partial r} + \frac{u_\theta}{r}\frac{\partial u_r}{\partial \theta} - \frac{u_\theta^2}{r} + u_z\frac{\partial u_r}{\partial z}\right) =$$

$$= -\frac{\partial p}{\partial r} + \left[\frac{1}{r}\frac{\partial}{\partial r}(r\tau_{rr}) + \frac{1}{r}\frac{\partial \tau_{r\theta}}{\partial \theta} - \frac{\tau_{\theta\theta}}{r} + \frac{\partial \tau_{rz}}{\partial z}\right] + \rho g_r$$

$\theta-$component :

$$\rho\left(\frac{\partial u_\theta}{\partial t} + u_r\frac{\partial u_\theta}{\partial r} + \frac{u_\theta}{r}\frac{\partial u_\theta}{\partial \theta} + \frac{u_r u_\theta}{r} + u_z\frac{\partial u_\theta}{\partial z}\right) =$$

$$= -\frac{1}{r}\frac{\partial p}{\partial \theta} + \left[\frac{1}{r^2}\frac{\partial}{\partial r}(r^2\tau_{r\theta}) + \frac{1}{r}\frac{\partial \tau_{\theta\theta}}{\partial \theta} + \frac{\partial \tau_{\theta z}}{\partial z}\right] + \rho g_\theta$$

$z-$component :

$$\rho\left(\frac{\partial u_z}{\partial t} + u_r\frac{\partial u_z}{\partial r} + \frac{u_\theta}{r}\frac{\partial u_z}{\partial \theta} + u_z\frac{\partial u_z}{\partial z}\right) =$$

$$= -\frac{\partial p}{\partial z} + \left[\frac{1}{r}\frac{\partial}{\partial r}(r\tau_{rz}) + \frac{1}{r}\frac{\partial \tau_{\theta z}}{\partial \theta} + \frac{\partial \tau_{zz}}{\partial z}\right] + \rho g_z$$

Table 3.3. *The equations of motion for incompressible fluids in cylindrical coordinates.*

Continuity equation

$$\frac{1}{r^2}\frac{\partial}{\partial r}\left(r^2 u_r\right) + \frac{1}{r\sin\theta}\frac{\partial}{\partial\theta}\left(u_\theta \sin\theta\right) + \frac{1}{r\sin\theta}\frac{\partial u_\phi}{\partial\phi} = 0$$

Momentum equation

$r-$component :

$$\rho\left(\frac{\partial u_r}{\partial t} + u_r\frac{\partial u_r}{\partial r} + \frac{u_\theta}{r}\frac{\partial u_r}{\partial\theta} + \frac{u_\phi}{r\sin\theta}\frac{\partial u_r}{\partial\theta} - \frac{u_\theta^2 + u_\phi^2}{r}\right) = -\frac{\partial p}{\partial r}$$

$$+ \left[\frac{1}{r^2}\frac{\partial}{\partial r}\left(r^2\tau_{rr}\right) + \frac{1}{r\sin\theta}\frac{\partial}{\partial\theta}\left(\tau_{r\theta}\sin\theta\right) + \frac{1}{r\sin\theta}\frac{\partial\tau_{r\phi}}{\partial\phi} - \frac{\tau_{\theta\theta} + \tau_{\phi\phi}}{r}\right] + \rho g_r$$

$\theta-$component :

$$\rho\left(\frac{\partial u_\theta}{\partial t} + u_r\frac{\partial u_\theta}{\partial r} + \frac{u_\theta}{r}\frac{\partial u_\theta}{\partial\theta} + \frac{u_\phi}{r\sin\theta}\frac{\partial u_\theta}{\partial\theta} + \frac{u_r u_\theta}{r} - \frac{u_\phi^2\cot\theta}{r}\right) = -\frac{1}{r}\frac{\partial p}{\partial\theta}$$

$$+ \left[\frac{1}{r^2}\frac{\partial}{\partial r}\left(r^2\tau_{r\theta}\right) + \frac{1}{r\sin\theta}\frac{\partial}{\partial\theta}\left(\tau_{\theta\theta}\sin\theta\right) + \frac{1}{r\sin\theta}\frac{\partial\tau_{\theta\phi}}{\partial\phi} + \frac{\tau_{r\theta}}{r} - \frac{\cot\theta}{r}\tau_{\phi\phi}\right] + \rho g_\theta$$

$\phi-$component :

$$\rho\left(\frac{\partial u_\phi}{\partial t} + u_r\frac{\partial u_\phi}{\partial r} + \frac{u_\theta}{r}\frac{\partial u_\phi}{\partial\theta} + \frac{u_\phi}{r\sin\theta}\frac{\partial u_\phi}{\partial\phi} + \frac{u_\phi u_r}{r} + \frac{u_\theta u_\phi}{r}\cot\theta\right) =$$

$$= -\frac{1}{r\sin\theta}\frac{\partial p}{\partial\phi} + \left[\frac{1}{r^2}\frac{\partial}{\partial r}\left(r^2\tau_{r\phi}\right) + \frac{1}{r}\frac{\partial\tau_{\theta\phi}}{\partial\theta} + \frac{1}{r\sin\theta}\frac{\partial\tau_{\phi\phi}}{\partial\phi} + \frac{\tau_{r\phi}}{r} + 2\frac{\cot\theta}{r}\tau_{\theta\phi}\right] + \rho g_\phi$$

Table 3.4. *The equations of motion for incompressible fluids in spherical coordinates.*

symmetry of the pipe. It is obvious then that $u_r=u_\theta=0$; since the flow is axisymmetric, $\partial u_z/\partial\theta=0$. The continuity equation from Table 3.3 then yields $\partial u_z/\partial z=0$. Therefore, the axial velocity is only a function of r, $u_z=u_z(r)$. Using $g_z = -g\sin\phi$, the z-component of the momentum equation becomes

$$0 = -\frac{\partial p}{\partial z} + \frac{1}{r}\frac{\partial}{\partial r}(r\tau_{rz}) + \frac{\partial\tau_{zz}}{\partial z} + \rho g_z . \tag{3.37}$$

The above microscopic, differential equation has a form similar to the macroscopic one (final result of Example 3.2.2). As discussed in Chapter 5, Eq. (3.37) can be solved for the unknown velocity profile, $u_z(r)$, given an appropriate *constitutive equation* that relates velocity to viscous stresses.   □

3.4   Problems

3.1. Repeat Example 3.2.1 for the conservation of linear momentum. Assume that the control volume travels with the fluid, i.e., it is a *material volume.*

3.2. Derive the equation of change of mechanical energy under the conditions of Example 3.2.2.

3.3. Prove that the velocity in the surrounding liquid at distance $r > R(t)$ of the growing bubble of Example 3.2.3 is

$$u_r = \left(\frac{\rho_L - \rho_G}{\rho_L}\right)\frac{R^2(t)}{r^2}\frac{dR(t)}{dt} ,$$

using as a control volume either
(a) a fixed sphere of radius $r > R(t)$, or
(b) a sphere of constant mass with radius $r > R(t)$
that contains the growing bubble and the adjacent part of the liquid.

3.4. Starting from the macroscopic mechanical energy equation, Eq. (3.18), show how the corresponding differential one, Eq. (3.29), is obtained. Explain the physical significance of each of the terms in Eq. (3.29). Repeat for Eqs. (3.17) and (3.30), and Eqs. (3.16) and (3.35).

3.5. For a three-dimensional source at the origin, the radial velocity $\mathbf{u}$ is given by

$$\mathbf{u} = \frac{k}{r^2}\,\mathbf{e}_r ,$$

where k is a constant. This expression represents the Eulerian description of the flow. Determine the Lagrangian description of this velocity field. Show that the flow is dynamically admissible.

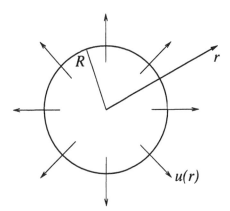

Figure 3.3. *Radial flow from a porous sphere.*

3.6. Analyze the *purely radial flow* of water through a porous sphere of radius R_0 by first identifying and then simplifying the appropriate equations of motion.

3.7. What are the appropriate conservation equations for steady, isothermal, compressible flow in a pipe?

3.8. The *momentum equation* for Newtonian liquid is

$$\rho\left(\frac{\partial u}{\partial t} + \mathbf{u} \cdot \nabla \mathbf{u}\right) = -\nabla p + \eta \nabla^2 \mathbf{u} + \rho \mathbf{g}\,.$$

Assuming that the liquid is incompressible, and by using vector-vector, vector-tensor, and differential operations, show how to derive the following equations:

(a) *Conservation of vorticity,* $\boldsymbol{\omega} = \nabla \times \mathbf{u}$

(b) *Kinetic energy change,* $E_k = 1/2(\mathbf{u} \cdot \mathbf{u})\rho$

(c) *Conservation of angular momentum,* $\mathbf{J}_\theta = \mathbf{r} \times \mathbf{J} = \mathbf{r} \times \rho \mathbf{u}$

Explain the physical significance of the terms in each equation.

3.9. *Incompressibility paradox* [7]. Here is a proof that the only velocity field that satisfies incompressibility is a zero velocity! Starting with

$$\nabla \cdot \mathbf{u} = 0\,, \tag{3.38}$$

where $\mathbf{u}$ is the velocity field, and using the divergence theorem, we find that

$$\int_S \mathbf{n} \cdot \mathbf{u}\, dS = \int_V \nabla \cdot \mathbf{u}\, dV = 0\,. \tag{3.39}$$

As a result of Eq. (3.38), there is a stream function, $\mathbf{A}$, such that

$$\mathbf{u} = \nabla \times \mathbf{A},$$

and, therefore, Eq. (3.39) implies that

$$\int_S \mathbf{n} \cdot (\nabla \times \mathbf{A}) \, dS = 0.$$

Using Stokes' theorem we get,

$$\oint_C (\mathbf{A} \cdot \mathbf{t}) \, d\ell = \int_S \mathbf{n} \cdot (\nabla \times \mathbf{A}) \, dS = 0.$$

The circulation of $\mathbf{A}$ is path-independent and, therefore, there exists a scalar function, ψ, such that

$$\mathbf{A} = \nabla \psi,$$

and

$$\mathbf{u} = \nabla \times \mathbf{A} = \nabla \times \nabla \psi = \mathbf{0}.$$

What went wrong in this derivation?

3.10. *Conservative force and work* [8]. A conservative force, $\mathbf{F}$, is such that

$$\mathbf{F} = -\nabla \phi,$$

where ϕ is a scalar field, called potential.
(a) Show that any work done by a conservative force is path-independent.
(b) Show that the sum of the potential and the kinetic energy of a system under only conservative force action is constant.
(c) Consider a sphere moving along an inclined surface in a uniform gravity field. Identify the developed forces, characterize them as conservative or not, and evaluate the work done by them during a translation $d\mathbf{r}$. Show that the system is not conservative. Under what conditions does the system approach a conservative one?

3.5   References

1. R.B. Bird, W.E. Stewart, and E.N. Lightfoot, *Transport Phenomena*, John Wiley & Sons, New York, 1960.

2. L.E. Scriven, *Intermediate Fluid Mechanics Lectures*, University of Minnesota, 1980.

3. R.L. Panton, *Incompressible Flow*, John Wiley & Sons, New York, 1984.

4. F. Cajori, *Sir Isaac Newton's Mathematical Principles*, University of California Press, Berkeley, 1946.

5. R.H. Kadlec, *Hydrodynamics of Wetland Treatment Systems, Constructed Wetlands for Wastewater Treatment*, Lewis Publishers, Chelsea, Michigan, 1989.

6. H.A. Stone, "A simple derivation of the time-dependent convective-diffusion equation for surfactant transport along a deforming interface," *Phys. Fluids A.* **2**, 111 (1990).

7. H.M. Schey, *Div, Grad, Curl and All That*, W.W. Norton & Company, Inc., New York, 1973.

8. R.R. Long, *Engineering Science Mechanics*, Prentice-Hall, Englewood Cliffs, NJ, 1963.

Chapter 4

STATIC EQUILIBRIUM OF FLUIDS AND INTERFACES

Fluids considered as continuum media are in *static equilibrium* when, *independently* of any stationary or moving frame of reference, there is no relative motion between any of their parts. Since, by definition, a fluid cannot support shear stresses without deforming continuously, static equilibrium is characterized by the *absence of shear stresses* or any other mechanism that gives rise to relative motion. Consequently, *velocity gradients* in static equilibrium do not exist. According to these definitions, a fluid is under static equilibrium even if it is subjected to a *rigid-body translation* and/or *rotation*, since these types of bulk motion do not involve relative motion between parts of the fluid. In fact, for these motions there is a reference frame moving and/or rotating with the velocity of the rigid-body motion such that the velocity of any part of the liquid with respect to the frame of reference is zero. Therefore, the only velocity and acceleration that can exist under static equilibrium are uniform and common to all parts of the fluid.

The lack of velocity gradients in static equilibrium implies that the only stress present is an *isotropic pressure* that is normal to fluid surfaces of any orientation. The pressure develops due to *body forces*, such as gravity and centrifugal forces, that counterbalance *contact forces*. The equilibrium between these forces is expressed by the *hydrostatic equation*. In static equilibrium, the state of stress is characterized by a *diagonal stress tensor* with components identical to the negative value of the pressure. Moreover, the mechanical pressure is identical to the *thermodynamic pressure* due to random molecular motions and collisions. Under flow conditions, the two are different from each other.

Consider two stratified *immiscible liquids* of different densities ρ_A and ρ_B, with one fluid on top of the other under the influence of gravity, or next to each other in a centrifugal field. At the area of contact, the density changes continuously from ρ_A to ρ_B over a short distance so that a discrete *macroscopic interface* develops [1].

Figure 4.1 shows the microscopic and macroscopic transition from one liquid to another.

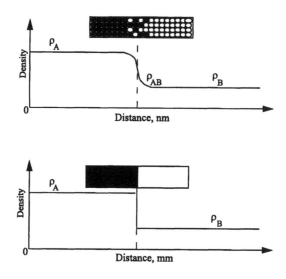

Figure 4.1. *On a macroscopic scale, in nm, the sharp continuous microscopic transition from ρ_A to ρ_B appears as a mathematical discontinuity.*

Due to anisotropic interactions of molecules adjacent to interfaces or free surfaces –interfaces between liquids and gases– the resulting state of stress and deformation are different from those of the bulk liquid. This is true for both static equilibrium and flow conditions. In static equilibrium, the state of the stress is modeled by the *Young–Laplace equation*, which relates the pressure discontinuity across the interface to *surface tension* and the *curvature* of the interface. The Young–Laplace equation combines with the hydrostatic equations of each phase to form the *generalized Laplace equation* for the interface configuration. The generalized Laplace equation then allows the determination of the curvature of the interface as a function of the associated body forces, surface tension, and the densities of the two phases.

This chapter combines both the mechanics of static equilibrium for single phases, and the mechanics of interfaces that are common boundaries to single phases. In most phenomena and applications, the two coexist. Typical examples are: density stratification of fluids; droplets and bubbles in equilibrium; wetting and static contact lines and angles; capillary climbing or dipping; buoyancy across interfaces; free surfaces of liquids trapped by solid substrate(s); thin film flows, spreading, and leveling of liquids on substrates; etc. Most of these phenomena are analyzed below either in examples or in problems at the end of the chapter.

4.1   Mechanics of Static Equilibrium

The most common body force on a control volume under static equilibrium is the force due to gravity,

$$\mathbf{F}_g = \mathbf{g} \, , \tag{4.1}$$

where $\mathbf{F}_g$ is the gravity force per unit mass, and $\mathbf{g}$ is the gravitational acceleration vector. Occasionally, we may also have electromagnetic forces of the form

$$\mathbf{F}_q = k \frac{Qq}{r^2} \mathbf{e}_r \, , \tag{4.2}$$

where k is a material constant, Q is the charge of the source of the force, q is the charge density of the static system, r is the distance between the two, and $\mathbf{e}_r$ is the force direction.

The only contact force acting along the boundaries of the system is a normal pressure force which is identical to the hydrostatic pressure, p^H, according to

$$\mathbf{F}_c = \mathbf{n} \cdot \mathbf{T} = \mathbf{n} \cdot (-p^H \mathbf{I}) = -p^H \mathbf{n} = -p \, \mathbf{n} \, , \tag{4.3}$$

where $\mathbf{n}$ is the unit normal vector to the boundary and $\mathbf{T}$ is the total stress tensor. The system can be in rigid-body translation and/or rotation with acceleration $\mathbf{a}$, which is also equivalent to an inertia force per unit mass of

$$\mathbf{F}_I = \mathbf{a} \, . \tag{4.4}$$

In the absence of convective momentum, the momentum or force balance for a system of fixed volume V, bounded by surface S, and moving with uniform velocity $\mathbf{u}$, is expressed as

$$\frac{d}{dt} \int_V \rho \mathbf{u} \, dV = \int_V \rho \mathbf{F}_b \, dV + \int_S \mathbf{F}_c \, dS \, , \tag{4.5}$$

where $\mathbf{F}_b = \mathbf{F}_g + \mathbf{F}_q$ and $\mathbf{F}_c = \mathbf{n} \cdot \mathbf{T} = -p\mathbf{n}$. For a constant volume V, application of the Gauss theorem, reduces Eq. (4.5) to

$$\int_V \left[\frac{\partial}{\partial t}(\rho \mathbf{u}) - \rho \mathbf{F}_b + \nabla p \right] dV = \mathbf{0} \, . \tag{4.6}$$

For an arbitrary control volume and constant density,

$$\mathbf{a} - \mathbf{F}_b + \frac{1}{\rho} \nabla p = \mathbf{0} \, , \tag{4.7}$$

which is the *hydrostatic equation*. Equation (4.7) relates the pressure distribution to density, acceleration, and the body force under hydrostatic conditions. As shown below, Eq. (4.7) can be solved easily by expressing all terms in gradient form.

The gravitational force according to *Newton's law of universal gravity* is

$$\mathbf{F}_g = \frac{GM_e}{r^2} \mathbf{e}_r , \tag{4.8}$$

where $G = 6.67 \times 10^{-8} dyn \cdot cm^2/gm^2$, M_e is Earth's mass, and r is the distance from Earth's center to the center of mass of the system. Since both M_e and r are large, Eq. (4.8) can be written as

$$\mathbf{F}_g = \frac{GM_e}{r_0^2} \nabla(\mathbf{r} - \mathbf{r}_0) = g_0 \, \mathbf{e}_r , \tag{4.9}$$

where r_0 is the local Earth's radius, and g_0 is the *local gravitational acceleration* (approximately 9.81 m/sec^2). The gravitational body force can then be cast in the form

$$\mathbf{F}_g = g_0 \nabla(\mathbf{r} - \mathbf{r}_0) \simeq -g_0 \nabla z \simeq -g \nabla z , \tag{4.10}$$

where z measures the local vertical distance from Earth's surface. Density is often a dependent variable, expressed as a function of the thermodynamic pressure, and temperature. Equations of state relate density to other thermodynamic properties, i.e., to pressure and temperature,

$$\rho = \rho(p, T) . \tag{4.11}$$

Density may change with pressure, depending on the *isothermal compressibility*, defined by

$$\beta \equiv \left[\frac{\partial(\ln \rho)}{\partial p} \right]_T , \tag{4.12}$$

and with temperature, depending on the *coefficient of thermal expansion*,

$$\alpha \equiv - \left[\frac{\partial(\ln \rho)}{\partial T} \right]_p . \tag{4.13}$$

When changes in pressure and temperature are small, a commonly used equation of state is

$$\rho = \rho_0 \left[1 + \beta(p - p_0) - \alpha(T - T_0) \right] , \tag{4.14}$$

where ρ_0, p_0, T_0 are respectively reference values for density, pressure and temperature.

Equation (4.14) includes the behavior of *incompressible liquids* in the limiting case of

$$\alpha = \beta = 0 \,. \tag{4.15}$$

Real and *ideal gases* may also be approximated by the ideal gas law when the *compressibility factor* Z is near unity as

$$\rho = \left(\frac{M}{ZRT}\right) p \,, \tag{4.16}$$

where M is the molecular weight, and R is the universal gas constant. (The approximation, of course, is exact for ideal gases where $Z=1$.) In isothermal processes, the instantaneous density is proportional to p, while in isentropic processes it is proportional to p^γ, where $\gamma \equiv C_p/C_v$ is the ratio of the specific heats at constant pressure and volume. In general polytropic processes, density is proportional to p^n, where n is a constant.

Equations (4.14) to (4.16) express density as a function of both pressure and temperature. Fluids for which the temperature can be neglected or eliminated so that density is a function of the pressure alone,

$$\rho = \rho(p) \,, \tag{4.17}$$

are called *barotropic*.

A way to eliminate the temperature dependence is to express both the pressure and temperature in terms of a unique new variable, for instance, the elevation in atmospheric air, or to describe the way the two vary during a process, e.g., by means of Eq. (4.16). In case of *barotropic gases* and *incompressible fluids*, the density-pressure term of the hydrostatic equation can be cast in gradient form as

$$\frac{1}{\rho} \nabla p = \begin{cases} \nabla\left(\frac{p}{\rho}\right) \,, & \text{incompressible fluid} \\[2mm] c\,\nabla(p^m) \,, & \text{barotropic gas} \end{cases} \tag{4.18}$$

where c and m are appropriate constants.

The *rigid-body translational acceleration* can be written as

$$\mathbf{a}_l = \nabla(\mathbf{a}_l \cdot \mathbf{r}) \,. \tag{4.19}$$

Similarly, the *rigid-body rotational acceleration* in uniform rotation is expressed as

$$\mathbf{a}_r = \frac{d}{dt}(\boldsymbol{\omega} \times \mathbf{r}) = \boldsymbol{\omega} \times \mathbf{u} = \boldsymbol{\omega} \times \boldsymbol{\omega} \times \mathbf{r} = -\frac{1}{2}\nabla(\boldsymbol{\omega} \times \mathbf{r})^2 \,. \tag{4.20}$$

Therefore, the acceleration can be cast in the form

$$\mathbf{a} = \nabla \Phi , \quad \Phi = \mathbf{a}_t \cdot \mathbf{r} - \frac{1}{2} (\boldsymbol{\omega} \times \mathbf{r})^2 . \tag{4.21}$$

If $\mathbf{a}$ is viewed as an inertia force, Φ can be interpreted as a kinetic energy related to the work done by the system due to changes in its position, $\mathbf{r}$. In fact, for a uniform circular motion

$$\Phi = -\frac{1}{2} (\boldsymbol{\omega} \times \mathbf{r})^2 = -\frac{1}{2} \omega^2 R^2 = -\frac{1}{2} \mathbf{u}^2 , \tag{4.22}$$

which is exactly the kinetic energy per unit mass. The same is true for linear motion where $d\mathbf{u} = \mathbf{a} \, dt$ and $d\mathbf{r} = \mathbf{u} \, dt$. Therefore, inertia forces can be viewed as the gradient of the kinetic energy potential.

By casting all terms in gradient form, Eq. (4.7) is expressed as

$$\nabla (\Phi + gz) + \frac{1}{\rho} \nabla p = \mathbf{0} . \tag{4.23}$$

By integrating the above equation between two arbitrary points 1 and 2, we get

$$\int_{p_1}^{p_2} \frac{dp}{\rho(p)} + g (z_2 - z_1) + a_t (x_2 - x_1) - \frac{\omega^2}{2} (R_2^2 - R_1^2) = 0 . \tag{4.24}$$

In Eq. (4.24), the pressure difference between any two points of a fluid in static equilibrium is given in terms of the elevation difference $(z_2 - z_1)$, the distance in the direction of the acceleration $(x_2 - x_1)$, the radii difference from the axis of rotation $(R_2^2 - R_1^2)$, the gravitational acceleration g, the uniform angular velocity ω, the translational acceleration a_t, and the density distribution between the points.

The pressure term can be integrated in case of
(a) *incompressible fluids* to

$$\int_{p_1}^{p_2} \frac{dp}{\rho} = \frac{p_2 - p_1}{\rho} ; \tag{4.25}$$

(b) *ideal gases under isothermal conditions* to

$$\int_{p_1}^{p_2} \frac{dp}{\rho} = \frac{RT}{M} \ln \frac{p_2}{p_1} ; \tag{4.26}$$

(c) *isentropic* or *polytropic ideal gases* to

$$\int_{p_1}^{p_2} \frac{dp}{\rho} = C \, (p_1^n - p_2^n) . \tag{4.27}$$

Equation (4.24) generalizes the steady *Bernoulli equation* for static incompressible liquids,

$$\frac{p_2 - p_1}{\rho} + g(z_2 - z_1) + \frac{u_2^2 - u_1^2}{2} = 0 \,, \qquad (4.28)$$

to include the effects of rigid-body motion as well. Indeed, for the last two terms of Eq. (4.24), we get

$$a_t(x_2 - x_1) - \frac{\omega^2}{2}(R_2^2 - R_1^2) = \frac{\Delta u_t^2 + \Delta u_r^2}{2} = \frac{\Delta u^2}{2} = \frac{u_2^2 - u_1^2}{2} \,. \qquad (4.29)$$

This similarity shows that gravity forces are gradients of the potential energy, inertia forces are gradients of the kinetic energy, and pressure forces are gradients of the pressure or strain energy.

Under relative flow and deformation, Eqs. (4.24) and (4.28) are generalized *along streamlines* to

$$\frac{p_2 - p_1}{\rho} + g(z_2 - z_1) + \frac{u_2^2 - u_1^2}{2} + \int_1^2 (\boldsymbol{\tau} : \nabla\mathbf{u})\, ds = 0 \,. \qquad (4.30)$$

The last term represents loss of mechanical energy to heat by *viscous dissipation* along the streamline.

Example 4.1.1 below highlights the application of the hydrostatic equation to an engineering problem, dealing with forces on bodies submerged in fluids. Example 4.1.2 highlights the derivation of the well known *Archimedes' principle of buoyancy* [4]: "bodies in fluids are subjected to buoyancy forces equal to the weight of the displaced fluid."

Example 4.1.1. Force on a submerged surface

Find the resultant force vector on the hemispherical cavity with radius R shown in Fig. 4.2. Assume that the center of the cavity is at a depth $h > R$, below the free surface of a liquid of density ρ.

Solution:
The pressure at a point on the hemisphere is given by $p = \rho g(h - R\sin\theta)$. The force on an infinitesimal area dS is then

$$d\mathbf{F} = -\mathbf{n}p\, dS = -\mathbf{e}_r p\, dS = -\mathbf{e}_r p\, R^2 \cos\theta d\theta d\phi \qquad \Longrightarrow$$

$$d\mathbf{F} = -(\sin\theta\mathbf{j} + \cos\theta\,\cos\phi\mathbf{i} + \cos\theta\,\sin\phi\mathbf{k})\,[\rho g(h - R\sin\theta)]R^2\,\cos\theta d\theta d\phi \,. \qquad (4.31)$$

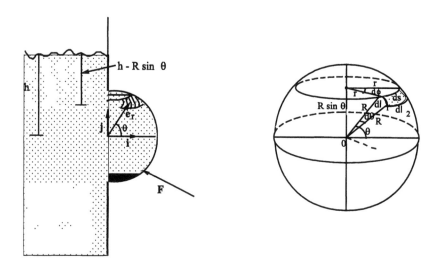

Figure 4.2. *Force on a submerged surface.*

Due to symmetry, the total force is

$$\mathbf{F} = -\int_{-\pi/2}^{\pi/2}\int_{-\pi/2}^{\pi/2}[(\ \ )\mathbf{i} + (\ \ )\mathbf{j}]d\theta d\phi = -\left(2\rho g R^2 \int_{-\pi/2}^{\pi/2}(h - R\sin\theta)\cos^2\theta d\theta\right)\mathbf{i}$$

$$-\left(\pi\rho g R^2 \int_{-\pi/2}^{\pi/2}(h - R\sin\theta)\sin\theta\cos\theta d\theta\right)\mathbf{j} \quad\Longrightarrow$$

$$\mathbf{F} = -(\pi\rho g R^2 h)\mathbf{i} + \left(\frac{2\pi\rho g R^3}{3}\right)\mathbf{j}. \tag{4.32}$$

The magnitude of the force is

$$F = \sqrt{F_x^2 + F_y^2} = \pi\rho g R^2\sqrt{h^2 + \frac{4}{9}R^2} = \pi\rho g R^3\sqrt{\frac{h^2}{R^2} + \frac{4}{9}}; \tag{4.33}$$

for its direction, we get

$$\phi_F = \arctan\frac{F_y}{F_x} = \arctan\frac{2R}{3h} < 0. \tag{4.34}$$

Hence, the force is directed downwards and inwards. As $R/h \to 0$, the force becomes horizontal. $\qquad\square$

Example 4.1.2. Archimedes' principle of buoyancy

A solid of volume V_s and density ρ_s is submerged in a stationary liquid of density ρ_L (Fig. 4.3). Show that the buoyancy force on the solid is identical to the weight of the liquid displaced by the solid.

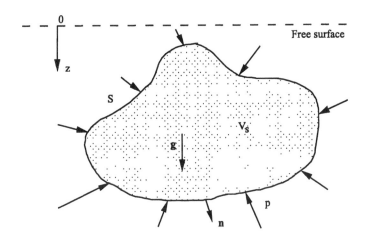

Figure 4.3. *Buoyancy force.*

Solution:
The contact force along the surface S is given by

$$\mathbf{F}_b = \int_S (-p\mathbf{n})\, dS = -\rho_L g \int_S z\mathbf{n}\, dS = -\rho_L g \int_V \nabla z\, dV = -\rho_L g \int_V \mathbf{k}\, dV \implies$$

$$\mathbf{F}_b = (-\rho_L g V_s)\, \mathbf{k}\,. \tag{4.35}$$

The last term is the buoyancy force directed upwards with magnitude equal to the weight of the displaced liquid. The solid will equilibrate under an external force $\mathbf{F}_e$, such that

$$\mathbf{F}_e = -\int_{V_s} \rho_s \mathbf{g}\, dV - \int_S (-p\mathbf{n})\, dS = -\rho_s \mathbf{g} V_s - (-\rho_L g V_s)\mathbf{k} \implies$$

$$\mathbf{F}_e = V_s g\, (\rho_L - \rho_s)\, \mathbf{k}\,. \tag{4.36}$$

Thus, the external force is directed either upwards or downwards depending on the density difference, $(\rho_L - \rho_s)$. In the case of $\rho_L = \rho_s$, the solid will equilibrate without any external force applied. □

Example 4.1.3. Archimedes' principle generalized to two fluids

Repeat Example 4.1.2 for a solid in equilibrium across a planar interface of two immiscible liquids of densities ρ_A and ρ_B, with $\rho_A > \rho_B$, as shown Fig. 4.4.

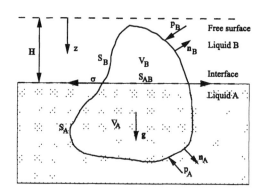

Figure 4.4. *Buoyancy force on a stationary body at the interface of two immiscible fluids.*

Solution:
The buoyancy force is

$$
\begin{aligned}
\mathbf{F}_b &= \int_{S_A+S_{AB}} -p\mathbf{n}\,dS + \int_{S_B+S_{AB}} -p\mathbf{n}\,dS = -\int_{V_A}(\nabla p)dV - \int_{V_B}(\nabla p)\,dV \\
&= -\int_{V_A} g\nabla[\rho_B H + \rho_A(z-H)]\,dV - \int_{V_B} g\nabla(\rho_B z)\,dV \\
&= -g\int_{V_A}\rho_A\nabla z\,dV - g\int_{V_B}\rho_B\nabla z\,dV \quad\Longrightarrow
\end{aligned}
$$

$$
\mathbf{F}_b = -g\,(\rho_A V_A - \rho_B V_B)\,\mathbf{k}. \tag{4.37}
$$

Therefore, the Archimedes' principle applies to stratified fluids as well, where each part of the solid is subjected to a buoyancy force equal to the weight of the displaced liquid. For the solid to equilibrate at the interface, an external force, $\mathbf{F}_e$, must be applied such that

$$
\mathbf{F}_e = -\mathbf{F}_b - \mathbf{F}_g = g(\rho_A V_A + \rho_B V_B)\mathbf{k} - \rho_s g(V_A + V_B)\mathbf{k} \quad\Longrightarrow
$$

$$
\mathbf{F}_e = g\,[\rho_A V_A + \rho_B V_B - \rho_s(V_A + V_B)]\,\mathbf{k}. \tag{4.38}
$$

The force is upwards, zero, or downwards depending on the value of the ratio

$$R = \frac{\rho_A V_A + \rho_B V_B}{\rho_s (V_A + V_B)} . \tag{4.39}$$

Notice that, in the case of a single liquid of density $\rho_A = \rho_B = \rho_L$,

$$R = \frac{\rho_L}{\rho_s} , \tag{4.40}$$

in agreement with Eq. (4.36) of Example 4.1.2. Note also that for a planar interface, there is a strictly horizontal surface tension force away from the body in all directions which, however, does not alter the vertical forces.   □

Example 4.1.4. Archimedes' principle in rigid-body motions

Derive Archimedes' principle of buoyancy for a solid of density ρ_s and volume V_s submerged in a liquid of density ρ_L which translates with velocity $\mathbf{U}$, and acceleration $\mathbf{a}$ (Fig 4.5). Assume no relative motion between solid and liquid.

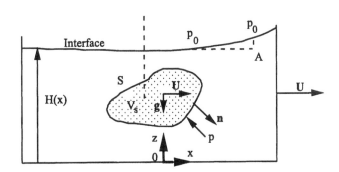

Figure 4.5. *Buoyancy under rigid-body translation.*

Solution:
The buoyancy force is given by

$$\mathbf{F}_b = \int_S (-p\mathbf{n}) \, dS = -\int_{V_s} \nabla p \, dV = -\int_{V_s} \nabla \left[p_0 + \rho_L g (H(x) - z) \right] \, dV$$

$$= \rho_L g \int_{V_s} \mathbf{k} \, dV - \rho_L g \int_{V_s} \frac{dH}{dx} \mathbf{i} \, dV = (\rho_L g V_s) \mathbf{k} - \rho_L g \int_{V_s} \frac{a}{g} \mathbf{i} \, dV \implies$$

$$\mathbf{F}_b = (\rho_L g V_s) \, \mathbf{k} - (\rho_L a V_s) \, \mathbf{i} . \tag{4.41}$$

Thus, the magnitude of the buoyancy force is

$$|\mathbf{F}_b| = \rho_L V_s \sqrt{g^2 + a^2} \,, \tag{4.42}$$

and its direction is given by

$$\tan\theta = -\frac{g}{a} \,. \tag{4.43}$$

To hold the solid body in place, an external force, $\mathbf{F}_e$, is required such that

$$\mathbf{F}_e = -\int_{V_s} \rho_s \, \mathbf{g} \, dV + \int_S \mathbf{n} \, p \, dS + \int_{V_s} \rho_s \, \mathbf{a} \, dV \,, \tag{4.44}$$

which reduces to

$$\mathbf{F}_e = \int_{V_s} (\rho_L - \rho_s)\,(\mathbf{g} - \mathbf{a})\,dV \,. \tag{4.45}$$

$\square$

4.2 Mechanics of Fluid Interfaces

A force balance on the interface S of two immiscible fluids A and B (Fig. 4.6) gives

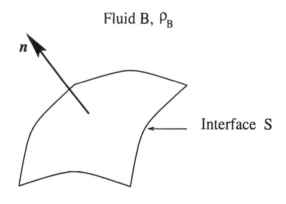

Fluid B, ρ_B

n

Interface S

Fluid A, ρ_A

Figure 4.6. *Interface of two immiscible fluids.*

$$\mathbf{n} \cdot (\mathbf{T}_B - \mathbf{T}_A) + \nabla_{II}\sigma + \mathbf{n}2H\sigma + \gamma\,(\mathbf{F}_g - \mathbf{a}) = \mathbf{0} \,, \tag{4.46}$$

where $\mathbf{n}$ is the unit normal vector to the interface pointing from liquid B to liquid A, σ is surface tension, γ is the surface density, $\mathbf{F}_g$ is the body force per unit mass, and $\mathbf{a}$ is the acceleration vector. Note that the gradient operator is defined in terms of local coordinates $(\mathbf{n}, \mathbf{t})$, i.e.,

$$\nabla_{II} = \mathbf{t} \, \frac{\partial}{\partial t} \, (\cdot) + \mathbf{n} \, \frac{\partial}{\partial n} \, (\cdot) \; . \tag{4.47}$$

The surface tension gradient, $\nabla_{II}\sigma$, which is, in general, present with surfactants at the vicinity of alternating curvature, and with nonisothermal interfaces, is responsible for *shear stress discontinuities* which may often initiate flow in thin films. In the absence of surfactant and temperature gradients, surface tension gradient is zero,

$$\nabla_{II}\sigma = \mathbf{0} \; . \tag{4.48}$$

Additionally, since the surface density is negligible, $\gamma = 0$, the equation reduces to the vector equation

$$\mathbf{n} \cdot (\mathbf{T}_B - \mathbf{T}_A) + \mathbf{n} \, 2H\sigma = \mathbf{0} \; . \tag{4.49}$$

The two components of the above equation are the *normal stress interface condition*,

$$\mathbf{n} \cdot [\mathbf{n} \cdot (\mathbf{T}_B - \mathbf{T}_A) + \mathbf{n}2H\sigma] = (p^B - p^A) + (\tau_{nn}^A - \tau_{nn}^B) - 2H\sigma = 0 \; , \tag{4.50}$$

and the *shear stress interface condition*,

$$\mathbf{t} \cdot [\mathbf{n} \cdot (\mathbf{T}_B - \mathbf{T}_A) + \mathbf{n} \, 2H\sigma] = \tau_{nt}^B - \tau_{nt}^A = 0 \; , \tag{4.51}$$

where τ_{nn}^i and τ_{nt}^i are normal and tangential shear stresses to the interface from the ith-fluid, i.e., they are stress components with respect to a natural coordinate system.

Equations (4.50) and (4.51) include the special case of *a free surface* when $\tau_{ij}^A = 0$, and $p^A = p_{gas}$. The mean curvature, $2H$, of a surface is necessary in order to account for the role of surface tension that gives rise to *normal stress discontinuities*.

The shape and curvature of surfaces are studied within the context of *differential geometry* [6]. Elements of the theory on surfaces, combined with surface tension mechanics, give rise to *capillary interfacial phenomena*, some of which are summarized below [7].

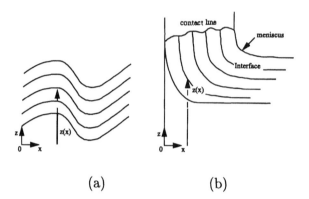

(a)   (b)

Figure 4.7. *Cylindrically symmetric fluid surfaces: (a) surface wave; (b) wall wetting and climbing by fluid.*

4.2.1   Interfaces in Static Equilibrium

Under no flow conditions and, therefore, zero viscous stresses, Eq. (4.49) simplifies to the Young–Laplace equation of capillarity that governs the configuration of interfaces under gravity and surface tension. Most classical and modern capillarity theories deal with interfaces that are *two-dimensional, cylindrical,* or *axisymmetric* (Fig. 4.7). Cylindrical means translational symmetry with constant curvature along straight lines known as the generators of the cylinder. Axisymmetric means rotational symmetry where the surface is generated by rotating a rigid curve around a fixed axis. In both cases, the position and the mean curvature of the interface are expressed in terms of a single surface coordinate and involve only ordinary derivatives. Therefore, the Young–Laplace equation [8],

$$\Delta p = 2H\sigma \,,$$

reduces to a nonlinear, second-order ordinary differential equation that can be solved either analytically or numerically. For other interface shapes, the Young–Laplace equation remains a second-order elliptic, nonlinear partial differential equation not amenable to analytical solution, and it is therefore solved numerically.

For interfaces and free surfaces with general configuration, the mean curvature $2H$ can be expressed as

$$2H\mathbf{n} = \frac{d\mathbf{t}}{ds} \,, \tag{4.52}$$

where $\mathbf{t}$ and $\mathbf{n}$ are the tangent and normal vectors, respectively, and s is the arc length.

For a *cylindrically symmetric surface*, e.g., the surface wave shown in Fig. 4.7, described by

$$z = z(x) \, , \tag{4.53}$$

or, equivalently, by

$$f(x, z) = z - z(x) \, , \tag{4.54}$$

the mean curvature is

$$2H = \frac{z_{xx}}{\sqrt{(1 + z_x^2)^3}} \, . \tag{4.55}$$

For a *rotationally symmetric surface*, e.g., liquid droplet on top of or hanging from a surface, as shown in Fig. 4.8, described by

$$z = z(r) \, , \tag{4.56}$$

or, equivalently, by

$$f(z, r) = z - z(r) \, , \tag{4.57}$$

the axisymmetric version of Eq. (4.52) yields

$$2H = \frac{z_r}{r(1 + z_r^2)^{1/2}} + \frac{z_{rr}}{(1 + z_r^2)^{3/2}} = \frac{1}{2} \frac{d}{dr} \left(\frac{r z_r}{\sqrt{1 + z_r^2}} \right) \, . \tag{4.58}$$

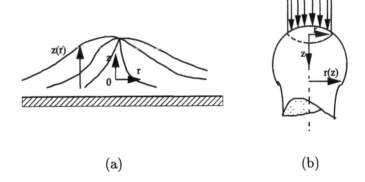

(a)   (b)

Figure 4.8. *Rotationally symmetric surfaces: (a) surface of droplet or liquid spreading on substrate, described by z=z(r); (b) swelling of liquid jet, described by r=r(z).*

The same surface can alternatively be described by

$$r = r(z) \, , \tag{4.59}$$

in which case,

$$2H = \frac{1}{r(1+r_z^2)^{1/2}} - \frac{r_{zz}}{(1+r_z^2)^{3/2}}.$$ (4.60)

The differential equation of interfacial statics is deduced easily from Eq. (4.46),

$$\mathbf{n}\,(p_A - p_B) + \mathbf{n}\,2H\sigma + \nabla_{II}\sigma + \gamma\,(\mathbf{F}_g - \mathbf{a}) = \mathbf{0}\,.$$ (4.61)

The difference $(p_B - p_A)$ is the pressure jump between the two bulk fluids. Thin films and membranes can be modeled as mathematical surfaces for which the same equation applies.

The component of Eq. (4.61) normal to the interface is

$$(p_A - p_B) + 2H\sigma + \gamma\mathbf{n}\cdot(\mathbf{F} - \mathbf{a}) = 0\,.$$ (4.62)

Its projection onto the tangent plane to the interface is

$$\nabla_{II}\sigma + \gamma\mathbf{P}_{II}\cdot(\mathbf{F} - \mathbf{a}) = \mathbf{0}\,,$$ (4.63)

where $\mathbf{P}_{II}$ is the surface projection tensor. Equation (4.63) reveals that film tension σ in any free–hanging film cannot be uniform since tension gradients must exist to offset the tangential component of the force due to gravity. These equations apply directly to soap films hanging in air, to lipid films supported in aqueous solutions, and to other mobile films. The general problem of determining the shape of fluid interfaces of uniform tension is relevant in a number of practical fields, including measurement of surface tension, wetting and spreading of liquids, application of thin films and coating, metal welding, bubbles and droplets, etc. [9, 10].

If the fluids on either side of an interface are incompressible, and body forces and accelerations are conservative, the equations of mechanical equilibrium of the bulk phases are

$$\nabla\,[p_A + \rho_A(\Phi + \Psi)] = \mathbf{0} \quad \text{and} \quad \nabla\,[p_B + \rho_B(\Phi + \Psi)] = \mathbf{0}\,,$$ (4.64)

where Φ and Ψ are, respectively, the gravitational and kinetic energy potentials. Subtracting one from the other gives

$$\nabla[p_A - p_B + (\rho_A - \rho_B)(\Phi + \Psi)] = \mathbf{0}\,,$$ (4.65)

which is integrated to

$$p_A - p_B + (\rho_A - \rho_B)(\Phi + \Psi) = (\rho_A - \rho_B)(\Phi_0 + \Psi_0) = \text{constant}\,.$$ (4.66)

The reference potential, $\Phi_0 + \Psi_0$, is taken where $p_A = p_B$, which is true at locations with planar interface. Equation (4.66) combines with Eq. (4.62) to yield the *generalized Laplace equation of capillarity*, along the interface

$$\mathbf{n}(\rho_B - \rho_A)(\Phi - \Phi_0 + \Psi - \Psi_0) = \mathbf{n}\, 2H\sigma + \nabla_{II}\sigma + \gamma\nabla(\Phi + \Psi) . \qquad (4.67)$$

In the common case of $\nabla_{II}\sigma = \mathbf{0}$ and $\gamma = 0$, this equation reduces to

$$2H = \frac{(\rho_B - \rho_A)(\Phi - \Phi_0 + \Psi - \Psi_0)}{\sigma} . \qquad (4.68)$$

If $\Psi = \Psi_0 = 0$ (absence of acceleration) and $\Phi = gz$, then

$$2H = \frac{(\rho_B - \rho_A)g(z - z_0)}{\sigma} = \frac{p_B - p_A}{\sigma} . \qquad (4.69)$$

This is the Young–Laplace equation for static interfaces. For planar interfaces, where $\Phi = \Phi_0$ and $\Psi = \Psi_0$, Eq. (4.69) reduces to

$$2H = 0 . \qquad (4.70)$$

Similar expressions can be derived for interfaces of known constant curvature, such as spheres and cylinders for which

$$2H = \frac{1}{R_1} + \frac{1}{R_2} = \frac{2}{R} , \qquad (4.71)$$

and

$$2H = \frac{1}{R_1} + \frac{1}{\infty} = \frac{1}{R} , \qquad (4.72)$$

respectively.

For interface configurations with variable curvature, the equation of capillarity is a nonlinear differential equation, which can be solved numerically to obtain the shape of the interface, e.g, the shapes of sessile and pendant drops and bubbles, static menisci, and downward and upward fluid spikes on substrates.

Example 4.2.1. Measurement of surface tension

The *Wilhelmy plate* is a widely used method to measure surface tension [11]. A plate of known dimensions S, L, and h, and density ρ_s is being pulled from a liquid of density ρ_B and surface tension σ in contact with air of density ρ_A, Fig. 4.9.

(a) What is the measured force $F(\sigma)$?

(b) The datum of force, F_0, is the force when the plate is entirely submerged in

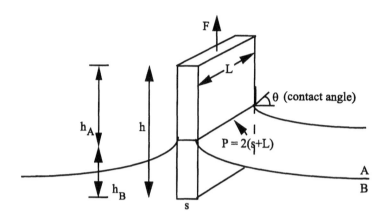

Figure 4.9. *The Wilhelmy plate for measuring surface tension.*

Phase A. An improved method is to position the plate in such a way that surface tension can be measured without knowing the densities ρ_A and ρ_B. What is the appropriate positioning?

Solution:
The net force exerted by Fluid A on the submerged part is

$$F_A = -\rho_A \, gh_A SL \,, \tag{4.73}$$

while net force exerted by Fluid B on the remaining part is

$$F_B = \rho_B \, gh_B SL \,. \tag{4.74}$$

The surface tension force on the plate is

$$F_\sigma = -\sigma P \cos \theta \,. \tag{4.75}$$

The weight of plate is

$$W = -\rho_s g V = -\rho_s g(h_A + h_B)SL \,. \tag{4.76}$$

The total force balance thus gives

$$F_A + F_B + F_\sigma + W = 0 \,. \tag{4.77}$$

Therefore,

$$F = gSL\left[h_A(\rho_S + \rho_A) + h_B(\rho_S - \rho_B)\right] + \sigma P \cos\theta \ . \tag{4.78}$$

The datum of force measured with the plate entirely submerged in Phase A is

$$F_0 = gSL(h_A + h_B)\left[\rho_S - \rho_A\right] \ . \tag{4.79}$$

Therefore,

$$F - F_0 = gSLh_B(\rho_A - \rho_B) + \sigma P \cos\theta \ . \tag{4.80}$$

To make $F - F_0 = \sigma P \cos\theta$, we must have $h_B = 0$, i.e., the bottom of the plate must be lined up with the level of the free surface.   □

Example 4.2.2. Capillary rise on vertical wall

The Young–Laplace equation for a translationally symmetric meniscus in the presence of gravity reduces to

$$\frac{\sigma}{R} = -g\,\Delta(\rho z) \ . \tag{4.81}$$

From differential geometry,

$$\frac{1}{R} = \frac{d\phi}{ds} \ , \tag{4.82}$$

where ϕ is the local inclination, and s is the arc length (Fig. 4.10). Calculate the shape of the resulting static meniscus.

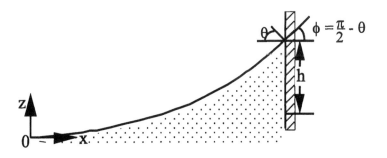

Figure 4.10. *Height of capillary climbing.*

Solution:
From differential geometry we have

$$\frac{dz}{ds} = \sin\phi \quad \text{and} \quad \frac{dx}{ds} = \cos\phi \ . \tag{4.83}$$

Therefore,

$$\frac{d\phi}{ds} = \frac{d\phi}{dz}\frac{dz}{ds} = \frac{d\phi}{dz}\sin\phi = -\frac{g\Delta\rho}{\sigma}z\,. \tag{4.84}$$

Integration gives

$$\cos\phi = -\frac{g\Delta\rho}{2\sigma}z^2 + c\,. \tag{4.85}$$

If we let $\cos\phi = 1$ at $z = 0$, then

$$\cos\phi - 1 = -2\sin^2(\phi/2) = -\frac{g\Delta\rho}{2\sigma}z^2\,, \tag{4.86}$$

which yields

$$z = \pm 2\sqrt{\sigma/g\Delta\rho}\,\sin(\phi/2)\,. \tag{4.87}$$

Height of rise on a vertical plane wall

The meniscus intersects the wall at a contact angle θ and a height h above the free surface. Therefore,

$$h = 2\sqrt{\frac{\sigma}{g\Delta\rho}}\,\sin(\pi/4 - \theta/2)\,. \tag{4.88}$$

If $\theta < \dfrac{\pi}{2}$, h is positive; if $\theta > \dfrac{\pi}{2}$, h is negative.

Alternatively, capillary climbing is solved by utilizing the mean curvature expressions given by Eqs. (4.52) to (4.60). For this case, the curvature is given by Eq. (4.55), with which Eq. (4.69) becomes

$$\frac{z_{xx}}{(1 + z_x^2)^{3/2}} = 2H = \frac{g\Delta\rho}{\sigma}\,. \tag{4.89}$$

Integration gives

$$-\frac{2}{(1 + z_x)^{1/2}} = g\frac{\Delta\rho}{\sigma}\,x + c\,, \tag{4.90}$$

subject to the boundary condition,

$$z_x(x = 0) = \tan\phi = 0\,. \tag{4.91}$$

The rest of the solution is left as an exercise for the reader.   □

Example 4.2.3. Interfacial tension by sessile droplet

The Young–Laplace equation relates interfacial tension to the local mean curvature of an interface and to the pressure jump across the interface. Consider the droplet of Liquid A shown in Fig 4.11 in contact with a solid surface and submerged in a Liquid B.

(a) If interfacial tension over a meniscus in the presence of a known body force field is uniform, explain how to determine the interfacial tension from measurements of the curvature without actually measuring the pressure jump.

(b) Show that, in static equilibrium, the interfacial tension over a meniscus must be uniform, i.e., $\nabla_{II}\sigma = 0$.

(c) In a standard sessile drop method of determining interfacial tension, only the dimensions h and d of an axisymmetric drop and the density difference $(\rho_B - \rho_A)$ are measured. Explain how these measurements replace direct measurement of local curvature (Fig. 4.11).

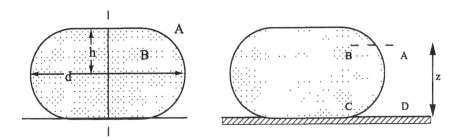

Figure 4.11. *Axisymmetric sessile drop.*

Solution:
Consider the points marked as A, B, C, and D in Fig. 4.11.

(a) Across a meniscus we have

$$p_B - p_A = 2H\sigma .\qquad(4.92)$$

From hydrostatics,

$$\left.\begin{array}{c} p_D - p_A = \rho_A\, gz \\ p_C - p_B = \rho_B\, gz \\ p_C = p_D \end{array}\right\} \quad\Longrightarrow\quad p_B - p_A = -(\rho_B - \rho_A)\, gz .\qquad(4.93)$$

Combining Eqs. (4.92) and (4.93), we obtain

$$\sigma = \frac{gz}{2H}\left(\rho_A - \rho_B\right).\qquad(4.94)$$

(b) Under the assumption of negligible interface mass ($\gamma{=}0$), Eq. (4.61) becomes

$$\mathbf{n}\,(p_B - p_A) + \mathbf{n}\,2H\sigma + \nabla_{II}\sigma = \mathbf{0}\,. \tag{4.95}$$

For the normal component of the above equation, we get

$$\mathbf{n}\cdot\mathbf{n}\,(p_B - p_A) + \mathbf{n}\cdot\mathbf{n}\,2H\sigma + \mathbf{n}\cdot\nabla_{II}\sigma = 0 \quad\Longrightarrow$$

$$(p_B - p_A) + 2H\sigma + \mathbf{n}\cdot\nabla_{II}\sigma = 0\,. \tag{4.96}$$

Since

$$\mathbf{n}\cdot\nabla_{II}\sigma = 0\,, \tag{4.97}$$

Eq. (4.96) reduces to

$$p_B - p_A + 2H\sigma = 0\,. \tag{4.98}$$

Substitution of this result into Eq. (4.95) gives

$$\nabla_{II}\sigma = \mathbf{0}\,. \tag{4.99}$$

(c) The surface of the droplet is given by $z{=}z(x,y)$, as shown in Fig. 4.12. From Eq. (4.60), we get

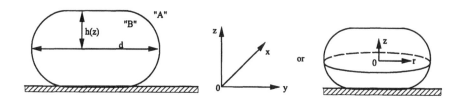

Figure 4.12. *Indirect curvature measurement by sessile droplet.*

$$2H = \frac{z_{xx}\left(1 + z_{yy}^2\right) - 2z_x z_y z_{xy} + 2z_{yy}(1 + z_x^2)}{(1 + z_x^2 + z_y^2)^{3/2}}\,. \tag{4.100}$$

The solution to this equation contains two constants which can be determined by applying the boundary conditions,

$$x = y = 0, \quad z = z_{max} \quad\text{or}\quad \partial z/\partial r = 0\,. \tag{4.101}$$

The description is complete by the additional condition

$$r = \pm\frac{d}{2}, \quad z_r \to \infty, \quad z = z_{max} - h\,, \tag{4.102}$$

which determines h. $\square$

4.3 Problems

4.1 A balloon is said to be in the "taut state" if the gas within is at a pressure just above the ambient pressure and thus completely distends the bag of the balloon. Otherwise, the balloon is said to be in a "limp state." In the taut state, any further increase in pressure causes gas to be released through a relief valve until the inside and outside pressures are again equal. Therefore, the balloon remains at essentially constant volume.

Consider a research balloon carrying a total load of M kg. The balloon is motionless at an equilibrium state at an altitude of h meters $(T = 0^0 C)$ in an adiabatically stratified atmosphere. What will be the effect of throwing a sack of sand of mass m overboard:
(a) if the balloon is initially in the taut state, and
(b) if the balloon is initially in the limp state?

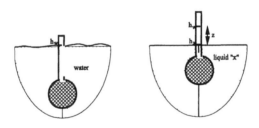

Figure 4.13. *Schematic of a pycnometer.*

4.2. The *pycnometer* shown in Fig. 4.13 is the most commonly used device for measuring density [12]. The heavy spherical head forces the pycnometer to submerge to a depth h_w in water. The same pycnometer, submerged in a liquid X of unknown density ρ_x, reads depth h_x.
(a) Find a working equation that estimates the unknown density ρ_x in terms of ρ_w and $z = h_w - h_x$.
(b) The sugar concentration of a natural juice, say grape juice, alters its density according to the expression

$$\rho_x = (1 - y)\,\rho_w + y\rho_s\,,$$

where y is the mole fraction of the dissolved sugar of density $\rho_s = 1.3 \; gr/cm^3$.

Derive the working equation of y vs. z. (The price at which wineries buy grapes from producers is largely based on y!)

4.3. Consider a cubic container of 2 m side that contains 0.8 m^3 of water at rest, in contact with still air. Calculate the resulting pressure distribution and the free surface profile for the following rigid-body motions of the container:
(a) No motion at all
(b) Vertical upward motion at speed $u=2$ m/sec
(c) Horizontal motion at speed $u=2$ m/sec, parallel to itself
(d) Horizontal motion at speed $u=2$ m/sec, in the direction of one of its diagonals
(e) Diagonal motion upwards at speed $u=2$ m/sec and angle 45°, with the container always parallel to itself
(f) Rotation at 10 rpm

4.4. *Static equilibrium of a rotating meniscus.* If surface-tension effects are negligible, what is the equilibrium shape of the interface between equal volumes of two liquids, A and B ($\rho_B > \rho_A$), contained in an open cylindrical vessel rotating about its axis, oriented vertically at the earth's surface?

What is the shape of the free surface of Liquid A? Where does the free surface intersect the wall of the container? Where does the liquid-liquid interface intersect the wall of the container? What is the maximum volume of liquid (equal volumes of A and B) that can be contained by the vessel rotating at a given angular velocity?

4.5. My son Charis (Papanastasiou), an elementary school, fourth grade beginner, came from school one day excited by a science experiment demonstrated in class by his teacher. A couple of raisins and baking soda were placed in a glass in which water and few drops of vinegar were added. This resulted in formation of gas bubbles. Some of these bubbles ascended to the free surface and ruptured, while others were deposited on the walls and onto the raisins. After two to five minutes the raisins covered with attached, nearly hemispherical gas bubbles, ascended from the bottom to the free surface, then sank to the bottom and remained there for a while before repeating the same motion.
(a) Explain in detail the physics involved in each stage of the experiment and justify the use of raisins with the soda/vinegar liquid.
(b) To quantify the phenomenon, assume spherical raisins of radius $R=0.8$ cm and density $\rho=1.1$ gr/cm^3 in a soda/vinegar foaming solution of depth $H=8$ cm and density $\rho_s=1$ gr/cm^3 at the time of the periodic motion. The average diameter of the deposited gas bubbles is 1 mm. Based on these, calculate the number of gas bubbles deposited at the inception of the motion of the raisin.
(c) Study the periodic motion by assuming that (i) all gas bubbles during the ascent remain attached while growing in size due to the diminishing external hydrostatic

pressure; (ii) at the free surface as many bubbles are ruptured as required to sink the raisin; (iii) during sinking, gas bubbles shrink in size due to the increasing hydrostatic pressure; and (iv) the raisins have zero velocity at the bottom and at the free surface.

4.6. *Water density stratification* [12]. The temperature of the water underneath the frozen surface of a lake varies linearly from 0°C to 4°C at a depth of 3 m. Describe the motion of a spherical ice piece of radius $R=1$ ft and density $\rho=0.999$ gr/cm^3 dropped with impact velocity of $u=5$ m/sec at the top. Neglect any temperature and size variations of the ice sphere. What would be the qualitative effect by taking such variations into account? Neglect also any viscous drag forces opposing the motion of the sphere.

4.7. *Self-gravitating fluid* [13]. In *gaseous stars* the gravitational attraction of distant parts provides the body force on fluid volumes. The density of such self-gravitating fluid is related to the induced gravitational potential by

$$\nabla^2 \phi = 4\pi g\rho \, .$$

(a) What is the resulting equation of static equilibrium?
(b) Show that, in case of spherically symmetric density and pressure distribution, this equation becomes

$$\frac{d}{dr}\left(\frac{r^2}{\rho}\frac{dp}{dr}\right) = -4\pi gr^2\rho \, .$$

(c) Under what conditions can a solution to the above equation be found? Solve the equation for uniform density and for uniform pressure throughout.

4.8. *Hydrostatics and capillarity.* The long cylinder, shown in Fig. 4.14, is filled with oil on top of water of given physical characteristics (density, viscosity, surface tension). A perfectly spherical air bubble of uniform pressure is trapped at the bottom. If the ambient pressure is p_0, what is the pressure inside the air bubble? Plot the pressure distribution along the axis of the cylinder.

4.9. Consider droplets of Fluid A of density ρ_A and radius R in another fluid of density ρ_B and interfacial tension σ_{AB}, under conditions of static equilibrium at rest or at rigid-body motion. Among these conditions and physical properties, what are the most favorable for ideal spherical droplet shape? Consider gravity and/or centrifugal fields or the absence of them, for example in space. Or, consider externally applied pressure or vacuum (without droplet evaporation) and combinations of them. Consider also the influence of vertical surface tension variation induced by temperature gradient.

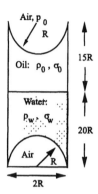

Figure 4.14. *Equilibrium of different fluids with surface tension.*

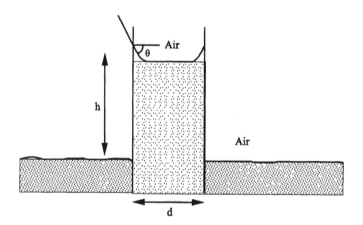

Figure 4.15. *Capillary rise.*

4.10. *Capillary force between parallel plates.* Consider two solid surfaces parallel to each other, separated by a small distance d. Suppose the gap is partly filled with liquid in the form of a captive drop with perimeter P and contact area on each surface A. Prove that the total capillary force tending to draw the two plates together is given by

$$F = \sigma P \sin\theta + \frac{\sigma(2A)\cos\theta}{d},$$

where θ is the contact angle, regarded as uniform along both contact lines (in which case the captive drop should be circular).

4.11. *Capillary rise.* Express the height h over the level of the pool liquid which the liquid climbs inside a capillary of diameter d. The liquid has density ρ_L and surface tension σ, and the contact angle is θ (Fig. 4.15).

4.12. *Hydrostatics and surface tension* [12]. Consider an infinitely long horizontal liquid container with the cross section shown in Fig. 4.16 in contact with stationary air of pressure $p=0$.

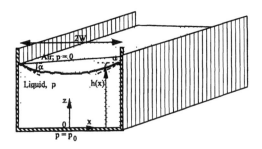

Figure 4.16. *Free surface in slightly inclined channel.*

The surface tension σ of the liquid is small such that the curvature of the cylindrically symmetric free surface is

$$2H \approx \frac{d^2 h}{dx^2} \, ,$$

and the contact angle, a, between the liquid and the side walls is small such that,

$$a \approx \sin a \approx \tan a \, .$$

(a) Derive the equation that describes the free surface shape by combining the Young–Laplace equation across the free surface with the hydrostatic equation within the liquid, given that the pressure at the bottom, $z=0$, is uniform, $p(z = 0)=p_0$.

(b) Nondimensionalize the equation and identify the resulting dimensionless numbers.

(c) Show that in the case of $\sigma=0$, a planar free surface is obtained.

(d) Find the shape of the free surfaces when $\sigma \neq 0$.

(e) Show that the solution for (d) yields the result in (c) in the limit of $\sigma \to 0$.

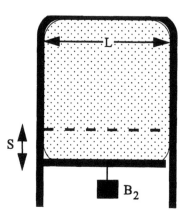

Figure 4.17. *Work done by surface tension forces.*

4.13. *Surface tension measurement.* Figure 4.17 shows a method for measuring surface tension. The method is based on a Π-shaped metallic wire equipped with a freely sliding wire where weights can be attached. The wire is introduced into soap water where a thin soap film is formed. The resultant surface tension force is larger than the weight B_1 of the sliding wire in a vertical arrangement. With additional weight (B_2) the film is stretched to a new position at a distance S downstream.
(a) By considering the work done by the attached weight against the surface tension forces, show that

$$\sigma = \frac{B_2}{2L},$$

where the factor 2 accounts for the two film-free surfaces.
(b) By considering the hydrostatic pressure distribution, show that the thickness of the film cannot be uniform throughout. What is the vertical thickness profile?

4.4   References

1. S.A. Rice and P. Gray, *The Statistical Mechanics of Simple Liquids*, Wiley & Sons, New York, 1965.

2. F. Cajori, *Sir Isaac Newton's Mathematical Principles*, University of California Press, Berkeley, 1946.

3. T. Carmady and H. Kobus, *Hydrodynamics by Daniel Bernoulli and Hydraulics by Joham Bernoulli*, Dover Publications, Inc., New York, 1968.

4. T.L. Heath, *The Works of Archimedes*, Dover Publications, Cambridge University Press, 1897; also reprint by Dover Publications, New York.

5. L.E. Scriven, *Intermediate Fluid Mechanics Lectures*, University of Minnesota, Minneapolis, 1981.

6. M.M. Lipshutz, *Theory and Problems of Differential Geometry*, Schaum's Outline Series, McGraw-Hill, New York, 1969.

7. L.E. Scriven, *Interfacial Phenomena Lectures*, University of Minnesota, Minneapolis, 1982.

8. L.E. Scriven and C.V. Sterling, "The Marangoni effects: Cause of and resistance to surface moments," *Nature* **187**, 186 (1960).

9. V.G. Levich, *Physicochemical Hydrodynamics*, Prentice-Hall, Englewood Cliffs, NJ, 1962.

10. J.T. Davies and E.K. Rideal, *Interfacial Phenomena*, Academic Press, London, 1963.

11. *Krüss Instruments for Rheology and Surface Chemistry*, Krüss USA, Charlotte, NC, 1990.

12. T.C. Papanastasiou, *Applied Fluid Mechanics*, Prentice-Hall, Englewood Cliffs, 1994.

13. G.K. Batchelor, *An Introduction to Fluid Dynamics*, Cambridge University Press, Cambridge, 1979.

Chapter 5

THE NAVIER–STOKES EQUATIONS

In a general isothermal flow the primary unknowns include the pressure, p, and the components of the velocity vector, $\mathbf{u}$, which are functions of the spatial coordinates and time. In problems involving an unknown boundary, such as a free surface or an interface, the location h of this boundary is usually determined by the *kinematic equation*. Other flow variables, such as the nine components of the stress tensor, the residence time, the streamlines, etc., can be evaluated *a posteriori* once the primary unknowns have been calculated.

It is a standard mathematical rule that, in order to determine a number of unknowns, as many equations that contain these unknowns must be solved. Therefore, the five equations – continuity, the three momentum components, and the kinematic equation– can be solved for the five unknowns (p, u_x, u_y, u_z, h) only if the stress components τ_{ij}, in these equations are expressed in terms of the primary unknowns. *Constitutive relations*, i.e., relations between stress and strain or rate of strain, do exactly that [1]. These relations must account for all events during the motion of fluid particles that contribute to the local stress. The measurable effects of these events, i.e., the strain and rate of strain, can be quantified, for example, by the *Rivlin–Ericksen strain tensors* [2]

$$\mathbf{A}_n \;=\; \frac{d^n}{d\tau^n}\left[\mathbf{G}_t\left(\tau\right)\right]\big|_{\tau=t}\,, \tag{5.1}$$

where $\mathbf{G}_t(\tau)$ is the *Green relative-strain tensor*. The stress tensor, $\mathbf{T}$, can then be approximated as a functional of the form

$$\mathbf{T} \;=\; -p\,\mathbf{I} + f(\mathbf{A}_1, \mathbf{A}_2, \ldots, \mathbf{A}_k)\,. \tag{5.2}$$

Since the Rivlin–Ericksen tensors are directly related to the rate of strain tensor, $\mathbf{D}$, and its substantial derivatives, Eq. (5.2) can be cast in the form

$$\mathbf{T} \;=\; -p\,\mathbf{I} + f(\mathbf{D}, \dot{\mathbf{D}}, \ddot{\mathbf{D}}, \ldots)\,, \tag{5.3}$$

165

where, the dots indicate differentiation with respect to time. The number of tensors needed to approximate the stress well is proportional to the memory of the fluid, i.e., the ability of the fluid to remember and return to its undeformed state once the gradients driving the flow are removed. This memory arises from the elastic properties of the involved molecules which, when stretched, compressed, or twisted, develop internal forces that resist deformation and tend to spontaneously return to their undeformed or unstressed state.

The *zero-order fluid* is defined as a fluid at rest where the molecules move in Brownian fashion, which gives rise to a thermodynamic pressure (stress) proportional to density (strain). The resulting stress is determined fully by the zero-order Rivlin–Ericksen tensor,

$$\mathbf{A}_0 \equiv \mathbf{I}, \tag{5.4}$$

and, since the only stress at rest is the isotropic pressure, p, the resulting constitutive equation for the zero-order fluid is

$$\mathbf{T} = -p\mathbf{I}. \tag{5.5}$$

Newtonian or *first-order fluids* have small, stiff molecules and exhibit no memory. The local stress is entirely due to the local deformation which excludes any strain– which incorporates history effects and any rate of strain derivatives that violate the localization of the rate of strain. Thus, the constitutive equation for Newtonian fluids is

$$\mathbf{T} = -p\,\mathbf{I} + \alpha_1 \mathbf{D}, \tag{5.6}$$

where, $\alpha_1 = 2\eta$, η being the viscosity of the fluid. For Newtonian fluids, η is independent of the rate of strain.

Including higher-order derivatives leads to *fluids with memory*, the simplest of which is the *second-order fluid* described by

$$\mathbf{T} = -p\,\mathbf{I} + \alpha_1 \mathbf{D} + a_2 \mathbf{D}^2 + a_3 \dot{\mathbf{D}}. \tag{5.7}$$

In the above expressions, the parameters α_i are material constants or functions. As the elasticity and memory of the fluid increases, progressively more terms are required to approximate the stress. These liquids are best approximated by *integral constitutive equations* of the form [3]

$$\mathbf{T} = -p\,\mathbf{I} + \int_{\infty}^{t} \mathbf{M}\left(t - t'\right) \left[H\left(I, II\right) \mathbf{C}_t^{-1}\left(t'\right) \right] dt', \tag{5.8}$$

where $\mathbf{M}\left(t - t'\right)$ is a time-dependent memory function, and $\mathbf{C}_t^{-1}\left(t'\right)$ is the *Finger tensor*. The term in square brackets is a known strain-dependent kernel function in which H is a function of the first and second invariants I and II, defined as $I = tr\left(\mathbf{C}_t^{-1}(t')\right)$ and $II = tr\left(\mathbf{C}_t(t')\right)$. Constitutive equations for *fluids with memory* (i.e., *viscoelastic fluids*) are beyond the scope of this book and will not be discussed further.

5.1 The Newtonian Liquid

The stress tensor for Newtonian liquids,

$$\mathbf{T} = -p\,\mathbf{I} + 2\eta\mathbf{D} = -p\,\mathbf{I} + \eta[\nabla\mathbf{u} + (\nabla\mathbf{u})^T]\,, \tag{5.9}$$

includes the isotropic mechanical pressure which, under static equilibrium, is identical to the thermodynamic pressure. The mechanical pressure at a point is the average value of the total normal force on three mutually perpendicular surfaces. Furthermore, volume expansion or contraction of compressible fluids is included in the rate-of-strain tensor, which contributes to the normal stress differently from the viscous contribution. This contribution is also isotropic and is, therefore, equivalent to pressure.

Equation (5.9) originates from the equation proposed by Stokes [4] in 1845, i.e.,

$$\mathbf{T} = -(p - \eta_v\nabla\cdot\mathbf{u})\,\mathbf{I} + \eta\left[\nabla\mathbf{u} + (\nabla\mathbf{u})^T - \frac{2}{3}\nabla\cdot\mathbf{u}\right]\,, \tag{5.10}$$

where η_v is the *bulk viscosity* related to the viscosity, η, by

$$\eta_v = \lambda + \frac{2}{3}\eta\,. \tag{5.11}$$

According to Stokes' hypothesis, the *second viscosity coefficient*, λ, is taken to make $\eta_v = 0$. The constitutive equation resulting from this assumption is the Newton-Poisson law of viscosity,

$$\mathbf{T} = -p\,\mathbf{I} + \lambda(\nabla\cdot\mathbf{u})\mathbf{I} + \eta[\nabla\mathbf{u} + (\nabla\mathbf{u})^T]\,. \tag{5.12}$$

This relation is appropriate for low molecular weight fluids under laminar flow conditions [6]. Traditionally, the total stress tensor is expressed as a combination of an isotropic pressure and viscous contributions, i.e., $\mathbf{T} = -p\,\mathbf{I} + \boldsymbol{\tau}$.

For turbulent flow, $\boldsymbol{\tau}$ is expressed as

$$\boldsymbol{\tau} = \boldsymbol{\tau}^{(\ell)} + \boldsymbol{\tau}^{(R)}\,, \tag{5.13}$$

where $\boldsymbol{\tau}^{(\ell)}$ is the laminar stress tensor corresponding to time-averaged turbulent kinematics, and $\boldsymbol{\tau}^{(R)}$ is the turbulent *Reynolds stress tensor* [6] which is a function of turbulent velocity fluctuations.

Substitution of Eq. (5.12) into the momentum stress equation results in the Navier–Poisson equation

$$\rho\left(\frac{\partial \mathbf{u}}{\partial t} + \mathbf{u} \cdot \nabla \mathbf{u}\right) = -\nabla p + (\lambda + \eta)\nabla(\nabla \cdot \mathbf{u}) + \eta \nabla^2 \mathbf{u} + \rho\, \mathbf{g}\,, \qquad (5.14)$$

for viscous, compressible, laminar flow of a Newtonian liquid. For incompressible flow (i.e., $\nabla \cdot \mathbf{u}{=}0$), the above equation reduces to the Navier–Stokes equation

$$\rho\left(\frac{\partial u}{\partial t} + \mathbf{u} \cdot \nabla \mathbf{u}\right) = -\nabla p + \eta \nabla^2 \mathbf{u} + \rho \mathbf{g}\,. \qquad (5.15)$$

Historically, the development of the Navier–Stokes equation can be considered as being based on *Euler equation* [7]

$$\rho\left(\frac{\partial \mathbf{u}}{\partial t} + \mathbf{u} \cdot \nabla \mathbf{u}\right) = -\nabla p + \rho \mathbf{g}\,, \qquad (5.16)$$

which is valid for inviscid flow. Navier extended the Euler equation by adding a stress contribution due to forces between molecules in motion. About the same time, Cauchy presented his equation [9],

$$\rho\left(\frac{\partial \mathbf{u}}{\partial t} + \mathbf{u} \cdot \nabla \mathbf{u}\right) = \nabla \cdot \mathbf{T} + \rho \mathbf{g}\,. \qquad (5.17)$$

For simple, fully developed, laminar pipe flow, the above equations predict the well-known expression for volumetric flow rate, Q,

$$Q = \frac{\pi R^4}{8\eta}\left(\frac{\Delta p}{\Delta L}\right), \qquad (5.18)$$

where $\Delta P/\Delta L$ is the constant pressure gradient across the pipe. This result was validated experimentally by Hagen and Poiseuille in capillary flow in tubes (see Chapter 6).

The Navier–Stokes equations were also rederived by Maxwell using kinetic theory [10]. Most of these developments can be found in Lamb's *Treatise on the Mathematical Theory of Motion of Fluids* [11], and in *Hydrodynamics* by Basset [12]. An interesting overview is also given by Whitaker [13].

Zero-order liquids were examined in Chapter 4 with the discussion on static equilibrium. Inviscid fluids, which are idealizations of real fluids, can be considered as Newtonian liquids of zero viscosity. This chapter is restricted to incompressible Newtonian liquids, which follow Newton's law of viscosity for incompressible liquids,

$$\mathbf{T} = -p\,\mathbf{I} + \eta[\nabla \mathbf{u} + (\nabla \mathbf{u})^T]\,. \qquad (5.19)$$

The viscosity, η, is, in general, a function of temperature and concentration and a weak function of pressure. Unless otherwise stated, the viscosity here will be treated as constant. The viscosity of gases can be approximated by molecular dynamics, based on the assumption that the primary source of shear stress is the microscopic transfer of momentum by random molecular motion, as

$$\eta = \frac{2}{3d^2} \sqrt{\frac{MkT}{\pi^3}} , \qquad (5.20)$$

where d and M are, respectively, the molecular diameter and mass, k is the Boltzmann constant, and T is the temperature [1]. Thus, the viscosity of gases increases with temperature and molecular weight, decreases with molecular size, and is independent of pressure. The viscosity of liquids is difficult to model by molecular dynamics. Experiments, however, show that the viscosity is virtually independent of pressure and decreases with temperature.

Equation (5.19) is a tensor equation equivalent to nine scalar equations corresponding to the components T_{ij}, according to

$$T_{ij} = -p\,\delta_{ij} + \eta \left(\frac{\partial u_i}{\partial x_j} + \frac{\partial u_j}{\partial x_i} \right), \quad i,j = 1,2,3. \qquad (5.21)$$

By virtue of the fact that the moment of momentum of a material volume is zero (which requires cancellation of shear stress on adjacent perpendicular surfaces), the viscous stress tensor is symmetric, i.e., $\tau_{ij} = \tau_{ji}$. Therefore, there are only six independent stress components. The components of the Newtonian constitutive relation in Cartesian, cylindrical, and spherical coordinates are tabulated in Table 5.1

Example 5.1.1

Find the x-component of the Navier–Stokes equation in Cartesian coordinates for an incompressible fluid.

Solution:
From Table 3.2, we have

$$\rho \left(\frac{\partial u_x}{\partial t} + u_x \frac{\partial u_x}{\partial x} + u_y \frac{\partial u_x}{\partial y} + u_z \frac{\partial u_x}{\partial z} \right) = -\frac{\partial p}{\partial x} + \left[\frac{\partial \tau_{xx}}{\partial x} + \frac{\partial \tau_{yx}}{\partial y} + \frac{\partial \tau_{zx}}{\partial z} \right] + \rho g_x .$$

Substituting the three stress components as given by Table 5.1,

$$\tau_{xx} = 2\eta \frac{\partial u_x}{\partial x}, \quad \tau_{yx} = \eta \left(\frac{\partial u_x}{\partial y} + \frac{\partial u_y}{\partial x} \right), \quad \tau_{zx} = \eta \left(\frac{\partial u_x}{\partial z} + \frac{\partial u_z}{\partial x} \right),$$

Cartesian coordinates (x, y, z)	Cylindrical coordinates (r, θ, z)
$\tau_{xx} = 2\eta \dfrac{\partial u_x}{\partial x}$	$\tau_{rr} = 2\eta \dfrac{\partial u_r}{\partial r}$
$\tau_{yy} = 2\eta \dfrac{\partial u_y}{\partial y}$	$\tau_{\theta\theta} = 2\eta \left(\dfrac{1}{r} \dfrac{\partial u_\theta}{\partial \theta} + \dfrac{u_r}{r} \right)$
$\tau_{zz} = 2\eta \dfrac{\partial u_z}{\partial z}$	$\tau_{zz} = 2\eta \dfrac{\partial u_z}{\partial z}$
$\tau_{xy} = \tau_{yx} = \eta \left(\dfrac{\partial u_x}{\partial y} + \dfrac{\partial u_y}{\partial x} \right)$	$\tau_{r\theta} = \tau_{\theta r} = \eta \left[r \dfrac{\partial}{\partial r} \left(\dfrac{u_\theta}{r} \right) + \dfrac{1}{r} \dfrac{\partial u_r}{\partial \theta} \right]$
$\tau_{xz} = \tau_{zx} = \eta \left(\dfrac{\partial u_x}{\partial z} + \dfrac{\partial u_z}{\partial x} \right)$	$\tau_{rz} = \tau_{zr} = \eta \left[\dfrac{\partial u_z}{\partial r} + \dfrac{\partial u_r}{\partial z} \right]$
$\tau_{yz} = \tau_{zy} = \eta \left(\dfrac{\partial u_y}{\partial z} + \dfrac{\partial u_z}{\partial y} \right)$	$\tau_{z\theta} = \tau_{\theta z} = \eta \left[\dfrac{\partial u_\theta}{\partial z} + \dfrac{1}{r} \dfrac{\partial u_z}{\partial \theta} \right]$

Spherical coordinates (r, θ, ϕ)

$$\tau_{rr} = 2\eta \frac{\partial u_r}{\partial r}$$

$$\tau_{\theta\theta} = 2\eta \left[\frac{1}{r} \frac{\partial u_\theta}{\partial \theta} + \frac{u_r}{r} \right]$$

$$\tau_{\phi\phi} = 2\eta \left[\frac{1}{r \sin\theta} \frac{\partial u_\phi}{\partial \phi} + \frac{u_r}{r} + \frac{u_\theta \cot\theta}{r} \right]$$

$$\tau_{r\theta} = \tau_{\theta r} = \eta \left[r \frac{\partial}{\partial r} \left(\frac{u_\theta}{r} \right) + \frac{1}{r} \frac{\partial u_r}{\partial \theta} \right]$$

$$\tau_{r\phi} = \tau_{\phi r} = \eta \left[\frac{1}{r \sin\theta} \frac{\partial u_r}{\partial \phi} + r \frac{\partial}{\partial r} \left(\frac{u_\phi}{r} \right) \right]$$

$$\tau_{\theta\phi} = \tau_{\phi\theta} = \eta \left[\frac{\sin\theta}{r} \frac{\partial}{\partial \theta} \left(\frac{u_\theta}{\sin\theta} \right) + \frac{1}{r \sin\theta} \frac{\partial u_\theta}{\partial \phi} \right]$$

Table 5.1. *Components of the viscous stress tensor τ for incompressible Newtonian fluid in various coordinate systems.*

we obtain,

$$\rho\left(\frac{\partial u_x}{\partial t} + u_x\frac{\partial u_x}{\partial x} + u_y\frac{\partial u_x}{\partial y} + u_z\frac{\partial u_x}{\partial z}\right) =$$

$$= -\frac{\partial p}{\partial x} + \eta\left(2\frac{\partial^2 u_x}{\partial x^2} + \frac{\partial^2 u_y}{\partial y^2} + \frac{\partial^2 u_y}{\partial x\partial y} + \frac{\partial^2 u_x}{\partial z^2} + \frac{\partial^2 u_z}{\partial z\partial x}\right) + \rho g_x$$

$$= -\frac{\partial p}{\partial x} + \eta\left(\frac{\partial^2 u_x}{\partial x^2} + \frac{\partial^2 u_x}{\partial y^2} + \frac{\partial^2 u_x}{\partial z^2}\right) + \rho g_x + \eta\frac{\partial}{\partial x}\left(\frac{\partial u_x}{\partial x} + \frac{\partial u_y}{\partial y} + \frac{\partial u_z}{\partial z}\right).$$

Due to mass conservation, the last term is identically equal to zero. Therefore,

$$\rho\left(\frac{\partial u_x}{\partial t} + u_x\frac{\partial u_x}{\partial x} + u_y\frac{\partial u_x}{\partial y} + u_z\frac{\partial u_x}{\partial z}\right) = -\frac{\partial p}{\partial x} + \eta\left(\frac{\partial^2 u_x}{\partial x^2} + \frac{\partial^2 u_x}{\partial y^2} + \frac{\partial^2 u_x}{\partial z^2}\right) + \rho g_x.$$

$\square$

The equations of motion for incompressible Newtonian fluids in Cartesian, cylindrical, and spherical coordinates are tabulated in Tables 5.2 to 5.4.

5.2 Alternative Forms of the Navier–Stokes Equations

The momentum equation,

$$\rho\frac{D\mathbf{u}}{Dt} = \rho\left(\frac{\partial\mathbf{u}}{\partial t} + \mathbf{u}\cdot\nabla\mathbf{u}\right) = -\nabla p + \eta\nabla^2\mathbf{u} + \rho\mathbf{g}, \tag{5.22}$$

states that the rate of change of momentum per unit volume is caused by pressure, viscous, and gravity forces.

Fluids of vanishingly small viscosity, $\eta\approx 0$, are called *inviscid*. Their motion is described by the *Euler equation*, which is easily obtained by neglecting the viscous term in the Navier–Stokes equation [7],

$$\rho\left(\frac{\partial\mathbf{u}}{\partial t} + \mathbf{u}\cdot\nabla\mathbf{u}\right) = -\nabla p + \rho\mathbf{g}. \tag{5.23}$$

The Euler equation also holds for irrotational flow (i.e., $\boldsymbol{\omega}=\mathbf{0}$) of incompressible liquids (i.e., $\nabla\cdot\mathbf{u}=0$) with nonzero viscosity. Under these conditions, according to the identities

$$\nabla^2\mathbf{u} = \nabla(\nabla\cdot\mathbf{u}) - \nabla\times(\nabla\times\mathbf{u}) = \mathbf{0} - \nabla\times\boldsymbol{\omega} = \mathbf{0}, \tag{5.24}$$

Continuity equation

$$\frac{\partial u_x}{\partial x} + \frac{\partial u_y}{\partial y} + \frac{\partial u_z}{\partial z} = 0$$

Momentum equation

x−component :

$$\rho\left(\frac{\partial u_x}{\partial t} + u_x\frac{\partial u_x}{\partial x} + u_y\frac{\partial u_x}{\partial y} + u_z\frac{\partial u_x}{\partial z}\right) =$$

$$= -\frac{\partial p}{\partial x} + \eta\left(\frac{\partial^2 u_x}{\partial x^2} + \frac{\partial^2 u_x}{\partial y^2} + \frac{\partial^2 u_x}{\partial z^2}\right) + \rho g_x$$

y−component :

$$\rho\left(\frac{\partial u_y}{\partial t} + u_x\frac{\partial u_y}{\partial x} + u_y\frac{\partial u_y}{\partial y} + u_z\frac{\partial u_y}{\partial z}\right) =$$

$$= -\frac{\partial p}{\partial y} + \eta\left(\frac{\partial^2 u_y}{\partial x^2} + \frac{\partial^2 u_y}{\partial y^2} + \frac{\partial^2 u_y}{\partial z^2}\right) + \rho g_y$$

z−component :

$$\rho\left(\frac{\partial u_z}{\partial t} + u_x\frac{\partial u_z}{\partial x} + u_y\frac{\partial u_z}{\partial y} + u_z\frac{\partial u_z}{\partial z}\right) =$$

$$= -\frac{\partial p}{\partial z} + \eta\left(\frac{\partial^2 u_z}{\partial x^2} + \frac{\partial^2 u_z}{\partial y^2} + \frac{\partial^2 u_z}{\partial z^2}\right) + \rho g_z$$

Table 5.2. *The equations of motion for incompressible Newtonian fluid in Cartesian coordinates* (x, y, z).

Continuity equation

$$\frac{1}{r}\frac{\partial}{\partial r}(ru_r) + \frac{1}{r}\frac{\partial u_\theta}{\partial \theta} + \frac{\partial u_z}{\partial z} = 0$$

Momentum equation

$r-$component :

$$\rho\left(\frac{\partial u_r}{\partial t} + u_r\frac{\partial u_r}{\partial r} + \frac{u_\theta}{r}\frac{\partial u_r}{\partial \theta} - \frac{u_\theta^2}{r} + u_z\frac{\partial u_z}{\partial z}\right) =$$

$$= -\frac{\partial p}{\partial r} + \eta\left[\frac{\partial}{\partial r}\left(\frac{1}{r}\frac{\partial}{\partial r}(ru_r)\right) + \frac{1}{r^2}\frac{\partial^2 u_r}{\partial \theta^2} - \frac{2}{r^2}\frac{\partial u_\theta}{\partial \theta} + \frac{\partial^2 u_r}{\partial z^2}\right] + \rho g_r$$

$\theta-$component :

$$\rho\left(\frac{\partial u_\theta}{\partial t} + u_r\frac{\partial u_\theta}{\partial r} + \frac{u_\theta}{r}\frac{\partial u_\theta}{\partial \theta} + \frac{u_r u_\theta}{r} + u_z\frac{\partial u_\theta}{\partial z}\right) =$$

$$= -\frac{1}{r}\frac{\partial p}{\partial \theta} + \eta\left[\frac{\partial}{\partial r}\left(\frac{1}{r}\frac{\partial}{\partial r}(ru_\theta)\right) + \frac{1}{r^2}\frac{\partial^2 u_\theta}{\partial \theta^2} + \frac{2}{r^2}\frac{\partial u_r}{\partial \theta} + \frac{\partial^2 u_\theta}{\partial z^2}\right] + \rho g_\theta$$

$z-$component :

$$\rho\left(\frac{\partial u_z}{\partial t} + u_r\frac{\partial u_z}{\partial r} + \frac{u_\theta}{r}\frac{\partial u_z}{\partial \theta} + u_z\frac{\partial u_z}{\partial z}\right) -$$

$$= -\frac{\partial p}{\partial z} + \eta\left[\frac{1}{r}\frac{\partial}{\partial r}\left(r\frac{\partial u_z}{\partial r}\right) + \frac{1}{r^2}\frac{\partial^2 u_z}{\partial \theta^2} + \frac{\partial^2 u_z}{\partial z^2}\right] + \rho g_z$$

Table 5.3. *The equations of motion for incompressible Newtonian fluid in cylindrical coordinates* (r, θ, z).

Continuity equation

$$\frac{1}{r^2}\frac{\partial}{\partial r}(r^2 u_r) + \frac{1}{r\sin\theta}\frac{\partial}{\partial\theta}(u_\theta\sin\theta) + \frac{1}{r\sin\theta}\frac{\partial u_\phi}{\partial\phi} = 0$$

Momentum equation

$r-$component :

$$\rho\left(\frac{\partial u_r}{\partial t} + u_r\frac{\partial u_r}{\partial r} + \frac{u_\theta}{r}\frac{\partial u_r}{\partial\theta} + \frac{u_\phi}{r\sin\theta}\frac{\partial u_r}{\partial\phi} - \frac{u_\theta^2 + u_\phi^2}{r}\right) =$$

$$= -\frac{\partial p}{\partial r} + \eta\left[\nabla^2 u_r - \frac{2}{r^2}u_r - \frac{2}{r^2}\frac{\partial u_\theta}{\partial\theta} - \frac{2}{r^2}u_\theta\cot\theta - \frac{2}{r^2\sin\theta}\frac{\partial u_\phi}{\partial\phi}\right] + \rho g_r$$

$\theta-$component :

$$\rho\left(\frac{\partial u_\theta}{\partial t} + u_r\frac{\partial u_\theta}{\partial r} + \frac{u_\theta}{r}\frac{\partial u_\theta}{\partial\theta} + \frac{u_\phi}{r\sin\theta}\frac{\partial u_\theta}{\partial\phi} + \frac{u_r u_\theta}{r} - \frac{u_\phi^2\cot\theta}{r}\right) =$$

$$= -\frac{1}{r}\frac{\partial p}{\partial\theta} + \eta\left[\nabla^2 u_\theta + \frac{2}{r^2}\frac{\partial u_r}{\partial\theta} - \frac{u_\theta}{r^2\sin^2\theta} - \frac{2\cos\theta}{r^2\sin^2\theta}\frac{\partial u_\phi}{\partial\phi}\right] + \rho g_\theta$$

$\phi-$component :

$$\rho\left(\frac{\partial u_\phi}{\partial t} + u_r\frac{\partial u_\phi}{\partial r} + \frac{u_\theta}{r}\frac{\partial u_\phi}{\partial\theta} + \frac{u_\phi}{r\sin\theta}\frac{\partial u_\phi}{\partial\phi} + \frac{u_\phi u_r}{r} + \frac{u_\theta u_\phi}{r}\cot\theta\right) =$$

$$= -\frac{1}{r\sin\theta}\frac{\partial p}{\partial\phi} + \eta\left[\nabla^2 u_\phi - \frac{u_\phi}{r^2\sin^2\theta} + \frac{2}{r^2\sin\theta}\frac{\partial u_r}{\partial\phi} + \frac{2\cos\theta}{r^2\sin^2\theta}\frac{\partial u_\theta}{\partial\phi}\right] + \rho g_\phi$$

where

$$\nabla^2 u_i = \frac{1}{r^2}\frac{\partial}{\partial r}\left(r^2\frac{\partial u_i}{\partial r}\right) + \frac{1}{r^2\sin\theta}\frac{\partial}{\partial\theta}\left(\sin\theta\frac{\partial u_i}{\partial\theta}\right) + \frac{1}{r^2\sin^2\theta}\frac{\partial^2 u_i}{\partial\phi^2}$$

Table 5.4. *The equations of motion for incompressible Newtonian fluid in spherical coordinates* (r, θ, ϕ).

the viscous contribution, $\nabla^2\mathbf{u}$, vanishes. Now, by using the identity

$$\mathbf{u}\cdot\nabla\mathbf{u} = \nabla\left(\frac{\mathbf{u}^2}{2}\right) - \mathbf{u}\times(\nabla\times\mathbf{u}) = \frac{\nabla\mathbf{u}^2}{2},\qquad(5.25)$$

and the expression $\mathbf{g}=-g\nabla z$, the Euler equation simplifies to

$$\rho\left(\frac{\partial\mathbf{u}}{\partial t} + \frac{\nabla\mathbf{u}^2}{2}\right) = \nabla\left(-p-\rho gz\right).\qquad(5.26)$$

The integration of the Euler equation between two points, 1 and 2, along a streamline yields the *Bernoulli equation* [8]:

$$\int_1^2\left[\rho\left(\frac{\partial\mathbf{u}}{\partial t}\right) + \nabla(\rho\frac{\mathbf{u}^2}{2}+p+\rho gz)\right] d\ell = 0.\qquad(5.27)$$

In steady-state, the above equation becomes

$$\left(\frac{\mathbf{u}^2}{2} + \frac{p}{\rho} + gz\right)_2 - \left(\frac{\mathbf{u}^2}{2} + \frac{p}{\rho} + gz\right)_1 = 0.\qquad(5.28)$$

The *generalized Bernoulli equation* for a viscous incompressible flow is given by

$$\left(\rho\frac{\mathbf{u}^2}{2}+p+\rho gz\right)_1 - \left(\rho\frac{\mathbf{u}^2}{2}+p+\rho gz\right)_2 = \int_1^2(\nabla\cdot\boldsymbol{\tau})\,d\ell.\qquad(5.29)$$

When the viscosity is large, and viscous forces dominate the flow, i.e., when

$$\rho\left|\frac{D\mathbf{u}}{Dt}\right| \ll \eta\left|\nabla^2\mathbf{u}\right|,\qquad(5.30)$$

the flow is called *creeping*. For creeping flow, the momentum equation reduces to the *Stokes equation*,

$$-\nabla p + \eta\nabla^2\mathbf{u} + \rho\mathbf{g} = \mathbf{0}.\qquad(5.31)$$

Conservation equations of secondary field variables, such as the vorticity vector, can be obtained by taking the curl of the momentum equation. For compressible Newtonian liquids, we get

$$\frac{\partial\boldsymbol{\omega}}{\partial t} + \mathbf{u}\cdot\nabla\boldsymbol{\omega} = \boldsymbol{\omega}\cdot\nabla\mathbf{u} - \boldsymbol{\omega}(\nabla\cdot\mathbf{u}) + \nu\,\nabla^2\boldsymbol{\omega}.\qquad(5.32)$$

The left-hand terms represent the time change and convection of vorticity, respectively. The terms in the right-hand side represent *intensification of vorticity* by *vortex stretching* and by volume expansion and *diffusion of vorticity*, with the kinematic viscosity acting as a diffusivity coefficient.

The Navier–Stokes equation can also be converted into the *equation of change of circulation*, Γ, defined by

$$\Gamma \equiv \oint_C \mathbf{u} \cdot d\mathbf{r} = \oint_C (\mathbf{u} \cdot \mathbf{t})\, dr , \qquad (5.33)$$

where C is a closed curve within the flow field (Chapter 1). As shown below, the concept of circulation is directly related to the normal component of the vorticity vector enclosed within a surface S and bounded by C. Consider, for instance, any surface S having as boundary the closed curve C. By invoking the Stokes theorem, we have

$$\int_S (\mathbf{n} \cdot \boldsymbol{\omega})\, dS = \int_S (\mathbf{n} \cdot (\nabla \times \mathbf{u}))\, dS = \oint_C (\mathbf{u} \cdot \mathbf{t})\, dr = \Gamma . \qquad (5.34)$$

In terms of Γ, the momentum equation is expressed as

$$\frac{D\Gamma}{Dt} = \int_C \mathbf{g} \cdot d\mathbf{r} - \int_C \frac{1}{\rho} \nabla p \cdot d\mathbf{r} + \int_C \frac{1}{\rho} (\nabla \cdot \boldsymbol{\tau}) \cdot d\mathbf{r} , \qquad (5.35)$$

which indicates that the rate of change of circulation is due to work done by body, pressure, and viscous forces. For *conservative forces* the body force contribution is zero, and for *barotropic fluids* the pressure term is zero. Therefore,

$$\frac{D\Gamma}{Dt} = \int_C \frac{1}{\rho} (\nabla \cdot \boldsymbol{\tau}) \cdot d\mathbf{r} . \qquad (5.36)$$

If, in addition, the fluid is inviscid,

$$\frac{D\Gamma}{Dt} = 0 , \qquad (5.37)$$

which is known as *Kelvin's circulation theorem*.

The *mechanical energy* equation is obtained by taking the dot product of the velocity vector with the momentum equation:

$$\rho \mathbf{u} \cdot \frac{\partial \mathbf{u}}{\partial t} + \rho \mathbf{u} \cdot [\mathbf{u} \cdot \nabla \mathbf{u}] = -\mathbf{u} \cdot \nabla p + \mathbf{u} \cdot \nabla \cdot \boldsymbol{\tau} + \rho \mathbf{u} \cdot \mathbf{g} . \qquad (5.38)$$

By invoking the vector identity of Eq. (5.25), the above equation reduces to

$$\rho \frac{D\mathbf{u}^2}{Dt} = \frac{\partial}{\partial t}\left(\frac{\rho u^2}{2}\right) + \mathbf{u} \cdot \nabla \left(\frac{\rho u^2}{2}\right) \qquad \Longrightarrow$$

$$\rho \frac{Du^2}{Dt} = p\nabla \cdot \mathbf{u} - \nabla \cdot (p\mathbf{u}) + \nabla \cdot (\boldsymbol{\tau} \cdot \mathbf{u}) - \boldsymbol{\tau} : \nabla \mathbf{u} + \rho(\mathbf{u} \cdot \mathbf{g}) \,, \qquad (5.39)$$

in agreement with Eq. (3.39).

Finally, the equation of conservation of *angular momentum*,

$$\mathbf{J}_\theta \equiv \mathbf{r} \times \mathbf{J} \,, \qquad (5.40)$$

where $\mathbf{J}$ is the linear momentum, and $\mathbf{r}$ is the position vector from the center of rotation, is obtained by taking the cross product of the position vector with the momentum equation. The resulting angular momentum conservation equation is

$$\rho \frac{D(\mathbf{r} \times \mathbf{u})}{Dt} = -\mathbf{r} \times \nabla p + \mathbf{r} \times \nabla \cdot \boldsymbol{\tau} + \rho \mathbf{r} \times \mathbf{g} \,. \qquad (5.41)$$

Therefore, the momentum equation, which can be viewed as *the application of Newton's law of motion to liquids*, appears to be the most important conservation equation, as most conservation laws can be derived from it.

Isothermal, incompressible, and Newtonian flow problems are analyzed by solving the continuity and momentum equations tabulated in Tables 5.2 to 5.4. Occasionally, for flows involving free surfaces or interfaces, the kinematic equation must also be considered along with the equations of motion in order to determine the locations of the free surfaces or interfaces. In bidirectional free-surface flow, for instance, the kinematic equation may be expressed as

$$\frac{\partial h}{\partial t} + u_x \frac{\partial h}{\partial x} = u_y \,, \qquad (5.42)$$

where $h-h(x,t)$ denotes the position of the free surface. Finally, note that the complete solution to the governing equations requires the specification of *boundary* and *initial conditions*. These are discussed below.

5.3 Boundary Conditions

In most cases, fluids interact with their surroundings through common boundaries. The mathematical formulation of these boundary interactions result in boundary conditions. Thus, boundary conditions are constraints that are imposed on the conservation equations in order to describe how the field under consideration conforms to its surroundings. Therefore, boundary conditions come from nature and are mathematical descriptions of the physics at the boundary. Once a system or a flow field is chosen, these conditions follow automatically. In fact, if boundary conditions are not obvious, then the boundaries of the system may not be natural, which may lead to an ill-posed mathematical problem. The boundary conditions may describe conditions along the boundary dealing with motion, external stresses, rate of mass and momentum flux, boundary values of field variables, as well as relations among them. When the solution involves the time evolution of flow fields, in addition to boundary conditions, *initial conditions* are also required.

The required number of boundary conditions is determined by the nature of the governing partial differential equations inasmuch as the physics may provide several forms of boundary conditions. In general, *elliptic equations* require boundary conditions on each portion of the boundary; *hyperbolic equations* require boundary conditions at upstream, but not downstream boundaries; and *parabolic equations* require initial conditions and boundary conditions everywhere except at downstream boundaries. The Navier-Stokes equation is hyperbolic at high Reynolds numbers and elliptic at low Reynolds numbers. The Euler equation is the upper limit of hyperpolicity and the Stokes equation is the lower limit of ellipticity.

In general, there are three kinds of boundary conditions:

(a) *First kind* or *Dirichlet* boundary condition: the value of a dependent variable, u, is imposed along $\mathbf{r}_s$

$$u(\mathbf{r}_s, t) = f(\mathbf{r}_s, t) \,, \tag{5.43}$$

where f is a known function. Typical Dirichlet boundary conditions are the no-slip boundary condition for the velocity (i.e., $\mathbf{u}_s=0$) and the specification of inlet and/or outlet values for the velocity.

(b) *Second kind* or *Neumann* boundary condition: the normal derivative of the dependent variable is specified

$$\frac{\partial u}{\partial n} = g(\mathbf{r}_s, t) \,, \tag{5.44}$$

where g is a known function. Examples of Neumann boundary conditions are symmetry conditions and free-surface and interface stress conditions.

(c) *Third kind* or *Robin* boundary condition: the dependent variable and its normal derivative are related by the general expression

$$au + b\frac{\partial u}{\partial n} = c \,, \tag{5.45}$$

where a, b, and c are known functions. The slip boundary condition, and the free-surface and interface stress conditions are typical Robin conditions.

Example 5.3.1 demonstrates the application of boundary conditions to a general fluid mechanics problem that involves solid boundaries, a free surface, inlet and outlet boundaries, and symmetry boundaries.

Example 5.3.1. The extrudate-swell problem

The Navier–Stokes equation at low Reynolds numbers is a nonlinear partial differential equation of elliptic type. Therefore, boundary conditions are required at each portion of the boundary. Consider the planar *extrudate-swell problem* [14] shown in Fig. (5.1). A liquid under pressure exits from an orifice to form a jet. The boundary conditions are shown in Fig. 5.1 and discussed below.

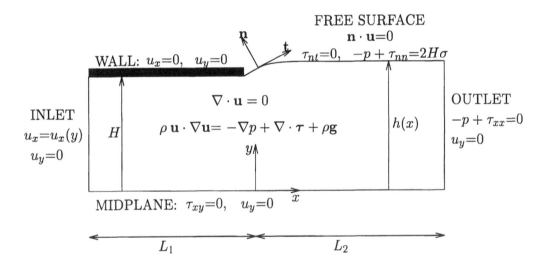

Figure 5.1. *Schematic of the planar extrudate-swell problem with governing equations and boundary conditions.*

(a) <u>Solid boundaries</u>

Assuming no-slip conditions, at $y=H$ and $0 \leq x < L_1$ we have $u_x=u_y=0$.

In case of a solid boundary moving with velocity $\mathbf{V}_w$, the boundary condition becomes $\mathbf{u}=\mathbf{V}_w$. The no-slip boundary condition is a consequence of the fact that the liquid wets or sticks to the boundary without penetration and, therefore, it is forced to move with the boundary.

(b) <u>Plane of symmetry</u>

The flow is symmetric with respect to the plane $y=0$. The proper boundary conditions along $y=0$ and $0 < x < L_1 + L_2$ are $u_y=\tau_{yx}=0$.

(c) Free surface

Free surfaces are described by two different boundary conditions: a kinematic and a dynamic boundary one.

The kinematic condition describes the motion of the free surface based on the observation that a fluid particle, which at an earlier time was at a free surface, will alway remain on the free surface. For the planar problem considered here, the height of the free surface $h(x,t)$ satisfies the kinematic condition

$$\frac{\partial h}{\partial t} + u_x \frac{\partial h}{\partial x} = u_y \, .$$

The dynamic condition describes the balance of forces along the free surface, expressed using the traction force per unit area vector, $\mathbf{f} = \mathbf{n} \cdot \mathbf{T}$, along the free surface, with $\mathbf{f}$ being the externally applied force. Along $y = h(x,t)$ and $L_1 \le x \le L_2$,

$$\tau_{nt} = f_t \quad \text{and} \quad -p + \tau_{nn} = f_n \, ,$$

where subscripts n and t refer to components along the normal and tangential directions, respectively.

In case of no applied external force, $f_t = f_n = 0$, we have the well known no-traction force boundary condition. In case of significant forces due to surface tension, the boundary condition is given by

$$\tau_{nt} = \mathbf{t} \cdot \nabla_{||}\sigma \quad \text{and} \quad -p + \tau_{nn} = 2H\sigma \, ,$$

where σ is the surface tension and $2H$ is the mean curvature, defined here as

$$2H = \frac{\dfrac{d^2 h}{dx^2}}{\left[1 + \left(\dfrac{dh}{dx}\right)^2\right]^{1/2}} \, .$$

In the absence of temperature or concentration (of surfactant) gradients, $\nabla_{||}\sigma$ is taken to be zero.

(d) Inlet boundary

At the inlet, the flow is assumed to be fully developed channel flow of known velocity profile (Chapter 6). At $0 \le y \le H$ and $x=0$, $u_x = u_x(y)$ and $u_y = 0$.

(e) Outlet boundary

At the outlet, the flow is assumed to be plug. Therefore, along $x = L_1 + L_2$ and $0 < y < h(L_2)$, $u_y = 0$ and $-p + \tau_{xx} = 0$.   □

The velocity across the interface of two immiscible liquids, A and B, is continuous. Therefore,

$$\mathbf{u}^A = \mathbf{u}^B .$$

In the absence of surface tension gradients, the shear stress is also continuous,

$$\tau_{nt}^A = \tau_{nt}^B .$$

The total normal stress difference is balanced by the capillary pressure, therefore,

$$(-p^A + \tau_{nn}^A) - (-p^B + \tau_{nn}^B) = 2H\sigma ,$$

where σ is the interfacial tension.

The no-slip boundary condition was first introduced by Bernoulli to dispute Navier's (1827) original hypothesis of slip, i.e., of a finite velocity, u_{slip}, at a solid wall moving with velocity V_w. Stokes' hypothesis is formulated as

$$\tau_{xy} = \frac{\eta}{\delta}(u_w - V_w) = \frac{\eta}{\delta}u_{slip} . \tag{5.46}$$

Equation (5.46) assumes that the fluid slips along the wall due to a thin stagnant liquid film with thickness δ adjacent to the solid wall. This film allows for different velocities between the solid wall and the fluid (which *macroscopically* appears to adhere to the wall).

A *continuous derivation* of Eq. (5.46) is highlighted in Example 5.3.2. The validity of slip and no-slip boundary conditions is examined by molecular dynamics simulations where the motion of individual molecules near a solid wall under a uniform external acceleration is calculated [15]. Results from this approach agree with the general rules discussed in this section and illustrated by Fig. 5.2. In particular, these simulations show that the contact angle of a meniscus separating two immiscible fluids advancing "steadily" in Poiseuille-like flow changes slowly with time, and that the liquid that preferentially wets the walls forms a thin film along the walls when it moves or it is displaced by the other liquid [16]. Therefore, the no-slip condition appears to break down at the contact line due to a jet flowing back into the liquid, as shown in Fig. 5.2.

Example 5.3.2. Derivation of the slip boundary condition

We assume that the velocity u_x varies linearly with y in a thin layer of thickness, δ, as shown in Fig. 5.3,

$$u_x = u_w \left(1 - \frac{y}{\delta}\right) .$$

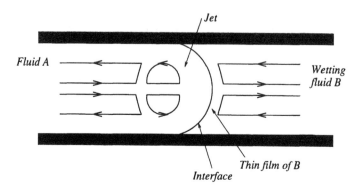

Figure 5.2. *Fluid dynamics in the vicinity of a moving contact angle, suggested by molecular simulations reported in [15].*

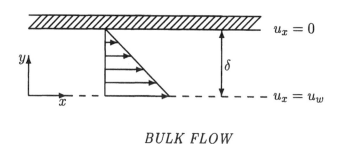

BULK FLOW

Figure 5.3. *Slip layer along a solid wall.*

At the wall ($y=\delta$), u_x is zero; at the other end of the layer ($y=0$), $u_x=u_w$ where u_w is the finite *slip velocity*.

If τ_w is the shear stress exerted by the fluid on the wall,

$$\tau_w = -\tau_{yx}\big|_{y=\delta} \quad \Longrightarrow \quad \tau_w = -\eta\frac{du_x}{dy} = \frac{\eta}{\delta}\,u_w\,.$$

Setting $\eta/\delta = \beta$, we get

$$\tau_w = \beta\,u_w\,.$$

The above slip equation includes the *no-slip boundary condition* as $\beta \to \infty$, and the *perfect-slip* case when $\beta=0$.   □

In transient flows, an initial state must be specified to initiate the involved nonzero time derivatives. Commonly, the primary unknowns are specified everywhere:

$$
\begin{aligned}
u_x(x, y, z, t = 0) &= u_x^o(x, y, z) \\
u_y(x, y, z, t = 0) &= u_y^o(x, y, z) \\
u_z(x, y, z, t = 0) &= u_z^o(x, y, z) \\
p(x, y, z, t = 0) &= p^o(x, y, z)
\end{aligned}
$$

where u_x^o, u_y^o, u_z^o, and p^o are prescribed distributions.

5.4   Problems

5.1. Show that the Euler and the Stokes equations are obtained from the Navier–Stokes equations in the limit of small and large Reynolds numbers. Write down flow situations where these limiting behaviors may apply.

5.2. *Eccentric Rheometer* [17]. Two large parallel disks of radius R and distance h apart are both rotating at constant angular velocity Ω. Their axes of rotation are displaced at distance a, where $a \ll h$. When a Newtonian oil was placed between the disks, the following velocity profile was observed in terms of a Cartesian coordinate system on the axis of the lower disk:

$$
\begin{aligned}
v_x &= -\Omega y + \frac{a}{h}\Omega z \,; \\
v_y &= \Omega x \,; \\
v_z &= 0 \,.
\end{aligned}
$$

(a) Verify that the above velocity profile satisfies the continuity, the momentum equations, and the vorticity equations.

(b) Is the oil experiencing any relative deformation? Is all motion a solid body rotation? Justify your answer.

(c) Neglecting body forces and surface tension, determine the forces acting on the lower disk.

(d) How could this device be used to determine the viscosity of the oil?

5.3. List the boundary conditions for flow down a vertical plate of a nonuniform liquid film under the action of surface tension and gravity, in contact with stationary air, at a given flow rate per unit plate width.

5.4. For a linearly elastic isotropic solid the constitutive equation is

$$
\tau = \lambda \nabla \cdot \mathbf{r} + \mu [\nabla \mathbf{r} + (\nabla \mathbf{r})^T] \,,
$$

where λ and μ are material parameters, and $\mathbf{r}$ is the displacement vector. Given that the inertia term, $\rho D\mathbf{u}/Dt$, is vanishingly small, derive the governing equation of motion (or of displacement).

5.5. Starting from the Navier–Stokes equation, show how one can arrive at the z-momentum component equation of Table 5.3. Then simplify this equation for flow in a horizontal annulus. State clearly your assumptions.

5.6. Identify and then simplify the appropriate equations, stating your assumptions in omitting terms, to analyze the following flow situations:
(a) Sink flow to a two-dimensional hole of diameter D at flow rate Q
(b) Source flow from a porous cylinder of radius R and length L at flow rate Q
(c) Flow around a growing bubble of radius $R(t)$ at rate $dR/dt = k$
(d) Flow in a horizontal pipe
(e) Flow in an inclined channel
(f) Film flow down a vertical wall
(g) Tornado, torsional flow
(h) Torsional flow between rotating concentric cylinders
What are the appropriate boundary conditions required to find the solution to the simplified equations?

5.7. Derive or identify the appropriate equations of motion of a compressible gas of vanishingly small viscosity. What are the corresponding equations for one-dimensional pipe flow of the gas? What are the appropriate boundary conditions? Can these equations be solved?

5.8. A liquid rests between two infinitely long and wide plates separated by a distance H. Suddenly the upper plate is set to motion under a constant external stress, τ. What are the appropriate initial and boundary conditions to this flow?

5.9. *Stress decomposition.* Split the following total stress tensor in two dimensions into an isotropic and an anisotropic part:

$$T = \begin{bmatrix} a/2 & 0 \\ 0 & -\frac{3a}{2} \end{bmatrix} = \begin{bmatrix} & \\ & \end{bmatrix} + \begin{bmatrix} & \\ & \end{bmatrix}.$$

(a) The two parts arise due to what?
(b) What are the principal directions and values of the total stress? How are they related to those of the pressure and the viscous stress?
(c) Find the velocity components with respect to the principal axes.
(d) Sketch the principal axes and the streamlines.
(e) A deformable small cube is introduced parallel to the streamlines. At what state of deformation and orientation exits the flow?

5.10. Formulate the appropriate boundary conditions to study a horizontal thin film flow in the presence of surface tension that varies linearly with the horizontal distance. Solve the equations for this (nearly) one-dimensional flow.

5.11. To address problems involving a moving front of unknown shape and location, such as the one shown in Fig. 5.4 for mold filling by a polymeric melt through a gate, typical of injection molding processes [18], the equations of motion are often formulated with respect to an observer moving with velocity **U**.

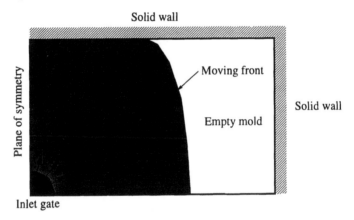

Figure 5.4. *Mold filling in injection molding.*

(a) Show that the resulting forms of the continuity, momentum, and energy equations are:

$$\nabla \cdot \mathbf{u} = 0 ,$$

$$\rho \left[\frac{\partial \mathbf{u}}{\partial t} + (\mathbf{u} - \mathbf{U}) \cdot \nabla \mathbf{u} \right] = \nabla \cdot (-p\,\mathbf{I} + \tau) ,$$

$$\rho C \left[\frac{\partial T}{\partial t} + (\mathbf{u} - \mathbf{U}) \cdot \nabla T \right] = k \nabla^2 T - \tau : \nabla \mathbf{u} ,$$

where **u** is the absolute velocity of the melt, and **U** is the velocity of the moving frame of reference.

(b) Show that the Eulerian equations are recovered for a stationary frame of reference, and the Lagrangian ones for a frame traveling with the liquid.

(c) If **u** is the velocity of the moving front, what are the appropriate boundary conditions at the solid walls, the lines of symmetry and along the moving front?

5.5 References

1. R.B. Bird, O. Hassager, R.C. Armstrong, and C.F. Curtiss, *Dynamics of Polymeric Liquids, Volume II, Kinetic Theory*, Wiley & Sons, Inc., New York, 1987.

2. L. E. Malvern, *Introduction to the Mechanics of a Continuous Medium*, Prentice Hall, Englewood Cliffs, New Jersey, 1969.

3. B. Bernstein, E.A. Kearsley, and L.J. Zappas, "A study of stress relaxation with finite strains," *Trans. Soc. Rheol.* **7**, 391 (1967).

4. G.G. Stokes, "On the theories of internal friction of fluids in motion, and of the equilibrium and motion of elastic solids", *Trans. Cambridge Phil. Soc.* **8**, 287 (1845).

5. C. Truesdell and W. Noll, "The classical field theories," *Handbuch der Physik, Vol. III*, Part 1, Springer-Verlag, 1960.

6. J.O. Hinze, *Turbulence*, McGraw-Hill, New York, 1959.

7. C. Truesdell, *Essays in the History of Mechanics*, Springer-Verlag, New York, 1968.

8. T. Carmady and H. Kobus, *Hydrodynamics by Daniel Bernoulli and Hydraulics by Johann Bernoulli*, Dover Publications, New York, 1968.

9. H. Rouse and S. Ince, *History of Hydraulics*, Dover Publications, New York, 1957.

10. J.C. Maxwell, "On the Dynamical Theory of Gases," *Phil. Trans. Roy. Soc.* **157**, 49 (1867).

11. H. Lamp, *Treatise on the Mathematical Theory of the Motion of Fluids*, Cambridge University Press, 1879.

12. A.B. Basset, *Hydrodynamics*, Dover Publications, New York, 1888.

13. S. Whitaker, *The Development of Fluid Mechanics in Chemical Engineering, in One Hundred Years of Chemical Engineering*, Kluwer Academic Publishers, Boston, 1989.

14. G. Georgiou, J. Wilkes, and T.C. Papanastasiou, "Laminar Jets at High Reynolds and High Surface Tension," *AIChE J.* **24**, No. 9, 1559-1562 (1988).

15. W.G. Hoover, in *Physics Today*, January 1984.

16. A. Khurana, "Numerical Simulations reveal fluid flows near solid boundaries," *Physics Today*, May 1988.

17. C.W. Macosko, *Rheological Measurements: Applications to Polymers, Suspensions and Processing*, University of Minnesota, 1989.

18. A.N. Alexandrou and A. Ahmed, "Injection molding using a generalized Eulerian–Lagrangian formulation," *Polymer Eng. Sci.* **33**, 1055-1064 (1994).

19. T.C. Papanastasiou, L.E. Scriven, and C.W. Macosko, "Bubble growth and collapse in viscoelastic liquids analyzed," *J. Non-Newtonian Fluid Mech.* **16**, 53 (1984).

Chapter 6

UNIDIRECTIONAL FLOWS

Isothermal, laminar, incompressible Newtonian flow is governed by a system of four scalar *partial differential equations* (PDEs); these are the continuity equation and the three components of the Navier–Stokes equation. The pressure and the three velocity components are the *primary unknowns* which are, in general, functions of time and of spatial coordinates. This system of PDEs is amenable to analytical solution for limited classes of flow. Even in the case of relatively simple flows in regular geometries, the nonlinearities introduced by the convective terms rule out the possibility of finding analytical solutions. This explains the extensive use of numerical methods in Fluid Mechanics [1]. *Computational Fluid Dynamics (CFD)* is certainly the fastest growing branch of fluid mechanics, largely as a result of the increasing availability and power of computers, and the parallel advancement of versatile numerical techniques.

In this chapter, we study certain classes of incompressible flows, in which the Navier–Stokes equations are simplified significantly to lead to analytical solutions. These classes concern *unidirectional* flows, that is, flows which have only one nonzero velocity component, u_i. Hence, the number of the primary unknowns is reduced to two: the velocity component, u_i, and pressure, p. In many flows of interest, the PDEs corresponding to the two unknown fields are decoupled. As a result, one can first find u_i by solving the corresponding component of the Navier–Stokes equation and then calculating the pressure. Another consequence of the unidirectionality assumption is that u_i is a function of at most two spatial variables and time. Therefore, in the worst case scenario of incompressible, unidirectional flow, one has to solve a PDE with three independent variables, one of which is time.

The number of independent variables is reduced to two in

(a) *transient one-dimensional* (1D) unidirectional flows in which u_i is a function of one spatial independent variable and time; and

(b) *steady two-dimensional* (2D) unidirectional flows in which u_i is a function of two spatial independent variables.

The resulting PDEs in the above two cases can often be solved using various techniques, such as the *separation of variables* [2] and *similarity methods* [3].

In *steady, one-dimensional unidirectional flows*, the number of independent variables is reduced to one. In these flows, the governing equation for the nonzero velocity component is just a *linear*, second-order *ordinary differential equation* (ODE) which can be solved easily using well-known formulas and techniques. Such flows are studied in the first three sections of this chapter. In particular, in Sections 1 and 2, we study flows in which the streamlines are straight lines, i.e., one-dimensional *rectilinear flows* with $u_x=u_x(y)$ and $u_y=u_z=0$ (Section 6.1), and *axisymmetric rectilinear flows* with $u_z=u_z(r)$ and $u_r=u_\theta=0$ (Section 6.2). In Section 6.3, we study *axisymmetric torsional* (or *swirling*) *flows* with $u_\theta=u_\theta(r)$ and $u_z=u_r=0$. In this case, the streamlines are circles centered at the axis of symmetry.

In Sections 6.4 and 6.5, we briefly discuss steady *radial flows* with *axial* and *spherical symmetry*, respectively. An interesting feature of radial flows is that the nonzero radial velocity component, $u_r=u_r(r)$, is determined from the continuity equation rather than from the radial component of the Navier–Stokes equation. In Section 6.6, we study transient, one-dimensional unidirectional flows. Finally, in Section 6.7, we consider examples of steady, two-dimensional unidirectional flows.

Unidirectional flows, although simple, are important in a diversity of fluid transferring and processing applications. As demonstrated in examples in the following sections, once the velocity and the pressure are known, the nonzero components of the stress tensor, such as the shear stress, as well as other useful macroscopic quantities, such as the volumetric flow rate and the shear force (or *drag*) on solid boundaries in contact with the fluid, can be easily determined.

Let us point out that analytical solutions can also be found for a limited class of two-dimensional *almost unidirectional* or *bidirectional* flows by means of the *potential function* and/or the *stream function*, as demonstrated in Chapters 8 to 10. Approximate solutions for limiting values of the involved parameters can be constructed by *asymptotic* and *perturbation analyses*, which are the topics of Chapters 7 and 9, with the most profound examples being the *lubrication*, *thin-film*, and *boundary-layer* approximations.

6.1 Steady, One-Dimensional Rectilinear Flows

Rectilinear flows, i.e., flows in which the streamlines are straight lines, are usually described in Cartesian coordinates, with one of the axes being parallel to the flow direction. If the flow is axisymmetric, a cylindrical coordinate system with the z-axis coinciding with the axis of symmetry of the flow is usually used.

Let us assume that a Cartesian coordinate system is chosen to describe a rectilinear flow, with the x-axis being parallel to the flow direction, as in Fig. 6.1, where the geometry of the flow in a channel of rectangular cross section is shown. Therefore, u_x is the only nonzero velocity component and

$$u_y = u_z = 0 . \tag{6.1}$$

From the continuity equation for incompressible flow,

$$\frac{\partial u_x}{\partial x} + \frac{\partial u_y}{\partial y} + \frac{\partial u_z}{\partial z} = 0 ,$$

we find that

$$\frac{\partial u_x}{\partial x} = 0 ,$$

which indicates that u_x does not change in the flow direction, i.e., u_x is independent of x:

$$u_x = u_x(y, z, t) . \tag{6.2}$$

Flows satisfying Eqs. (6.1) and (6.2) are called *fully developed*. Flows in tubes of constant cross section, such as the one shown in Fig. 6.1, can be considered fully developed if the tube is *sufficiently long* so that entry and exit effects can be neglected.

Due to Eqs. (6.1) and (6.2), the x-momentum equation,

$$\rho \left(\frac{\partial u_x}{\partial t} + u_x \frac{\partial u_x}{\partial x} + u_y \frac{\partial u_x}{\partial y} + u_z \frac{\partial u_x}{\partial z} \right) = -\frac{\partial p}{\partial x} + \eta \left(\frac{\partial^2 u_x}{\partial x^2} + \frac{\partial^2 u_x}{\partial y^2} + \frac{\partial^2 u_x}{\partial z^2} \right) + \rho g_x ,$$

is reduced to

$$\rho \frac{\partial u_x}{\partial t} = -\frac{\partial p}{\partial x} + \eta \left(\frac{\partial^2 u_x}{\partial y^2} + \frac{\partial^2 u_x}{\partial z^2} \right) + \rho g_x . \tag{6.3}$$

If now the flow is steady, then the time derivative in the x-momentum equation is zero, and Eq. (6.3) becomes

$$-\frac{\partial p}{\partial x} + \eta \left(\frac{\partial^2 u_x}{\partial y^2} + \frac{\partial^2 u_x}{\partial z^2} \right) + \rho g_x = 0 . \tag{6.4}$$

The last equation which describes any steady, two-dimensional rectilinear flow in the x-direction is studied in Section 6.5. In many unidirectional flows, it can be assumed that

$$\frac{\partial^2 u_x}{\partial y^2} \gg \frac{\partial^2 u_x}{\partial z^2} ,$$

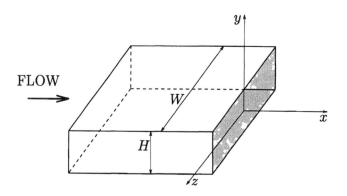

Figure 6.1. *Geometry of flow in a channel of rectangular cross section.*

and u_x can be treated as a function of y alone, i.e.,

$$u_x = u_x(y) \,. \tag{6.5}$$

With the latter assumption, the x-momentum equation is reduced to:

$$-\frac{\partial p}{\partial x} + \eta \frac{d^2 u_x}{dy^2} + \rho g_x = 0 \,. \tag{6.6}$$

The only nonzero component of the stress tensor is the shear stress τ_{yx},

$$\tau_{yx} = \eta \frac{du_x}{dy} \,, \tag{6.7}$$

in terms of which the x-momentum equation takes the form

$$-\frac{\partial p}{\partial x} + \frac{d\tau_{yx}}{dy} + \rho g_x = 0 \,. \tag{6.8}$$

Equation (6.6) is a linear, second-order ordinary differential equation and can be integrated directly if

$$\frac{\partial p}{\partial x} = \text{const} \,. \tag{6.9}$$

Its general solution is given by

$$u_x(y) = \frac{1}{2\eta} \left(\frac{\partial p}{\partial x} - \rho g_x \right) y^2 + c_1 y + c_2 \,. \tag{6.10}$$

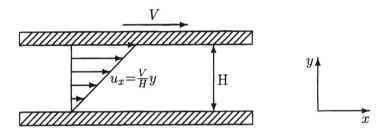

Figure 6.2. *Plane Couette flow.*

Therefore, the velocity profile is a parabola and involves two constants, c_1 and c_2, which are determined by applying appropriate boundary conditions for the particular flow. The shear stress, $\tau_{yx}=\tau_{xy}$, is linear, i.e.,

$$\tau_{yx} = \eta \frac{du_x}{dy} = \left(\frac{\partial p}{\partial x} - \rho g_x\right) y + \eta c_1 . \tag{6.11}$$

Note that the y- and z-momentum components do not involve the velocity u_x; since $u_y=u_z=0$, they degenerate to the hydrostatic pressure expressions

$$-\frac{\partial p}{\partial y} + \rho g_y = 0 \quad \text{and} \quad -\frac{\partial p}{\partial z} + \rho g_z = 0 . \tag{6.12}$$

Integrating Eqs. (6.9) and (6.12), we obtain the following expression for the pressure:

$$p = \frac{\partial p}{\partial x} x + \rho g_y y + \rho g_z z + c , \tag{6.13}$$

where c is a constant of integration which may be evaluated in any particular flow problem by specifying the value of the pressure at a point.

In Table 6.1, we tabulate the assumptions, the governing equations, and the general solution for steady, one-dimensional rectilinear flows in Cartesian coordinates. Important flows in this category are:

1. *Plane Couette flow,* i.e., fully-developed flow between parallel flat plates of infinite dimensions, driven by the steady motion of one of the plates. (Such a flow is called *shear-driven flow.*) The geometry of this flow is depicted in Fig. 6.2, where the upper wall is moving with constant speed V (so that it remains in the same plane) while the lower one is fixed. The pressure gradient is zero everywhere and the gravity term is neglected. This flow is studied in Example 1.6.1.

Assumptions:
$$u_y = u_z = 0, \qquad \frac{\partial u_x}{\partial z} = 0, \qquad \frac{\partial p}{\partial x} = \text{const.}$$

Continuity:

$$\frac{\partial u_x}{\partial x} = 0 \qquad \Longrightarrow \qquad u_x = u_x(y)$$

x-momentum:

$$-\frac{\partial p}{\partial x} + \eta \frac{d^2 u_x}{dy^2} + \rho g_x = 0$$

y-momentum:

$$-\frac{\partial p}{\partial y} + \rho g_y = 0$$

z-momentum:

$$-\frac{\partial p}{\partial z} + \rho g_z = 0$$

General solution:

$$u_x = \frac{1}{2\eta}\left(\frac{\partial p}{\partial x} - \rho g_x\right) y^2 + c_1 y + c_2$$

$$\tau_{yx} = \tau_{xy} = \left(\frac{\partial p}{\partial x} - \rho g_x\right) y + \eta c_1$$

$$p = \frac{\partial p}{\partial x} x + \rho g_y y + \rho g_z z + c$$

Table 6.1. *Governing equations and general solution for steady, one-dimensional rectilinear flows in Cartesian coordinates.*

2. *Fully-developed plane Poiseuille flow*, i.e., flow between parallel plates of infinite width and length, driven by a constant pressure gradient, imposed by a pushing or pulling device (a pump or vacuum, respectively) and/or gravity. This flow is an idealization of the flow in a channel of rectangular cross section, with the width W being much greater than the height H of the channel (see Fig. 6.1). Obviously, this idealization does not hold near the two lateral walls where the flow is two-dimensional. The geometry of the plane Poiseuille flow is depicted in Fig. 6.4. This flow is studied in Examples 6.1.2 to 6.1.5 for

different boundary conditions.

3. *Thin-film flow* down an inclined plane, driven by gravity (i.e., elevation differences), under the absence of surface tension. The pressure gradient is usually assumed to be everywhere zero. Such a flow is illustrated in Fig. 6.8, and is studied in Example 1.6.6.

All the above flows are rotational, with vorticity generation at the solid boundaries,

$$\boldsymbol{\omega} = \nabla \times \mathbf{u}|_w = \begin{vmatrix} \mathbf{i} & \mathbf{j} & \mathbf{k} \\ 0 & \frac{\partial}{\partial y} & 0 \\ u_x & 0 & 0 \end{vmatrix}_w = -\left(\frac{\partial u_x}{\partial y}\right)_w \mathbf{k} \neq \mathbf{0} \, .$$

The vorticity diffuses away from the wall and penetrates the main flow at a rate $\nu(d^2 u_x/dy^2)$. The extensional stretching or compression along streamlines is zero, i.e.,

$$\dot{\epsilon} = \frac{\partial u_x}{\partial x} = 0 \, .$$

Material lines connecting two moving fluid particles traveling along different streamlines both rotate and stretch, where stretching is induced by rotation. However, the principal directions of strain rotate with respect to those of vorticity. Therefore, strain is relaxed and the flow is weak.

Example 6.1.1. Plane Couette flow

Plane Couette flow,[1] named after Couette who introduced it in 1890 to measure viscosity, is fully-developed flow induced between two infinite parallel plates placed at a distance H apart, when one of them, say the upper one, is moving steadily with speed V relative to the other (Fig. 6.2). Assuming that the pressure gradient and the gravity in the x-direction are zero, the general solution for u_x is:

$$u_x = c_1 y + c_2 \, .$$

For the geometry depicted in Fig. 6.2, the boundary conditions are:

$$u_x = 0 \quad \text{at} \quad y = 0 \quad \text{(lower plate is stationary)};$$
$$u_x = V \quad \text{at} \quad y = H \quad \text{(upper plate is moving)}.$$

By means of the above two conditions, we find that $c_2 = 0$ and $c_1 = V/H$. Substituting the two constants into the general solution, yields

$$u_x = \frac{V}{H} y \, . \tag{6.14}$$

---

[1]Plane Couette flow is also known as *simple shear flow.*

The velocity u_x then varies linearly across the gap. The corresponding shear stress is constant,

$$\tau_{yx} = \eta \frac{V}{H} .$$
(6.15)

A number of macroscopic quantities, such as the volumetric flow rate and the shear stress at the wall, can be calculated. The volumetric flow rate per unit width is calculated by integrating u_x along the gap:

$$\frac{Q}{W} = \int_0^H u_x \, dy = \int_0^H \frac{V}{H} y \, dy \quad \Longrightarrow$$

$$\frac{Q}{W} = \frac{1}{2} HV .$$
(6.16)

The shear stress τ_w exerted by the fluid on the upper plate is

$$\tau_w = -\tau_{yx}|_{y=H} = -\eta \frac{V}{H} .$$
(6.17)

The minus sign accounts for the upper wall facing the negative y-direction of the chosen system of coordinates. The shear force per unit width required to move the upper plate is then

$$\frac{F}{W} = -\int_0^L \tau_w \, dx = \eta \frac{V}{H} L ,$$

where L is the length of the plate.

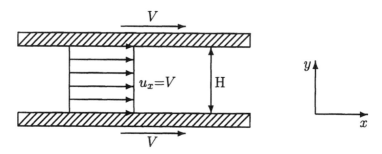

Figure 6.3. *Plug flow.*

Finally, let us consider the case where both plates move with the same speed V, as in Fig. 6.3. By invoking the boundary conditions

$$u_x(0) = u_x(H) = V ,$$

we find that $c_1=0$ and $c_2=V$ and, therefore,

$$u_x = V .$$

Thus, in this case, plane Couette flow degenerates into *plug flow*.   □

Example 6.1.2. Fully-developed plane Poiseuille flow

Plane Poiseuille flow, named after the channel experiments by Poiseuille in 1840, occurs when a liquid is forced between two stationary infinite flat plates under constant pressure gradient $\partial p/\partial x$ and zero gravity. The general steady-state solution is

$$u_x(y) = \frac{1}{2\eta}\frac{\partial p}{\partial x} y^2 + c_1 y + c_2 \tag{6.18}$$

and

$$\tau_{yx} = \frac{\partial p}{\partial x} y + \eta c_1 . \tag{6.19}$$

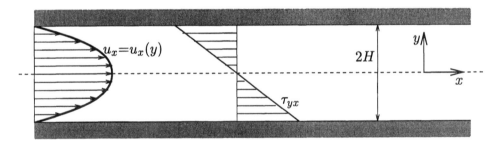

Figure 6.4. *Plane Poiseuille flow.*

By taking the origin of the Cartesian coordinates to be on the plane of symmetry of the flow, as in Fig. 6.4, and by assuming that the distance between the two plates is $2H$, the boundary conditions are:

$$\tau_{yx} = \eta \frac{du_x}{dy} = 0 \quad \text{at} \quad y = 0 \quad \text{(symmetry)} ;$$

$$u_x = 0 \quad \text{at} \quad y = H \quad \text{(stationary plate)} .$$

Note that the condition $u_x=0$ at $y=-H$ may be used instead of any of the above conditions. By invoking the boundary conditions at $y=0$ and H, we find that $c_1=0$ and

$$c_2 = -\frac{1}{2\eta}\frac{\partial p}{\partial x} H^2 .$$

The two constants are substituted into the general solution to obtain the following parabolic velocity profile,

$$u_x = -\frac{1}{2\eta} \frac{\partial p}{\partial x}(H^2 - y^2) \,. \tag{6.20}$$

If the pressure gradient is negative, then the flow is in the positive direction, as in Fig. 6.4. Obviously, the velocity u_x attains its maximum value at the centerline ($y=0$):

$$u_{x,max} = -\frac{1}{2\eta} \frac{\partial p}{\partial x} H^2 \,.$$

The volumetric flow rate per unit width is

$$\frac{Q}{W} = \int_{-H}^{H} u_x \, dy = 2 \int_{0}^{H} -\frac{1}{2\eta} \frac{\partial p}{\partial x}(H^2 - y^2) \, dy \quad \Longrightarrow$$

$$Q = -\frac{2}{3\eta} \frac{\partial p}{\partial x} H^3 W \,. \tag{6.21}$$

As expected, Eq. (6.21) indicates that the volumetric flow rate Q is proportional to the pressure gradient, $\partial p/\partial x$, and inversely proportional to the viscosity η. Note also that, since $\partial p/\partial x$ is negative, Q is positive. The average velocity, $\bar{u}_x$, in the channel is:

$$\bar{u}_x = \frac{Q}{WH} = -\frac{2}{3\eta} \frac{\partial p}{\partial x} H^2 \,.$$

The shear stress distribution is given by

$$\tau_{yx} = \frac{\partial p}{\partial x} y \,, \tag{6.22}$$

i.e., τ_{yx} varies linearly from $y=0$ to H, being zero at the centerline and attaining its maximum absolute value at the wall. The shear stress exerted by the fluid on the wall at $y=H$ is

$$\tau_w = -\tau_{yx}|_{y=H} = -\frac{\partial p}{\partial x} H \,.$$

$\square$

Example 6.1.3. Plane Poiseuille flow with slip

Consider again the fully-developed plane Poiseuille flow of the previous example, and assume that slip occurs along the two plates according to the slip law

$$\tau_w = \beta \, u_w \quad \text{at} \quad y = H \,,$$

where β is a material slip parameter, τ_w is the shear stress exerted by the fluid on the plate,

$$\tau_w = -\tau_{yx}\big|_{y=H} \, ,$$

and u_w is the *slip velocity*. Calculate the velocity distribution and the volume flow rate per unit width.

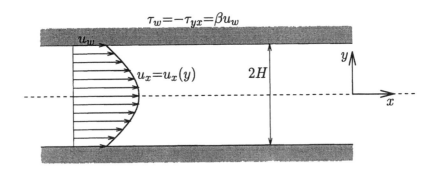

Figure 6.5. *Plane Poiseuille flow with slip.*

Solution:
We first note that the flow is still symmetric with respect to the centerline. In this case, the boundary conditions are:

$$\tau_{yx} = \eta \, \frac{du_x}{dy} = 0 \quad \text{at} \quad y = 0 \, ,$$

$$\tau_w = \beta \, u_w \quad \text{at} \quad y = H \, .$$

The condition at $y=0$ yields $c_1=0$. Consequently,

$$u_x = \frac{1}{2\eta} \, \frac{\partial p}{\partial x} \, y^2 + c_2 \, ,$$

and

$$\tau_{yx} = \frac{\partial p}{\partial x} \, y \quad \Longrightarrow \quad \tau_w = -\frac{\partial p}{\partial x} \, H \, .$$

Applying the condition at $y=H$, we obtain

$$u_w = \frac{1}{\beta}\tau_w \quad \Longrightarrow \quad u_x(H) = -\frac{1}{\beta} \, \frac{\partial p}{\partial x} \, H \quad \Longrightarrow \quad \frac{1}{2\eta} \, \frac{\partial p}{\partial x} \, H^2 + c_2 = -\frac{1}{\beta} \, \frac{\partial p}{\partial x} \, H \, .$$

Consequently,

$$c_2 = -\frac{1}{2\eta}\frac{\partial p}{\partial x}\left(H^2 + \frac{2\eta H}{\beta}\right) ,$$

and

$$u_x = -\frac{1}{2\eta}\frac{\partial p}{\partial x}\left(H^2 + \frac{2\eta H}{\beta} - y^2\right) . \tag{6.23}$$

Note that this expression reduces to the standard Poiseuille flow profile when $\beta \to \infty$. Since the slip velocity is inversely proportional to the slip coefficient β, the standard no-slip condition is recovered.

An alternative expression of the velocity distribution is

$$u_x = u_w - \frac{1}{2\eta}\frac{\partial p}{\partial x}\left(H^2 - y^2\right) ,$$

which indicates that u_x is just the superposition of the slip velocity u_w to the velocity distribution of the previous example.

For the volumetric flow rate per unit width, we obtain:

$$\frac{Q}{W} = 2\int_0^H u_x \, dy = 2u_w H - \frac{2}{3\eta}\frac{\partial p}{\partial x}H^3 \quad \Longrightarrow$$

$$Q = -\frac{2}{3\eta}\frac{\partial p}{\partial x}H^3\left(1 + \frac{3\eta}{\beta H}\right)W . \tag{6.24}$$

$\square$

Example 6.1.4. Plane Couette–Poiseuille flow

Consider again fully-developed plane Poiseuille flow with the upper plate moving with constant speed, V (Fig. 6.6). This flow is called *plane Couette-Poiseuille flow* or *general Couette flow*. In contrast to the previous two examples, this flow is not symmetric with respect to the centerline of the channel and, therefore, having the origin of the Cartesian coordinates on the centerline is not convenient. Therefore, the origin is moved to the lower plate.

The boundary conditions for this flow are:

$$u_x = 0 \quad \text{at} \quad y = 0 ,$$
$$u_x = V \quad \text{at} \quad y = a ,$$

where a is the distance between the two plates. Applying the two conditions, we get $c_2 = 0$ and

$$V = \frac{1}{2\eta}\frac{\partial p}{\partial x}a^2 + c_1 a \quad \Longrightarrow \quad c_1 = \frac{V}{a} - \frac{1}{2\eta}\frac{\partial p}{\partial x}a ,$$

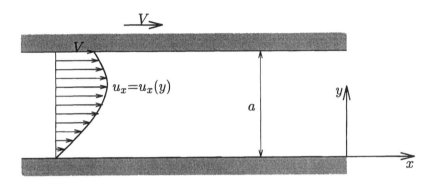

Figure 6.6. *Plane Poiseuille flow with the upper plate moving with constant speed.*

respectively. Therefore,

$$u_x = \frac{V}{a}y - \frac{1}{2\eta}\frac{\partial p}{\partial x}(ay - y^2).$$ (6.25)

The shear stress distribution is given by

$$\tau_{yx} = \eta\frac{V}{a} - \frac{1}{2}\frac{\partial p}{\partial x}(a - 2y).$$ (6.26)

It is a simple exercise to show that Eq. (6.25) reduces to the standard Poiseuille velocity profile for stationary plates given by Eq. (6.20). (Keep in mind that $a=2H$ and that the y-axis has been translated by a distance H.) If, instead, the pressure gradient is zero, the flow degenerates to the plane Couctte flow studied in Example 1.6.1, and the velocity distribution is linear. Hence, the solution in Eq. (6.25) is the sum of the solutions to the above two separate flow problems. This superposition of solutions is a result of the linearity of the governing Equation (6.6) and boundary conditions. Note also that Eq. (6.25) is valid not only when both the pressure gradient and the wall motion drive the fluid in the same direction, as in the present example, but also when they oppose each other. In the latter case, some reverse flow – in the negative x direction – can occur when $\partial p/\partial x > 0$.

Finally, let us find the point y^* where the velocity attains its maximum value. This point is a zero of the shear stress (or, equivalently, of the velocity derivative, du_x/dy):

$$0 = \eta\frac{V}{a} - \frac{1}{2}\frac{\partial p}{\partial x}(a - 2y^*) \quad\Longrightarrow\quad y^* = \frac{a}{2} + \frac{\eta V}{a\left(\frac{\partial p}{\partial x}\right)}.$$

The flow is symmetric with respect to the centerline if $y^*=a/2$, i.e., when $V=0$. The maximum velocity $u_{x,max}$ is determined by substituting y^* into Eq. (6.25).

□

Example 6.1.5. Poiseuille flow between inclined plates

Consider steady flow between two parallel inclined plates driven by both constant pressure gradient and gravity. The distance between the two plates is $2H$ and the chosen system of coordinates is shown in Fig. 6.7. The angle formed by the two plates and the horizontal direction is θ.

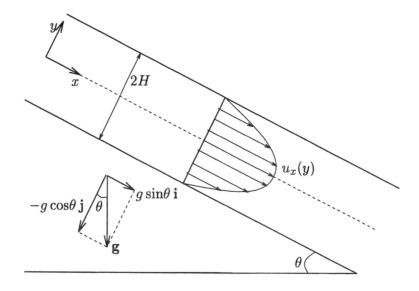

Figure 6.7. *Poiseuille flow between inclined plates.*

The general solution for u_x is given by Eq. (6.10):

$$u_x(y) = \frac{1}{2\eta} \left(\frac{\partial p}{\partial x} - \rho g_x \right) y^2 + c_1 y + c_2 \,.$$

Since,

$$g_x = g \sin\theta \,,$$

we get

$$u_x(y) = \frac{1}{2\eta} \left(\frac{\partial p}{\partial x} - \rho g \sin\theta \right) y^2 + c_1 y + c_2 \,.$$

Integration of this equation with respect to y and application of the boundary conditions, $du_x/dy=0$ at $y=0$ and $u_x=0$ at $y=H$, give

$$u_x(y) = \frac{1}{2\eta}\left(-\frac{\partial p}{\partial x} + \rho g \sin\theta\right)(H^2 - y^2).$$ (6.27)

The pressure is obtained from Eq. (6.13) as

$$p = \frac{\partial p}{\partial x} x + \rho g_y\, y + c \quad \Longrightarrow$$

$$p = \frac{\partial p}{\partial x} x + \rho g \cos\theta\, y + c$$ (6.28)

$\square$

Example 6.1.6. Thin-film flow

Consider a thin film of an incompressible Newtonian liquid flowing down an inclined plane (Fig. 6.8). The ambient air is assumed to be stationary and, therefore, the flow is driven by gravity alone. Assuming that the surface tension of the liquid is negligible, and that the film is of uniform thickness δ, calculate the velocity and the volumetric flow rate per unit width.

Solution:
The governing equation of the flow is

$$\eta\frac{d^2 u_x}{dy^2} + \rho g_x = 0 \quad \Longrightarrow \quad \eta\frac{d^2 u_x}{dy^2} = -\rho g \sin\theta,$$

with general solution

$$u_x = -\frac{\rho g \sin\theta}{\eta}\frac{y^2}{2} + c_1 y + c_2.$$

As for the boundary conditions, we have no slip along the solid boundary,

$$u_x = 0 \quad \text{at} \quad y = 0,$$

and no shearing at the free surface (the ambient air is stationary),

$$\tau_{yx} = \eta\frac{du_x}{dy} = 0 \quad \text{at} \quad y = \delta.$$

Applying the above two conditions, we find that $c_2=0$ and $c_1=\rho g \sin\theta/(\eta\delta)$, and thus

$$u_x = \frac{\rho g \sin\theta}{\eta}\left(\delta y - \frac{y^2}{2}\right).$$ (6.29)

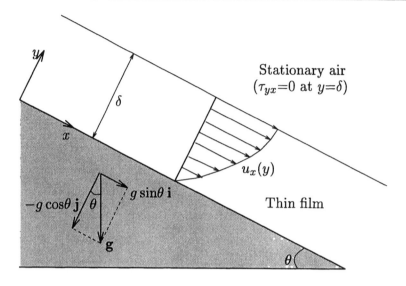

Figure 6.8. *Film flow down an inclined plane.*

The velocity profile is semiparabolic and attains its maximum value at the free surface,

$$u_{x,max} = u_x(\delta) = \frac{\rho g \sin\theta \, \delta^2}{2\eta} \, .$$

The volume flow rate per unit width is

$$\frac{Q}{W} = \int_0^\delta u_x \, dy = \frac{\rho g \sin\theta \, \delta^3}{3\eta} \, , \tag{6.30}$$

and the average velocity, $\bar{u}_x$, over a cross section of the film is given by

$$\bar{u}_x = \frac{Q}{W\delta} = \frac{\rho g \sin\theta \, \delta^2}{3\eta} \, .$$

Note that if the film is horizontal, then $\sin\theta = 0$ and u_x is zero, i.e., no flow occurs. If the film is vertical, then $\sin\theta = 1$, and

$$u_x = \frac{\rho g}{\eta} \left(\delta y - \frac{y^2}{2} \right) \tag{6.31}$$

and

$$\frac{Q}{W} = \frac{\rho g \delta^3}{3\eta} \,. \tag{6.32}$$

By virtue of Eq. (6.13), the pressure is given by

$$p = \rho g_y \, y + c = -\rho g \cos\theta \, y + c \,.$$

At the free surface, the pressure must be equal to the atmospheric pressure, p_0, so

$$p_0 = -\rho g \cos\theta \, \delta + c$$

and

$$p = p_0 + \rho g \, (\delta - y) \, \cos\theta \,. \tag{6.33}$$

$\square$

Example 6.1.7. Two-layer plane Couette flow

Two immiscible incompressible liquids A and B of densities ρ_A and ρ_B ($\rho_A > \rho_B$) and viscosities η_A and η_B flow between two parallel plates. The flow is induced by the motion of the upper plate which moves with speed V, while the lower plate is stationary (Fig. 6.9).

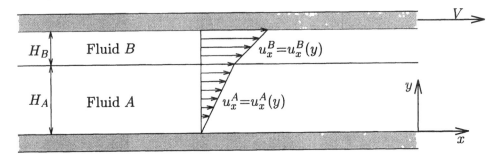

Figure 6.9. *Two-layer plane Couette flow.*

The velocity distributions in both layers obey Eq. (6.6) and are given by Eq. (6.10). Since the pressure gradient and gravity are both zero,

$$u_x^A = c_1^A y + c_2^A \,, \quad 0 \le y \le H_A \,,$$
$$u_x^B = c_1^B y + c_2^B \,, \quad H_A \le y \le H_A + H_B \,,$$

where c_1^A, c_2^A, c_1^B, and c_2^B are integration constants determined by conditions at the solid boundaries and the interface of the two layers. The no-slip boundary conditions at the two plates are applied first. At $y=0$, $u_x^A=0$; therefore,

$$c_2^A = 0 .$$

At $y=H_A + H_B$, $u_x^B=V$; therefore,

$$c_2^B = V - C_1^B (H_A + H_B) .$$

The two velocity distributions become

$$
\begin{aligned}
u_x^A &= c_1^A y , \quad 0 \leq y \leq H_A , \\
u_x^B &= V - c_1^B (H_A + H_B - y) , \quad H_A \leq y \leq H_A + H_B .
\end{aligned}
$$

At the interface $(y=H_A)$, we have two additional conditions: (a) the velocity distribution is continuous, i.e.,

$$u_x^A = u_x^B \quad \text{at} \quad y = H_A ;$$

(b) momentum transfer through the interface is continuous, i.e.,

$$\tau_{yx}^A = \tau_{yx}^B \quad \text{at} \quad y = H_A \quad \Longrightarrow$$

$$\eta_A \frac{du_x^A}{dy} = \eta_B \frac{du_x^B}{dy} \quad \text{at} \quad y = H_A .$$

From the interface conditions, we find that

$$c_1^A = \frac{\eta_B V}{\eta_A H_B + \eta_B H_A} \quad \text{and} \quad c_1^B = \frac{\eta_A V}{\eta_A H_B + \eta_B H_A} .$$

Hence, the velocity profiles in the two layers are

$$u_x^A = \frac{\eta_B V}{\eta_A H_B + \eta_B H_A} y , \quad 0 \leq y \leq H_A , \tag{6.34}$$

$$u_x^B = V - \frac{\eta_A V}{\eta_A H_B + \eta_B H_A} (H_A + H_B - y) , \quad H_A \leq y \leq H_A + H_B . \tag{6.35}$$

If the two liquids are of the same viscosity, $\eta_A=\eta_B=\eta$, then the two velocity profiles are the same, and the results simplify to the linear velocity profile for one-layer Couette flow,

$$u_x^A = u_x^B = \frac{V}{H_A + H_B} y .$$

$\square$

6.2   Steady, Axisymmetric Rectilinear Flows

Axisymmetric flows are conveniently studied in a cylindrical coordinate system, (r, θ, z), with the z-axis coinciding with the axis of symmetry of the flow. *Axisymmetry* means that there is no variation of the velocity with the angle θ,

$$\frac{\partial \mathbf{u}}{\partial \theta} = \mathbf{0} \,. \tag{6.36}$$

There are three important classes of axisymmetric *unidirectional* flows (i.e., flows in which only one of the three velocity components, u_r, u_θ, and u_z, is nonzero):

1. *Axisymmetric rectilinear flows*, in which only the axial velocity component, u_z, is nonzero. The streamlines are straight lines. Typical flows are fully-developed, pressure-driven flows in cylindrical tubes and annuli, and open film flows down cylinders or conical pipes.

2. *Axisymmetric torsional flows*, in which only the azimuthal velocity component, u_θ, is nonzero. The streamlines are circles centered on the axis of symmetry. These flows, studied in Section 6.3, are good prototypes of rigid-body rotation, flow in rotating mixing devices, and swirling flows, such as tornados.

3. *Axisymmetric radial flows*, in which only the radial velocity component, u_r, is nonzero. These flows, studied in Section 6.4, are typical models for radial flows through porous media, migration of oil towards drilling wells, and suction flows from porous pipes and annuli.

As already mentioned, in axisymmetric rectilinear flows,

$$u_r = u_\theta = 0 \,. \tag{6.37}$$

The continuity equation for incompressible flow,

$$\frac{1}{r}\frac{\partial}{\partial r}(r u_r) + \frac{1}{r}\frac{\partial u_\theta}{\partial \theta} + \frac{\partial u_z}{\partial z} = 0 \,,$$

becomes

$$\frac{\partial u_z}{\partial z} = 0 \,.$$

From the above equation and the axisymmetry condition (6.36), we deduce that

$$u_z = u_z(r, t) \,. \tag{6.38}$$

Due to Eqs. (6.36)-(6.38), the z-momentum equation,

$$\rho\left(\frac{\partial u_z}{\partial t} + u_r\frac{\partial u_z}{\partial r} + \frac{u_\theta}{r}\frac{\partial u_z}{\partial \theta} + u_z\frac{\partial u_z}{\partial z}\right) = -\frac{\partial p}{\partial z} + \eta\left[\frac{1}{r}\frac{\partial}{\partial r}\left(r\frac{\partial u_z}{\partial r}\right) + \frac{1}{r^2}\frac{\partial^2 u_z}{\partial \theta^2} + \frac{\partial^2 u_z}{\partial z^2}\right] + \rho g_z,$$

is simplified to

$$\rho\frac{\partial u_z}{\partial t} = -\frac{\partial p}{\partial z} + \eta\frac{1}{r}\frac{\partial}{\partial r}\left(r\frac{\partial u_z}{\partial r}\right) + \rho g_z. \tag{6.39}$$

For steady flow, $u_z=u_z(r)$ and Eq. (6.39) becomes an ordinary differential equation,

$$-\frac{\partial p}{\partial z} + \eta\frac{1}{r}\frac{d}{dr}\left(r\frac{du_z}{dr}\right) + \rho g_z = 0. \tag{6.40}$$

The only nonzero components of the stress tensor are the shear stresses τ_{rz} and τ_{zr},

$$\tau_{rz} = \tau_{zr} = \eta\frac{du_z}{dr}, \tag{6.41}$$

for which we have

$$-\frac{\partial p}{\partial z} + \frac{1}{r}\frac{d}{dr}(r\tau_{rz}) + \rho g_z = 0. \tag{6.42}$$

When the pressure gradient $\partial p/\partial z$ is constant, the general solution of Eq. (6.39) is

$$u_z = \frac{1}{4\eta}\left(\frac{\partial p}{\partial z} - \rho g_z\right)r^2 + c_1\ln r + c_2. \tag{6.43}$$

For τ_{rz}, we get

$$\tau_{rz} = \frac{1}{2}\left(\frac{\partial p}{\partial z} - \rho g_z\right)r + \eta\frac{c_1}{r}. \tag{6.44}$$

The constants c_1 and c_2 are determined from the boundary conditions of the flow. The assumptions, the governing equations, and the general solution for steady, axisymmetric rectilinear flows are summarized in Table 6.2.

Example 6.2.1. Hagen–Poiseuille flow

Fully-developed axisymmetric Poiseuille flow, or *Hagen-Poiseuille flow*, studied experimentally by Hagen in 1839 and Poiseuille in 1840, is the pressure-driven flow in infinitely long cylindrical tubes. The geometry of the flow is shown in Fig. 6.10.

Assuming that gravity is zero, the general solution for u_z is

$$u_z = \frac{1}{4\eta}\frac{\partial p}{\partial z}r^2 + c_1\ln r + c_2.$$

Assumptions:

$$u_r = u_\theta = 0, \qquad \frac{\partial u_z}{\partial \theta} = 0, \qquad \frac{\partial p}{\partial z} = \text{const.}$$

Continuity:

$$\frac{\partial u_z}{\partial z} = 0 \quad \Longrightarrow \quad u_z = u_z(r)$$

z-momentum:

$$-\frac{\partial p}{\partial z} + \eta \frac{1}{r} \frac{d}{dr}\left(r \frac{du_z}{dr}\right) + \rho g_z = 0$$

r-momentum:

$$-\frac{\partial p}{\partial r} + \rho g_r = 0$$

θ-momentum:

$$-\frac{1}{r}\frac{\partial p}{\partial \theta} + \rho g_\theta = 0$$

General solution:

$$u_z = \frac{1}{4\eta}\left(\frac{\partial p}{\partial z} - \rho g_z\right) r^2 + c_1 \ln r + c_2$$

$$\tau_{rz} = \tau_{zr} = \frac{1}{2}\left(\frac{\partial p}{\partial z} - \rho g_z\right) r + \eta \frac{c_1}{r}$$

$$p = \frac{\partial p}{\partial z} z + c(r, \theta)$$

$$[\, c(r, \theta) = \text{const. when } g_r = g_\theta = 0 \,]$$

Table 6.2. *Governing equations and general solution for steady, axisymmetric rectilinear flows.*

The constants c_1 and c_2 are determined by the boundary conditions of the flow. Along the axis of symmetry, the velocity u_z must be finite,

$$u_z \text{ finite} \quad \text{at} \quad r = 0\,.$$

Since the wall of the tube is stationary,

$$u_z = 0 \quad \text{at} \quad r = R\,.$$

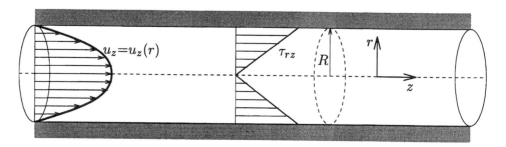

Figure 6.10. *Axisymmetric Poiseuille flow.*

By applying the two conditions, we get $c_1=0$ and

$$c_2 = -\frac{1}{4\eta}\frac{\partial p}{\partial z}R^2\,,$$

and, therefore,

$$u_z = -\frac{1}{4\eta}\frac{\partial p}{\partial z}\left(R^2 - r^2\right)\,, \qquad (6.45)$$

which represents a parabolic velocity profile (Fig. 6.10). The shear stress varies linearly with r,

$$\tau_{rz} = \frac{1}{2}\frac{\partial p}{\partial z}r\,,$$

and the shear stress exerted by the fluid on the wall is

$$\tau_w = -\tau_{rz}|_{r=R} = -\frac{1}{2}\frac{\partial p}{\partial z}R\,.$$

(Note that the contact area faces the negative r-direction.)

The maximum velocity occurs at $r=0$,

$$u_{z,max} = -\frac{1}{4\eta}\frac{\partial p}{\partial z}R^2\,.$$

For the volume flow rate, we get:

$$Q = \int_0^R u_z\, 2\pi r\, dr = -\frac{\pi}{2\eta}\frac{\partial p}{\partial z}\int_0^R (R^2 - r^2)r\, dr \qquad \Longrightarrow$$

$$Q = -\frac{\pi}{8\eta}\frac{\partial p}{\partial z}R^4\,. \qquad (6.46)$$

Note that, since the pressure gradient $\partial p/\partial z$ is negative, Q is positive. Equation (6.46) is the famous experimental result of Hagen and Poiseuille, also known as the *fourth-power law*. This basic equation is used to determine the viscosity from *capillary viscometer* data after taking into account the so-called *Bagley correction* for the inlet and exit pressure losses.

The average velocity, $\bar{u}_z$, in the tube is

$$\bar{u}_z = \frac{Q}{\pi R^2} = -\frac{1}{8\eta}\frac{\partial p}{\partial z}R^2 \,.$$

$\square$

Example 6.2.2. Fully-developed flow in an annulus

Consider fully-developed, pressure-driven flow of a Newtonian liquid in a sufficiently long annulus of radii R and κR, where $\kappa < 1$ (Fig. 6.11). For zero gravity, the general solution for the axial velocity u_z is

$$u_z = \frac{1}{4\eta}\frac{\partial p}{\partial z}r^2 + c_1 \ln r + c_2 \,.$$

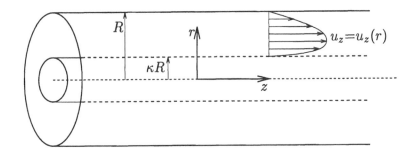

Figure 6.11. *Fully-developed flow in an annulus.*

Applying the boundary conditions,

$$u_z = 0 \quad \text{at} \quad r = \kappa R \,,$$
$$u_z = 0 \quad \text{at} \quad r = R \,,$$

we find that

$$c_1 = -\frac{1}{4\eta}\frac{\partial p}{\partial z}R^2\frac{1 - \kappa^2}{\ln(1/\kappa)}$$

and

$$c_2 = -\frac{1}{4\eta}\frac{\partial p}{\partial z}R^2 - c_1 \ln R.$$

Substituting c_1 and c_2 into the general solution we obtain:

$$u_z = -\frac{1}{4\eta}\frac{\partial p}{\partial z}R^2\left[1 - \left(\frac{r}{R}\right)^2 + \frac{1-\kappa^2}{\ln(1/\kappa)}\ln\frac{r}{R}\right]. \qquad (6.47)$$

The shear stress is given by

$$\tau_{rz} = \frac{1}{4}\frac{\partial p}{\partial z}R\left[2\left(\frac{r}{R}\right) - \frac{1-\kappa^2}{\ln(1/\kappa)}\left(\frac{R}{r}\right)\right]. \qquad (6.48)$$

The maximum velocity occurs at the point where $\tau_{rz}=0$ (which is equivalent to $du_z/dr=0$), i.e., at

$$r^* = R\left[\frac{1-\kappa^2}{2\ln(1/\kappa)}\right]^{1/2}.$$

Substituting into Eq. (6.47), we get

$$u_{z,max} = -\frac{1}{4\eta}\frac{\partial p}{\partial z}R^2\left\{1 - \frac{1-\kappa^2}{2\ln(1/\kappa)}\left[1 - \ln\frac{1-\kappa^2}{2\ln(1/\kappa)}\right]\right\}.$$

For the volume flow rate, we have

$$Q = \int_0^R u_z\, 2\pi r\, dr = -\frac{\pi}{2\eta}\frac{\partial p}{\partial z}R^2\int_0^R\left[1 - \left(\frac{r}{R}\right)^2 + \frac{1-\kappa^2}{\ln(1/\kappa)}\ln\frac{r}{R}\right]r\, dr \quad\Longrightarrow$$

$$Q = -\frac{\pi}{8\eta}\frac{\partial p}{\partial z}R^4\left[\left(1-\kappa^4\right) - \frac{(1-\kappa^2)^2}{\ln(1/\kappa)}\right]. \qquad (6.49)$$

The average velocity, $\bar{u}_z$, in the annulus is

$$\bar{u}_z = \frac{Q}{\pi R^2 - \pi(\kappa R)^2} = -\frac{1}{8\eta}\frac{\partial p}{\partial z}R^2\left[\left(1+\kappa^2\right) - \frac{(1-\kappa^2)}{\ln(1/\kappa)}\right].$$

$\square$

Example 6.2.3. Film flow down a vertical cylinder

A Newtonian liquid is falling vertically on the outside surface of an infinitely long cylinder of radius R, in the form of a thin, uniform axisymmetric film, in contact

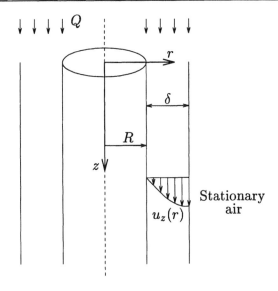

Figure 6.12. *Thin-film flow down a vertical cylinder.*

with stationary air (Fig. 6.12). If the volumetric flow rate of the film is Q, calculate its thickness δ. Assume that the flow is steady, and that surface tension is zero.

Solution:

Equation (6.43) applies with $\frac{\partial p}{\partial z} = 0$:

$$u_z = -\frac{1}{4\eta}\rho g_z\, r^2 + c_1 \ln r + c_2 \,.$$

Since the air is stationary, the shear stress on the free surface of the film is zero,

$$\tau_{rz} = \eta\frac{du_z}{dr} = 0 \quad \text{at} \quad r = R + \delta \quad \Longrightarrow \quad c_1 = \rho g\frac{(R+\delta)^2}{2\eta} \,.$$

At $r=R$, $u_z=0$; consequently,

$$c_2 = \frac{1}{4\eta}\rho g R^2 - c_1 \ln R \,.$$

Substituting into the general solution, we get

$$u_z = \frac{1}{4\eta}\rho g \left[R^2 - r^2 + 2(R+\delta)^2 \ln \frac{r}{R}\right] \,. \tag{6.50}$$

For the volume flow rate, Q, we have:

$$Q = \int_R^{R+\delta} u_z \, 2\pi r \, dr = \frac{\pi}{2\eta} \rho g \int_R^{R+\delta} \left[R^2 - r^2 + 2(R+\delta)^2 \, \ln \frac{r}{R} \right] r \, dr \, .$$

After integration and some algebraic manipulations, we find that

$$Q = \frac{\pi}{8\eta} \rho g R^4 \left\{ 4 \left(1 + \frac{\delta}{R} \right)^4 \ln \left(1 + \frac{\delta}{R} \right) - \frac{\delta}{R} \left(2 + \frac{\delta}{R} \right) \left[3 \left(1 + \frac{\delta}{R} \right)^2 - 1 \right] \right\} . \quad (6.51)$$

When the annular film is very thin, it can be approximated as a thin planar film. We will show that this is indeed the case by proving that for

$$\frac{\delta}{R} \ll 1 \, ,$$

Eq. (6.51) reduces to the expression found in Example 6.1.6 for a thin vertical planar film. Letting

$$\epsilon = \frac{\delta}{R}$$

leads to the following expression for Q,

$$Q = \frac{\pi}{8\eta} \rho g R^4 \left\{ 4 \left(1 + \epsilon \right)^4 \ln \left(1 + \epsilon \right) - \epsilon \left(2 + \epsilon \right) \left[3 \left(1 + \epsilon \right)^2 - 1 \right] \right\} .$$

Expanding $\ln(1 + \epsilon)$ into Taylor series, we get

$$\ln(1 + \epsilon) = \epsilon - \frac{\epsilon^2}{2} + \frac{\epsilon^3}{3} - \frac{\epsilon^4}{4} + O(\epsilon^5) \, .$$

Thus

$$(1 + \epsilon)^4 \ln(1 + \epsilon) = (1 + 4\epsilon + 6\epsilon^2 + 4\epsilon^3 + \epsilon^4) \left[\epsilon - \frac{\epsilon^2}{2} + \frac{\epsilon^3}{3} - \frac{\epsilon^4}{4} + O(\epsilon^5) \right]$$

$$= \epsilon + \frac{7}{2}\epsilon^2 + \frac{13}{3}\epsilon^3 + \frac{25}{12}\epsilon^4 + O(\epsilon^5) \, .$$

Consequently,

$$Q = \frac{\pi}{8\eta} \rho g R^4 \left\{ 4 \left[\epsilon + \frac{7}{2}\epsilon^2 + \frac{13}{3}\epsilon^3 + \frac{25}{12}\epsilon^4 + O(\epsilon^5) \right] - (4\epsilon + 14\epsilon^2 + 12\epsilon^3 + 3\epsilon^4) \right\} \, ,$$

or

$$Q = \frac{\pi}{8\eta} \rho g R^4 \left[\frac{16}{3}\epsilon^3 - \frac{11}{12}\epsilon^4 + O(\epsilon^5) \right] \, .$$

Keeping only the third-order term, we get

$$Q = \frac{\pi}{8\eta}\rho g R^4 \frac{16}{3}\left(\frac{\delta}{R}\right)^3 \implies \frac{Q}{2\pi R} = \frac{\rho g \delta^3}{3\eta}.$$

By setting $2\pi R$ equal to W, the last equation becomes identical to Eq. (6.32).   □

Example 6.2.4. Annular flow with the outer cylinder moving

Consider fully-developed flow of a Newtonian liquid between two coaxial cylinders of infinite length and radii R and κR, where $\kappa <1$. The outer cylinder is steadily translated parallel to its axis with speed V, whereas the inner cylinder is fixed (Fig. 6.13). For this problem, the pressure gradient and gravity are assumed to be negligible.

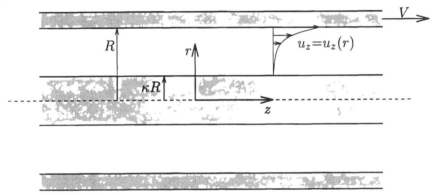

Figure 6.13. *Flow in an annulus driven by the motion of the outer cylinder.*

The general solution for the axial velocity u_z takes the form

$$u_z = c_1 \ln r + c_2 .$$

For $r=\kappa R$, $u_z=0$, and for $r=R$, $u_z=V$. Consequently,

$$c_1 = \frac{V}{\ln(1/\kappa)} \quad \text{and} \quad c_2 = -V\frac{\ln(\kappa R)}{\ln(1/\kappa)}.$$

Therefore, the velocity distribution is given by

$$u_z = V\frac{\ln\left(\frac{r}{\kappa R}\right)}{\ln(1/\kappa)} . \tag{6.52}$$

Let us now examine two limiting cases of this flow.

(a) For $\kappa \to 0$, the annular flow degenerates to flow in a tube. From Eq. (6.52), we have

$$u_z = \lim_{\kappa \to 0} V \, \frac{\ln\left(\frac{r}{\kappa R}\right)}{\ln(1/\kappa)} = V \lim_{\kappa \to 0} \left[1 + \frac{\ln \frac{r}{R}}{\ln(1/\kappa)}\right] = V .$$

In other words, we have plug flow (solid-body translation) in a tube.

(b) For $\kappa \to 1$, the annular flow is approximately a plane Couette flow. To demonstrate this, let

$$\epsilon = \frac{1}{\kappa} - 1 = \frac{1 - \kappa}{\kappa}$$

and

$$\Delta R = R - \kappa R = (1 - \kappa)R \quad \Longrightarrow \quad \kappa R = \frac{\Delta R}{\epsilon} .$$

Introducing Cartesian coordinates, (y, z), with the origin on the surface of the inner cylinder, we have

$$y = r - \kappa R \quad \Longrightarrow \quad \frac{r}{\kappa R} = 1 + \epsilon \frac{y}{\Delta R} .$$

Substituting into Eq. (6.52), we get

$$u_z = V \, \frac{\ln\left(1 + \epsilon \frac{y}{\Delta R}\right)}{\ln(1 + \epsilon)} . \tag{6.53}$$

Using L'Hôpital's rule, we find that

$$\lim_{\epsilon \to 0} V \, \frac{\ln\left(1 + \epsilon \frac{y}{\Delta R}\right)}{\ln(1 + \epsilon)} = \lim_{\epsilon \to 0} V \, \frac{y}{\Delta R} \frac{1 + \epsilon}{1 + \epsilon \frac{y}{\Delta R}} = V \frac{y}{\Delta R} .$$

Therefore, for small values of ϵ, that is for $\kappa \to 1$, we obtain a linear velocity distribution which corresponds to plane Couette flow between plates separated by a distance ΔR. $\square$

6.3   Steady, Axisymmetric Torsional Flows

In axisymmetric torsional flows, also referred to as *swirling flows*,

$$u_r = u_z = 0 , \tag{6.54}$$

and the streamlines are circles centered at the axis of symmetry. Such flows usually occur when rigid cylindrical boundaries (concentric to the symmetry axis of the flow) are rotating about their axes. Due to the axisymmetry condition, $\partial u_\theta / \partial \theta = 0$, the continuity equation for incompressible flow,

$$\frac{1}{r}\frac{\partial}{\partial r}(r u_r) + \frac{1}{r}\frac{\partial u_\theta}{\partial \theta} + \frac{\partial u_z}{\partial z} = 0 \,,$$

is automatically satisfied.

Assuming that the gravitational acceleration is parallel to the symmetry axis of the flow,

$$\mathbf{g} = -g\,\mathbf{e}_z \,, \tag{6.55}$$

the r- and z-momentum equations are simplified as follows,

$$\rho \frac{u_\theta^2}{r} = \frac{\partial p}{\partial r} \,, \tag{6.56}$$

$$\frac{\partial p}{\partial z} + \rho g = 0 \,. \tag{6.57}$$

Equation (6.56) suggests that the centrifugal force on an element of fluid balances the force produced by the radial pressure gradient. Equation (6.57) represents the standard hydrostatic expression. Note also that Eq. (6.56) provides an example in which the nonlinear convective terms are not vanishing. In the present case, however, this nonlinearity poses no difficulties in obtaining the analytical solution for u_θ. As explained below, u_θ is determined from the θ-momentum equation which is decoupled from Eq. (6.56).

By assuming that

$$\frac{\partial p}{\partial \theta} = 0$$

and by integrating Eq. (6.57), we get

$$p = -\rho g\,z + c(r,t) \,;$$

consequently, $\partial p / \partial r$ is not a function of z. Then, from Eq. (6.56) we deduce that

$$u_\theta = u_\theta(r,t) \,. \tag{6.58}$$

Due to the above assumptions, the θ-momentum equation reduces to

$$\rho \frac{\partial u_\theta}{\partial t} = \eta \frac{\partial}{\partial r}\left(\frac{1}{r}\frac{\partial}{\partial r}(r u_\theta)\right) \,. \tag{6.59}$$

For steady flow, we obtain the linear ordinary differential equation

$$\frac{d}{dr}\left(\frac{1}{r}\frac{d}{dr}(r u_\theta)\right) = 0 \,, \tag{6.60}$$

the general solution of which is

$$u_\theta = c_1 r + \frac{c_2}{r} .$$

(6.61)

The constants c_1 and c_2 are determined from the boundary conditions of the flow.

Assumptions:	
	$u_r = u_z = 0, \qquad \frac{\partial u_\theta}{\partial \theta}=0, \qquad \frac{\partial p}{\partial \theta}=0, \qquad \mathbf{g} = -g\, \mathbf{e}_z$
Continuity:	Satisfied identically
θ-momentum:	$\frac{d}{dr}\left(\frac{1}{r}\frac{d}{dr}(ru_\theta)\right) = 0$
z-momentum:	$\frac{\partial p}{\partial z} + \rho g = 0$
r-momentum:	$\rho\frac{u_\theta^2}{r} = \frac{\partial p}{\partial r} \qquad \Longrightarrow \qquad u_\theta = u_\theta(r)$
General solution:	$u_\theta = c_1 r + \frac{c_2}{r}$ $\tau_{r\theta} = \tau_{\theta r} = -2\eta\,\frac{c_2}{r^2}$ $p = \rho\left(\frac{c_1^2 r^2}{2} + 2c_1 c_2 \ln r - \frac{c_2^2}{2r^2}\right) - \rho g\, z + c$

Table **6.3.** *Governing equations and general solution for steady, axisymmetric torsional flows.*

The pressure distribution is determined by integrating Eqs. (6.56) and (6.57):

$$p = \int \frac{u_\theta^2}{r}\, dr - \rho g\, z \qquad \Longrightarrow$$

$$p = \rho \left(\frac{c_1^2 r^2}{2} + 2 c_1 c_2 \ln r - \frac{c_2^2}{2 r^2} \right) - \rho g \, z + c \,, \tag{6.62}$$

where c is a constant of integration, evaluated in any particular problem by specifying the value of the pressure at a reference point.

Note that, under the above assumptions, the only nonzero components of the stress tensor are the shear stresses,

$$\tau_{r\theta} = \tau_{\theta r} = \eta \, r \, \frac{d}{dr} \left(\frac{u_\theta}{r} \right) , \tag{6.63}$$

in terms of which the θ-momentum equation takes the form

$$\frac{d}{dr}(r^2 \tau_{r\theta}) = 0 \,. \tag{6.64}$$

The general solution for $\tau_{r\theta}$ is

$$\tau_{r\theta} = -2 \, \eta \, \frac{c_2}{r^2} \,. \tag{6.65}$$

The assumptions, the governing equations and the general solution for steady, axisymmetric torsional flows are summarized in Table 6.3.

Example 6.3.1. Steady flow between rotating cylinders

The flow between rotating coaxial cylinders is known as the *circular Couette flow*, and is the basis for Couette rotational-type viscometers. Consider the steady flow of an incompressible Newtonian liquid between two vertical coaxial cylinders of infinite length and radii R_1 and R_2, respectively, occurring when the two cylinders are rotating about their common axis with angular velocities Ω_1 and Ω_2, in the absence of gravity (Fig. 6.14).[2]

The general form of the angular velocity u_θ is given by Eq. (6.61),

$$u_\theta = c_1 \, r + \frac{c_2}{r} \,.$$

The boundary conditions,

$$u_\theta = \Omega_1 R_1 \quad \text{at} \quad r = R_1 \,,$$
$$u_\theta = \Omega_2 R_2 \quad \text{at} \quad r = R_2 \,,$$

---

[2]The time-dependent flow between rotating cylinders is much more interesting, especially the manner in which it destabilizes for large values of Ω_1, leading to the generation of axisymmetric *Taylor vortices* [4].

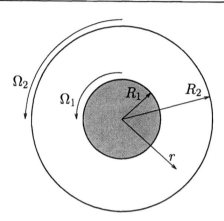

Figure 6.14. *Geometry of circular Couette flow.*

result in

$$c_1 = \frac{R_2^2\Omega_2 - R_1^2\Omega_1}{R_2^2 - R_1^2} \quad \text{and} \quad c_2 = -\frac{R_1^2 R_2^2}{R_2^2 - R_1^2}\,(\Omega_2 - \Omega_1)\,.$$

Therefore,

$$u_\theta = \frac{1}{R_2^2 - R_1^2}\left[(R_2^2\Omega_2 - R_1^2\Omega_1)\,r - R_1^2 R_2^2(\Omega_2 - \Omega_1)\,\frac{1}{r}\right]. \qquad (6.66)$$

Note that the viscosity does not appear in Eq. (6.66), because shearing between adjacent cylindrical shells of fluid is zero. This observation is analogous to that made for the plane Couette flow [Eq. (6.14)]. Also, from Eqs. (6.62) and (6.65), we get

$$p = \rho\frac{1}{(R_2^2 - R_1^2)^2}\left[\frac{1}{2}(R_2^2\Omega_2 - R_1^2\Omega_1)^2\,r^2 + 2R_1^2 R_2^2(R_2^2\Omega_2 - R_1^2\Omega_1)(\Omega_2 - \Omega_1)\ln r\right.$$

$$\left. - \frac{1}{2}R_1^4 R_2^4(\Omega_2 - \Omega_1)^2\,\frac{1}{r^2}\right] + c, \qquad (6.67)$$

and

$$\tau_{r\theta} = 2\eta\frac{R_1^2 R_2^2}{(R_2^2 - R_1^2)^2}\,(\Omega_2 - \Omega_1)\,\frac{1}{r^2}. \qquad (6.68)$$

Let us now examine the four special cases of flow between rotating cylinders illustrated in Fig. 6.15.

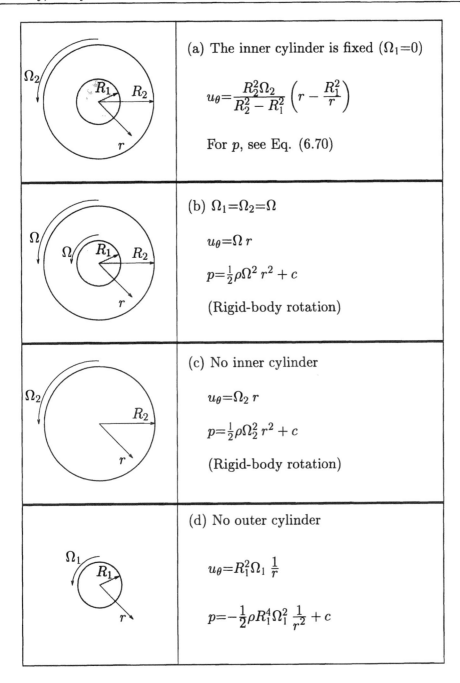

(a) The inner cylinder is fixed ($\Omega_1=0$)

$$u_\theta = \frac{R_2^2\Omega_2}{R_2^2 - R_1^2}\left(r - \frac{R_1^2}{r}\right)$$

For p, see Eq. (6.70)

(b) $\Omega_1=\Omega_2=\Omega$

$$u_\theta = \Omega\, r$$

$$p = \tfrac{1}{2}\rho\Omega^2\, r^2 + c$$

(Rigid-body rotation)

(c) No inner cylinder

$$u_\theta = \Omega_2\, r$$

$$p = \tfrac{1}{2}\rho\Omega_2^2\, r^2 + c$$

(Rigid-body rotation)

(d) No outer cylinder

$$u_\theta = R_1^2\Omega_1\, \frac{1}{r}$$

$$p = -\tfrac{1}{2}\rho R_1^4\Omega_1^2\, \frac{1}{r^2} + c$$

Figure 6.15. *Different cases of flow between rotating vertical coaxial cylinders of infinite height.*

(a) The inner cylinder is fixed, i.e., $\Omega_1 = 0$. In this case,

$$u_\theta = \frac{R_2^2 \Omega_2}{R_2^2 - R_1^2} \left(r - \frac{R_1^2}{r} \right) \tag{6.69}$$

and

$$p = \rho \frac{R_2^4 \Omega_2^2}{(R_2^2 - R_1^2)^2} \left(\frac{r^2}{2} + 2R_1^2 \ln r - \frac{R_1^4}{2r^2} \right) + c. \tag{6.70}$$

The constant c can be determined by setting $p = p_0$ at $r = R_1$; accordingly,

$$p = \rho \frac{R_2^4 \Omega_2^2}{(R_2^2 - R_1^2)^2} \left[\frac{r^2 - R_1^2}{2} + 2R_1^2 \ln \frac{r}{R_1} - \frac{R_1^4}{2} \left(\frac{1}{r^2} - \frac{1}{R_1^2} \right) \right] + p_0. \tag{6.71}$$

For the shear stress, $\tau_{r\theta}$, we get

$$\tau_{r\theta} = 2\eta \frac{R_1^2 R_2^2}{R_2^2 - R_1^2} \Omega_2 \frac{1}{r^2}. \tag{6.72}$$

The shear stress exerted by the liquid to the outer cylinder is

$$\tau_w = -\tau_{r\theta}|_{r=R_2} = -2\eta \frac{R_1^2}{R_2^2 - R_1^2} \Omega_2. \tag{6.73}$$

In viscosity measurements, one measures the torque T per unit height L, at the outer cylinder,

$$\frac{T}{L} = 2\pi R_2^2 (-\tau_w) \quad \Longrightarrow$$

$$\frac{T}{L} = 4\pi\eta \frac{R_1^2 R_2^2}{R_2^2 - R_1^2} \Omega_2. \tag{6.74}$$

The unknown viscosity of a liquid can be determined using the above relation.

When the gap between the two cylinders is very small, circular Couette flow can be approximated as a plane Couette flow. Indeed, letting $r = R_1 + \Delta r$, we get from Eq. (6.69)

$$u_\theta = \frac{R_2^2 \Omega_2}{R_2^2 - R_1^2} \frac{2 + \frac{\Delta r}{R_1}}{1 + \frac{\Delta r}{R_1}} \Delta r.$$

When $R_1 \to R_2$, $\Delta r / R_1 \ll 1$ and, therefore,

$$u_\theta = \frac{R_2 \Omega_2}{2(R_2 - R_1)} 2\Delta r = \frac{R_2 \Omega_2}{R_2 - R_1} \Delta r,$$

which is a linear velocity distribution corresponding to plane Couette flow between plates separated by a distance R_2-R_1, with the upper plate moving with velocity $R_2\Omega_2$.

(b) The two cylinders rotate with the same angular velocity, i.e.,

$$\Omega_1 = \Omega_2 = \Omega .$$

In thic case, $c_1=\Omega$ and $c_2=0$. Consequently,

$$u_\theta = \Omega r , \tag{6.75}$$

which corresponds to rigid-body rotation. This is also indicated by the zero tangential stress,

$$\tau_{r\theta} = -2\eta \frac{c_2}{r^2} = 0 .$$

For the pressure, we get

$$p = \frac{1}{2}\rho\Omega^2 r^2 + c . \tag{6.76}$$

(c) The inner cylinder is removed. In thic case, $c_1=\Omega_2$ and $c_2=0$, since u_θ (and $\tau_{r\theta}$) are finite at $r=0$. This flow is the limiting case of the previous one for $R_1 \to 0$,

$$u_\theta = \Omega_2 r , \quad \tau_{r\theta} = 0 \quad \text{and} \quad p = \frac{1}{2}\rho\Omega_2^2 r^2 + c .$$

(d) The outer cylinder is removed, i.e., the inner cylinder is rotating in an infinite pool of liquid. In this case, $u_\theta \to 0$ as $r \to \infty$, and, therefore, $c_1=0$. At $r=R_1$, $u_\theta=\Omega_1 R_1$ which gives

$$c_2 = R_1^2 \Omega_1 .$$

Consequently,

$$u_\theta = R_1^2\Omega_1 \frac{1}{r} , \tag{6.77}$$

$$\tau_{r\theta} = -2\eta \, R_1^2\Omega_1 \frac{1}{r^2} , \tag{6.78}$$

and

$$p = -\frac{1}{2}\rho \, R_1^4\Omega_1^2 \frac{1}{r^2} + c . \tag{6.79}$$

The shear stress exerted by the liquid to the cylinder is

$$\tau_w = \tau_{r\theta}|_{r=R_1} = -2\eta \, \Omega_1 . \tag{6.80}$$

The torque per unit height required to rotate the cylinder is

$$\frac{T}{L} \;=\; 2\pi R_1^2 \left(-\tau_w\right) \;=\; 4\pi\eta\, R_1^2 \Omega_1 \,. \tag{6.81}$$

$\square$

In the previous example, we studied flows between vertical coaxial cylinders of infinite height ignoring the gravitational acceleration. As indicated by Eq. (6.62), gravity has no influence on the velocity and affects only the pressure. In case of rotating liquids with a free surface, the gravity term should be included if the top part of the flow and the shape of the free surface are of interest. If surface tension effects are neglected, the pressure on the free surface is constant. Therefore, the locus of the free surface can be determined using Eq. (6.62).

Example 6.3.2. Shape of free surface in torsional flows

In this example, we study two different torsional flows with a free surface. First, we consider steady flow of a liquid contained in a large cylindrical container and agitated by a vertical rod of radius R that is coaxial to the container and rotates at angular velocity Ω. If the radius of the container is much larger than R, one may assume that the rod rotates in an infinite pool of liquid (Fig. 6.16).

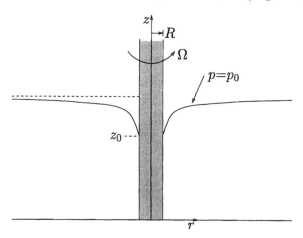

Figure 6.16. *Rotating rod in a pool of liquid.*

From the results of Example 6.3.1, we have $c_1=0$ and $c_2=\Omega R$. Therefore,

$$u_\theta \;=\; R^2\Omega\,\frac{1}{r}$$

and

$$p = -\frac{1}{2}\rho R^4 \Omega^2 \frac{1}{r^2} - \rho g\, z + c\,.$$

With the surface tension effects neglected, the pressure on the free surface is equal to the atmospheric pressure, p_0. To determine the constant c, we assume that the free surface contacts the rod at $z=z_0$. Thus, we obtain

$$c = p_0 + \frac{1}{2}\rho R^4 \Omega^2 \frac{1}{R^2} + \rho g\, z_0$$

and

$$p = \frac{1}{2}\rho R^4 \Omega^2 \left(\frac{1}{R^2} - \frac{1}{r^2}\right) - \rho g\,(z - z_0) + p_0\,. \tag{6.82}$$

Since the pressure is constant along the free surface, the equation of the latter is

$$0 = p - p_0 = \frac{1}{2}\rho R^4 \Omega^2 \left(\frac{1}{R^2} - \frac{1}{r^2}\right) - \rho g\,(z - z_0) \quad \Longrightarrow$$

$$z = z_0 + \frac{R^2 \Omega^2}{2g}\left(1 - \frac{R^2}{r^2}\right)\,. \tag{6.83}$$

The elevation of the free surface increases with the radial distance r and approaches asymptotically the value

$$z_\infty = z_0 + \frac{R^2 \Omega^2}{2g}\,.$$

This flow behavior, known as *rod dipping*, is a characteristic of generalized Newtonian liquids, whereas viscoelastic liquids exhibit *rod climbing* (i.e., they climb the rotating rod) [5].

Consider now steady flow of a liquid contained in a cylindrical container of radius R rotating at angular velocity Ω (Fig. 6.17). From Example 6.3.1, we know that this flow corresponds to rigid-body rotation, i.e.,

$$u_\theta = \Omega\, r\,.$$

The pressure is given by

$$p = \frac{1}{2}\rho\Omega^2 r^2 - \rho g\, z + c\,.$$

Letting z_0 be the elevation of the free surface at $r=0$, and p_0 be the atmospheric pressure, we get

$$c = p_0 + \rho g\, z_0\,,$$

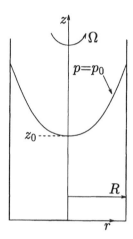

Figure 6.17. *Free surface of liquid in a rotating cylindrical container.*

and thus

$$p = \frac{1}{2}\rho\Omega^2\, r^2 - \rho g\,(z - z_0) + p_0\,. \tag{6.84}$$

The equation of the free surface is

$$0 = p - p_0 = \frac{1}{2}\rho\Omega^2\, r^2 - \rho g\,(z - z_0) \quad \Longrightarrow$$

$$z = z_0 + \frac{\Omega^2}{2g}\, r^2\,, \tag{6.85}$$

i.e., the free surface is a parabola. □

Example 6.3.3. Superposition of Poiseuille and Couette flows

Consider steady flow of a liquid in a cylindrical tube occurring when a constant pressure gradient $\partial p/\partial z$ is applied while the tube is rotating about its axis with constant angular velocity Ω (Fig. 6.18). This is obviously a *bidirectional* flow since the axial and azimuthal velocity components, u_z and u_θ, are nonzero.

The flow can be considered as a superposition of axisymmetric Poiseuille and circular Couette flows, for which we have:

$$u_z = u_z(r) = -\frac{1}{4\eta}\frac{\partial p}{\partial z}\,(R^2 - r^2) \quad \text{and} \quad u_\theta = u_\theta(r) = \Omega\, r\,.$$

This superposition is *dynamically admissible* since it does not violate the continuity equation, which is automatically satisfied.

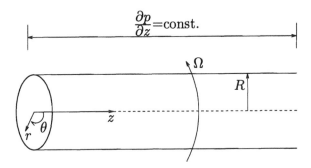

Figure 6.18. *Flow in a rotating tube under constant pressure gradient.*

Moreover, the governing equations of the flow, i.e., the z- and θ-momentum equations,

$$-\frac{\partial p}{\partial z} + \eta \frac{1}{r} \frac{\partial}{\partial r} \left(r \frac{\partial u_z}{\partial r} \right) = 0 \quad \text{and} \quad \frac{\partial}{\partial r} \left(\frac{1}{r} \frac{\partial}{\partial r}(r u_\theta) \right) = 0\,,$$

are linear and uncoupled. Hence, the velocity for this flow is given by

$$\mathbf{u} = u_z\,\mathbf{e}_z + u_\theta\,\mathbf{e}_\theta = -\frac{1}{4\eta} \frac{\partial p}{\partial z}\left(R^2 - r^2\right)\mathbf{e}_z + \Omega\,r\,\mathbf{e}_\theta\,, \tag{6.86}$$

which describes a *helical flow.*

The pressure is obtained by integrating the r-momentum equation,

$$\rho\,\frac{u_\theta^2}{r} = \frac{\partial p}{\partial r}\,,$$

taking into account that $\partial p / \partial z$ is constant. It turns out that

$$p = \frac{\partial p}{\partial z}\,z + \frac{1}{2}\rho\Omega^2\,r^2 + c\,, \tag{6.87}$$

which is simply the sum of the pressure distributions of the two superposed flows. It should be noted, however, that this might not be the case in superposition of other unidirectional flows. □

6.4 Steady, Axisymmetric Radial Flows

In axisymmetric radial flows,

$$u_z = u_\theta = 0 . \tag{6.88}$$

Evidently, the streamlines are straight lines perpendicular to the axis of symmetry (Fig. 6.19).

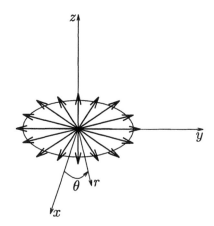

Figure 6.19. *Streamlines in axisymmetric radial flow.*

For the sake of simplicity, we will assume that u_r, in addition to being axisymmetric, does not depend on z. In other words, we assume that, in steady-state, u_r is only a function of r:

$$u_r = u_r(r) . \tag{6.89}$$

A characteristic of radial flows is that the nonvanishing radial velocity component is determined by the conservation of mass rather than by the r-component of the conservation of momentum equation. This implies that u_r is independent of the viscosity of the liquid. (More precisely, u_r is independent of the constitutive equation of the fluid.) Due to Eq. (6.88), the continuity equation is simplified to

$$\frac{\partial}{\partial r}(ru_r) = 0 , \tag{6.90}$$

which gives

$$u_r = \frac{c_1}{r} , \tag{6.91}$$

where c_1 is a constant. The velocity u_r can also be obtained from a macroscopic mass balance. If Q is the volumetric flow rate per unit height, L, then

$$Q = u_r\,(2\pi r L) \quad \Longrightarrow$$

$$u_r = \frac{Q}{2\pi L\, r} \, ,$$

(6.92)

which is identical to Eq. (6.91) for $c_1 = Q/(2\pi L)$.

Assumptions:

$$u_z = u_\theta = 0, \qquad u_r = u_r(r), \qquad \mathbf{g} = -g\, \mathbf{e}_z$$

Continuity:

$$\frac{d}{dr}(r u_r) = 0 \qquad \Longrightarrow \qquad u_r = \frac{c_1}{r}$$

r-momentum:

$$\rho\, u_r \frac{du_r}{dr} = -\frac{\partial p}{\partial r}$$

z-momentum:

$$\frac{\partial p}{\partial z} + \rho g = 0$$

θ-momentum:

$$\frac{\partial p}{\partial \theta} = 0 \qquad \Longrightarrow \qquad p = p(r, z)$$

General solution:

$$u_r = \frac{c_1}{r}$$

$$\tau_{rr} = -2\eta\, \frac{c_1}{r^2}, \qquad \tau_{\theta\theta} = 2\eta\, \frac{c_1}{r^2}$$

$$p = -\rho\frac{c_1^2}{2r^2} - \rho g\, z + c$$

Table 6.4. *Governing equations and general solution for steady, axisymmetric radial flows.*

Letting

$$\mathbf{g} = -g\, \mathbf{e}_z \, ,$$

(6.93)

the r-component of the Navier–Stokes equation is simplified to

$$\rho\, u_r \frac{du_r}{dr} = -\frac{\partial p}{\partial r} \, .$$

(6.94)

Note that the above equation contains a nonvanishing, nonlinear convective term. The z- and θ-components of the Navier–Stokes equation are reduced to the standard hydrostatic expression,

$$\frac{\partial p}{\partial z} + \rho g = 0 \,, \tag{6.95}$$

and to

$$\frac{\partial p}{\partial \theta} = 0 \,, \tag{6.96}$$

respectively. The latter equation dictates that $p=p(r,z)$. Integration of Eqs. (6.94) and (6.95) gives

$$\begin{aligned}
p(r,z) &= -\rho \int u_r \frac{du_r}{dr} \, dr - \rho g\, z + c \\
&= \rho\, c_1^2 \int \frac{1}{r^3} \, dr - \rho g\, z + c \qquad \Longrightarrow
\end{aligned}$$

$$p(r,z) = -\rho \frac{c_1^2}{2r^2} - \rho g\, z + c \,, \tag{6.97}$$

where the integration constant c is determined by specifying the value of the pressure at a point.

In axisymmetric radial flows, there are two nonvanishing stress components:

$$\tau_{rr} = 2\eta \frac{du_r}{dr} = -2\eta \frac{c_1}{r^2} \,; \tag{6.98}$$

$$\tau_{\theta\theta} = 2\eta \frac{u_r}{r} = 2\eta \frac{c_1}{r^2} \,. \tag{6.99}$$

The assumptions, the governing equations, and the general solution for steady, axisymmetric radial flows are summarized in Table 6.4.

6.5 Steady, Spherically Symmetric Radial Flows

In spherically symmetric radial flows, the fluid particles move towards or away from the center of solid, liquid, or gas spheres. Examples of such flows are flow around a gas bubble which grows or collapses in a liquid bath, flow towards a spherical sink, and flow away from a point source.

The analysis of spherically symmetric radial flows is similar to that of the axisymmetric ones. The assumptions and the results are tabulated in Table 6.5. Obviously,

Assumptions:	
	$$u_\theta = u_\phi = 0, \quad u_r = u_r(r), \quad \mathbf{g} = \mathbf{0}$$
Continuity:	
	$$\frac{d}{dr}(r^2 u_r) = 0 \quad \Longrightarrow \quad u_r = \frac{c_1}{r^2}$$
r-momentum:	
	$$\rho\, u_r \frac{du_r}{dr} = -\frac{\partial p}{\partial r}$$
θ-momentum:	
	$$\frac{\partial p}{\partial \theta} = 0$$
ϕ-momentum:	
	$$\frac{\partial p}{\partial \phi} = 0$$
General solution:	
	$$u_r = \frac{c_1}{r^2}$$
	$$\tau_{rr} = -4\eta\, \frac{c_1}{r^3}, \qquad \tau_{\theta\theta} = \tau_{\phi\phi} = 2\eta\, \frac{c_1}{r^3}$$
	$$p = -\rho\frac{c_1^2}{2r^4} + c$$

Table 6.5. *Governing equations and general solution for steady, spherically symmetric radial flows.*

spherical coordinates are the natural choice for the analysis. In steady-state, the radial velocity component is a function of the radial distance,

$$u_r = u_r(r), \tag{6.100}$$

while the other two velocity components are zero:

$$u_\theta = u_\phi = 0. \tag{6.101}$$

As in axisymmetric radial flows, u_r is determined from the continuity equation as

$$u_r = \frac{c_1}{r^2}, \tag{6.102}$$

or

$$u_r = \frac{Q}{4\pi r^2}, \tag{6.103}$$

where Q is the volumetric flow rate.

The pressure is given by

$$p(r) = -\rho \frac{c_1^2}{2r^4} + c. \tag{6.104}$$

(Note that in spherically symmetric flows, gravity is neglected.) Finally, there are now three nonvanishing stress components:

$$\tau_{rr} = 2\eta \frac{du_r}{dr} = -4\eta \frac{c_1}{r^3}; \tag{6.105}$$

$$\tau_{\theta\theta} = \tau_{\phi\phi} = 2\eta \frac{u_r}{r} = 2\eta \frac{c_1}{r^3}. \tag{6.106}$$

Example 6.5.1. Bubble growth in a Newtonian liquid

Boiling of a liquid often originates from small air bubbles which grow radially in the liquid. Consider a spherical bubble of radius $R(t)$ in a pool of liquid growing at a rate

$$\frac{dR}{dt} = k.$$

The velocity, u_r, and the pressure, p, can be calculated using Eqs. (6.102) and (6.104), respectively. At first, we calculate the constant c_1. At $r=R$, $u_r=dR/dt=k$ or

$$\frac{c_1}{R^2} = k \quad \Longrightarrow \quad c_1 = kR^2.$$

Substituting c_1 into Eqs. (6.102) and (6.104), we get

$$u_r = k \frac{R^2}{r^2}$$

and

$$p = -\rho k^2 \frac{R^4}{2r^4} + c.$$

Note that the pressure near the surface of the bubble may attain small or even negative values which favor evaporation of the liquid and expansion of the bubble.

□

6.6   Transient One-Dimensional Unidirectional Flows

In Sections 6.1 to 6.3, we studied three classes of steady-state unidirectional flows where the dependent variable, i.e., the nonzero velocity component, was assumed to be a function of a single spatial independent variable. The governing equation for such a flow is a linear, second-order, ordinary differential equation which is integrated to arrive at a general solution. The general solution contains two integration constants which are determined by the boundary conditions at the endpoints of the one-dimensional domain over which the analytical solution is sought.

In the present section, we consider one-dimensional, *transient*, unidirectional flows. Hence, the dependent variable is now a function of two independent variables, one of which is time, t. The governing equations for these flows are partial differential equations. In fact, we have already encountered some of these PDEs in Sections 6.1-6.3 while simplifying the corresponding components of the Navier–Stokes equation. For the sake of convenience, these are listed below.

(a) For transient one-dimensional rectilinear flow in Cartesian coordinates with $u_y = u_z = 0$ and $u_x = u_x(y, t)$,

$$\rho \frac{\partial u_x}{\partial t} = -\frac{\partial p}{\partial x} + \eta \frac{\partial^2 u_x}{\partial y^2} + \rho g_x . \qquad (6.107)$$

(b) For transient axisymmetric rectilinear flow with $u_r = u_\theta = 0$ and $u_z = u_z(r, t)$,

$$\rho \frac{\partial u_z}{\partial t} = \frac{\partial p}{\partial z} + \eta \frac{1}{r} \frac{\partial}{\partial r} \left(r \frac{\partial u_z}{\partial r} \right) + \rho g_z ,$$

or

$$\rho \frac{\partial u_z}{\partial t} = -\frac{\partial p}{\partial z} + \eta \left(\frac{\partial^2 u_z}{\partial r^2} + \frac{1}{r} \frac{\partial u_z}{\partial r} \right) + \rho g_z . \qquad (6.108)$$

(c) For transient axisymmetric torsional flow with $u_z = u_r = 0$ and $u_\theta = u_\theta(r, t)$,

$$\rho \frac{\partial u_\theta}{\partial t} = \eta \frac{\partial}{\partial r} \left(\frac{1}{r} \frac{\partial}{\partial r} (r u_\theta) \right) ,$$

or

$$\rho \frac{\partial u_\theta}{\partial t} = \eta \left(\frac{\partial^2 u_\theta}{\partial r^2} + \frac{1}{r} \frac{\partial u_\theta}{\partial r} - \frac{1}{r^2} u_\theta \right) . \qquad (6.109)$$

The above equations are all *parabolic* PDEs. For any particular flow, they are supplemented by appropriate boundary conditions at the two endpoints of the one-dimensional flow domain, and by an *initial condition* for the entire flow domain. Note that the pressure gradients in Eqs. (6.107) and (6.108) may be functions of time. These two equations are *inhomogeneous* due to the presence of the pressure gradient and gravity terms. The inhomogeneous terms can be eliminated by decomposing the dependent variable into a properly chosen steady-state component (satisfying the corresponding steady-state problem and the boundary conditions) and a transient one which satisfies the *homogeneous* problem. A similar decomposition is often used for transforming inhomogeneous boundary conditions into homogeneous ones. *Separation of variables* [2] and the *similarity solution* method [3,6] are the standard methods for solving Eq. (6.109) and the homogeneous counterparts of Eqs. (6.107) and (6.108).

In homogeneous problems admitting separable solutions, the dependent variable $u(x_i, t)$ is expressed in the form

$$u(x_i, t) = X(x_i) \, T(t) \, . \tag{6.110}$$

Substitution of the above expression into the governing equation leads to the equivalent problem of solving two ordinary differential equations with X and T as the dependent variables.

In similarity methods, the two independent variables, x_i and t, are combined into the *similarity variable*

$$\xi = \xi(x_i, t) \, . \tag{6.111}$$

If a similarity solution does exist, then the original partial differential equation for $u(x_i, t)$ is reduced to an ordinary differential equation for $u(\xi)$.

Similarity solutions exist for problems involving parabolic PDEs in two independent variables where external length and time scales are absent. A typical problem is flow of a semi-infinite fluid above a plate suddenly set in motion with a constant velocity (Example 6.6.1). Length and time scales do exist in transient plane Couette flow, and in flow of a semi-infinite fluid above a plate oscillating along its own plane. In the former flow, the length scale is the distance between the two plates, whereas in the latter case the length scale is the period of oscillations. These two flows are governed by Eq. (6.107), with the pressure-gradient and gravity terms neglected; they are solved in Examples 6.6.2 and 6.6.3 using separation of variables. In Example 6.6.4 we solve the problem of transient plane Poiseuille flow due to the sudden application of a constant pressure gradient.

Finally, in the last two examples we solve transient axisymmetric rectilinear and torsional flow problems governed, respectively, by Eqs. (6.108) and (6.109). In

Example 6.6.5, we consider transient axisymmetric Poiseuille flow, and in Example 6.6.6, we consider flow inside an infinite long cylinder which is suddenly rotated.

Example 6.6.1. Flow near a plate suddenly set in motion

Consider a semi-infinite incompressible Newtonian liquid of viscosity η and density ρ, bounded below by a plate at $y=0$ (Fig. 6.20). Initially, both the plate and the liquid are at rest. At time $t=0^+$, the plate starts moving in the x direction (i.e., along its plane) with constant speed V. Pressure gradient and gravity in the direction of the flow are zero. This flow problem was studied by Stokes in 1851 and is called *Rayleigh's problem* or *Stokes' first problem*.

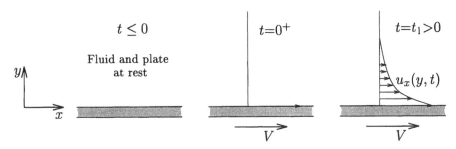

Figure 6.20. *Flow near a plate suddenly set in motion.*

The governing equation for $u_x(y,t)$ is homogeneous:

$$\frac{\partial u_x}{\partial t} = \nu \frac{\partial^2 u_x}{\partial y^2}, \tag{6.112}$$

where $\nu \equiv \eta/\rho$ is the kinematic viscosity. Mathematically, Eq. (6.112) is called the *heat* or *diffusion equation*. The boundary and initial conditions are:

$$\left. \begin{array}{llll} u_x = V & \text{at} & y = 0, \; t > 0 \\ u_x = 0 & \text{at} & y \to \infty, \; t \geq 0 \\ u_x = 0 & \text{at} & t = 0, \; 0 \leq y < \infty \end{array} \right\}. \tag{6.113}$$

The problem described by Eqs. (6.112) and (6.113) can be solved by Laplace transforms and by the similarity method. Here, we employ the latter, which is useful in solving some nonlinear problems arising in boundary layer theory (see Chapter 8). A solution with Laplace transforms can be found in Ref. [7].

Examining Eq. (6.112), we observe that if y and t are magnified k and k^2 times, respectively, Eq. (6.112), along with the boundary and initial conditions (6.113), will still be satisfied. This clearly suggests that u_x depends on a combination of y and t of the form $y/\sqrt{t}$. The same conclusion is reached by noting that the dimensionless velocity u_x/V must be a function of the remaining kinematic quantities of this flow problem: ν, t and y. From these three quantities, only one dimensionless group can be formed, $\xi = y/\sqrt{\nu t}$.

Let us, however, assume that the existence of a similarity solution and the proper combination of y and t are not known *a priori*, and assume that the solution is of the form

$$u_x(y, t) = V f(\xi), \tag{6.114}$$

where

$$\xi = a\,\frac{y}{t^n}, \quad \text{with} \quad n > 0. \tag{6.115}$$

Here $\xi(y, t)$ is the similarity variable, a is a constant to be determined later so that ξ is dimensionless, and n is a positive number to be chosen so that the original partial differential equation (6.112) can be transformed into an ordinary differential equation with f as the dependent variable and ξ as the independent one. Note that a precondition for the existence of a similarity solution is that ξ is of such a form that the original boundary and initial conditions are combined into two boundary conditions for the new dependent variable f. This is easily verified in the present flow. The boundary condition at $y=0$ is equivalent to

$$f = 1 \quad \text{at} \quad \xi = 0, \tag{6.116}$$

whereas the boundary condition at $y \to \infty$ and the initial condition collapse to a single boundary condition for f,

$$f = 0 \quad \text{at} \quad \xi \to \infty. \tag{6.117}$$

Differentiation of Eq. (6.114) using the chain rule gives

$$\frac{\partial u_x}{\partial t} = -V\,n\frac{ay}{t^{n+1}}\,f' = -V\,n\frac{\xi}{t}\,f',$$

$$\frac{\partial u_x}{\partial y} = V\,\frac{a}{t^n}\,f' \quad \text{and} \quad \frac{\partial^2 u_x}{\partial y^2} = V\,\frac{a^2}{t^{2n}}\,f'',$$

where primes denote differentiation with respect to ξ. Substitution of the above derivatives into Eq. (6.112) gives the following equation:

$$f'' + \frac{n\xi}{\nu a^2}\,t^{2n-1}\,f' = 0.$$

By setting $n=1/2$, time is eliminated and the above expression becomes a second-order, ordinary differential equation,

$$f'' + \frac{\xi}{2\nu a^2} f' = 0 \quad \text{with} \quad \xi = a \frac{y}{\sqrt{t}} .$$

Taking a equal to $1/\sqrt{\nu}$ makes the similarity variable dimensionless. For convenience in the solution of the differential equation, we set $a=1/(2\sqrt{\nu})$. Hence,

$$\xi = \frac{y}{2\sqrt{\nu t}} , \tag{6.118}$$

whereas the resulting ordinary differential equation is

$$f'' + 2\xi f' = 0 . \tag{6.119}$$

This equation is subject to the boundary conditions (6.116) and (6.117). By straight-forward integration, we obtain

$$f(\xi) = c_1 \int_0^\xi e^{-z^2} dz + c_2 ,$$

where z is a dummy variable of integration. At $\xi=0$, $f=1$; consequently, $c_2=1$. At $\xi \rightarrow \infty$, $f=0$; therefore,

$$c_1 \int_0^\infty e^{-z^2} dz + 1 = 0 \quad \text{or} \quad c_1 = -\frac{2}{\sqrt{\pi}} ,$$

and

$$f(\xi) = 1 - \frac{2}{\sqrt{\pi}} \int_0^\xi e^{-z^2} dz = 1 - \text{erf}(\xi) , \tag{6.120}$$

where erf is the *error function* defined as

$$\text{erf}(\xi) \equiv \frac{2}{\sqrt{\pi}} \int_0^\xi e^{-z^2} dz . \tag{6.121}$$

Values of the error function are tabulated in several math textbooks. It is a monotone increasing function with

$$erf(0) = 0 \quad \text{and} \quad \lim_{\xi \rightarrow \infty} \text{erf}(\xi) = 1 .$$

Note that the second expression was used when calculating the constant c_1. Substituting into Eq. (6.114), we obtain the solution

$$u_x(y, t) = V \left[1 - \text{erf} \left(\frac{y}{2\sqrt{\nu t}} \right) \right] . \tag{6.122}$$

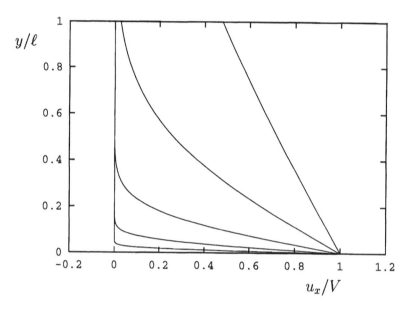

Figure 6.21. *Transient flow due to the sudden motion of a plate. Velocity profiles at $\nu t/\ell^2 = 0.0001$, 0.001, 0.01, 0.1, and 1, where ℓ is an arbitrary length scale.*

The evolution of $u_x(y, t)$ is illustrated in Fig. 6.21, where the velocity profiles are plotted at different values of $\nu t/\ell^2$, ℓ being an arbitrary length scale.

From Eq. (6.122), we observe that, for a fixed value of u_x/V, y varies as $2\sqrt{\nu t}$. A *boundary-layer thickness*, $\delta(t)$, can be defined as the distance from the moving plate at which $u_x/V = 0.01$. This happens when ξ is about 1.8, and thus

$$\delta(t) = 3.6 \sqrt{\nu t}.$$

The sudden motion of the plate generates vorticity since the velocity profile is discontinuous at the initial distance. The thickness $\delta(t)$ is the penetration of vorticity distance into regions of uniform velocity after a time t. Note that Eq. (6.112) can also be viewed as a vorticity diffusion equation. Indeed, since $\mathbf{u} = u_x(y, t)\mathbf{i}$,

$$\omega(y, t) = |\boldsymbol{\omega}| = |\nabla \times \mathbf{u}| = \frac{\partial u_x}{\partial y},$$

and Eq. (6.112) can be cast in the form

$$\frac{\partial}{\partial t} \int_0^y \omega \, dy = \nu \frac{\partial \omega}{\partial y},$$

or, equivalently,

$$\frac{\partial \omega}{\partial t} = \nu \frac{\partial^2 \omega}{\partial y^2}. \tag{6.123}$$

The above expression is a vorticity conservation equation and highlights the role of kinematic viscosity, which acts as a *vorticity diffusion* coefficient in a manner analogous to that of thermal diffusivity in heat diffusion.

The shear stress on the plate is given by

$$\tau_w = \tau_{yx}|_{y=0} = \eta \left.\frac{\partial u_x}{\partial y}\right|_{y=0} = -\eta V \left.\frac{\partial \mathrm{erf}(\xi)}{\partial \xi}\right|_{\xi=0} \left.\frac{\partial \xi}{\partial y}\right|_{y=0} = -\frac{\eta V}{\sqrt{\pi \nu t}}, \tag{6.124}$$

which suggests that the stress is singular at the instant the plate starts moving, and decreases as $1/\sqrt{t}$.

The physics of this example are similar to those of boundary layer flow, which is examined in detail in Chapter 8. In fact, the same similarity variable was invoked by Rayleigh to calculate skin friction over a plate moving with velocity V through a stationary liquid which leads to [8]

$$\tau_w = \frac{\eta V}{\sqrt{\pi \nu}} \sqrt{\frac{V}{x}},$$

by simply replacing t by x/V in Eq. (6.124). This situation arises in free stream flows overtaking submerged bodies, giving rise to boundary layers [9].

□

In the following example, we demonstrate the use of separation of variables by solving a transient plane Couette flow problem.

Example 6.6.2. Transient plane Couette flow

Consider a Newtonian liquid of density ρ and viscosity η bounded by two infinite parallel plates separated by a distance H, as shown in Fig. 6.22. The liquid and the two plates are initially at rest. At time $t=0^+$, the lower plate is suddenly brought to a steady velocity V in its own plane, while the upper plate is held stationary.

The governing equation is the same as in the previous example,

$$\frac{\partial u_x}{\partial t} = \nu \frac{\partial^2 u_x}{\partial y^2}, \tag{6.125}$$

with the following boundary and initial conditions:

$$\left.\begin{array}{lll} u_x = V & \text{at} & y = 0,\ t > 0 \\ u_x = 0 & \text{at} & y = H,\ t \geq 0 \\ u_x = 0 & \text{at} & t = 0,\ 0 \leq y \leq H \end{array}\right\}. \tag{6.126}$$

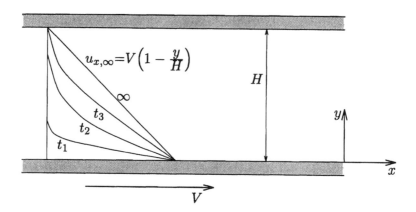

Figure 6.22. *Schematic of the evolution of the velocity in start-up plane Couette flow.*

Note that, while the governing equation is homogeneous, the boundary conditions are inhomogeneous. Therefore, separation of variables cannot be applied directly. We first have to transform the problem so that the governing equation and the two boundary conditions are homogeneous. This can be achieved by decomposing $u_x(y,t)$ into the steady plane Couette velocity profile, which is expected to prevail at large times, and a transient component:

$$u_x(y,t) \; = \; V\left(1 - \frac{y}{H}\right) - u_x'(y,t) \, . \tag{6.127}$$

Substituting into Eqs. (6.125) and (6.126), we obtain the following problem

$$\frac{\partial u_x'}{\partial t} \; = \; \nu \, \frac{\partial^2 u_x'}{\partial y^2} \, , \tag{6.128}$$

with

$$\left.\begin{array}{lll} u_x' = 0 & \text{at} \quad y = 0, \; t > 0 \\ u_x' = 0 & \text{at} \quad y = H, \; t \geq 0 \\ u_x' = V\left(1 - \frac{y}{H}\right) & \text{at} \quad t = 0, \; 0 \leq y \leq H \end{array}\right\} \, . \tag{6.129}$$

Note that the new boundary conditions are homogeneous, while the governing equation remains unchanged. Therefore, separation of variables can now be used. The first step is to express $u_x'(y,t)$ in the form

$$u_x'(y,t) \; = \; Y(y) \, T(t) \, . \tag{6.130}$$

Substituting into Eq. (6.128) and separating the functions Y and T, we get

$$\frac{1}{\nu T}\frac{dT}{dt} = \frac{1}{Y}\frac{d^2Y}{dY^2}.$$

The only way a function of t can be equal to a function of y is for both functions to be equal to the same constant. For convenience, we choose this constant to be $-\alpha^2/H^2$. (One advantage of this choice is that α is dimensionless.) We thus obtain two ordinary differential equations:

$$\frac{dT}{dt} + \frac{\nu\alpha^2}{H^2}T = 0, \tag{6.131}$$

$$\frac{d^2Y}{dy^2} + \frac{\alpha^2}{H^2}Y = 0. \tag{6.132}$$

The solution to Eq. (6.131) is

$$T = c_0\, e^{-\frac{\nu\alpha^2}{H^2}t}, \tag{6.133}$$

where c_0 is an integration constant to be determined.

Equation (6.132) is a homogeneous second-order ODE with constant coefficients, and its general solution is

$$Y(y) = c_1\, \sin(\frac{\alpha y}{H}) + c_2\, \cos(\frac{\alpha y}{H}). \tag{6.134}$$

The form of the general solution justifies the choice we made earlier for the constant $-\alpha^2/H^2$. The constants c_1 and c_2 are determined by the boundary conditions. Applying Eq. (6.130) to the boundary conditions at $y=0$ and H, we obtain

$$Y(0)\, T(t) = 0 \quad \text{and} \quad Y(H)\, T(t) = 0.$$

The case of $T(t)=0$ is excluded since this corresponds to the steady-state problem. Hence, we get the following boundary conditions for Y:

$$Y(0) = 0 \quad \text{and} \quad Y(H) = 0. \tag{6.135}$$

Note that in order to get the boundary conditions on Y, it is essential that the boundary conditions are homogeneous.

Applying the boundary condition at $y=0$, we get $c_2=0$. Thus,

$$Y(y) = c_1\, \sin(\frac{\alpha y}{H}). \tag{6.136}$$

Applying now the boundary condition at $y=H$, we get

$$\sin(\alpha) = 0 , \tag{6.137}$$

which has an infinite number of roots,

$$\alpha_k = k\pi , \quad k = 1, 2, \cdots \tag{6.138}$$

To each of these roots correspond solutions Y_k and T_k. These solutions are super-imposed by defining

$$u'_x(y, t) = \sum_{k=1}^{\infty} B_k \sin(\frac{\alpha_k y}{H}) e^{-\frac{\nu \alpha_k^2}{H^2} t} = \sum_{k=1}^{\infty} B_k \sin(\frac{k\pi y}{H}) e^{-\frac{k^2 \pi^2}{H^2} \nu t} , \tag{6.139}$$

where the constants $B_k = c_{0k} c_{1k}$ are determined from the initial condition. For $t=0$, we get

$$\sum_{k=1}^{\infty} B_k \sin(\frac{k\pi y}{H}) = V \left(1 - \frac{y}{H}\right) . \tag{6.140}$$

To isolate B_k, we will take advantage of the orthogonality property

$$\int_0^1 \sin(k\pi x) \sin(n\pi x) \, dx = \begin{cases} \frac{1}{2} , & k = n \\ 0 , & k \neq n \end{cases} . \tag{6.141}$$

By multiplying both sides of Eq. (6.140) by $\sin(n\pi y/H) \, dy$, and by integrating from 0 to H, we have:

$$\sum_{k=1}^{\infty} B_k \int_0^H \sin(\frac{k\pi y}{H}) \sin(\frac{n\pi y}{H}) \, dy = V \int_0^H \left(1 - \frac{y}{H}\right) \sin(\frac{n\pi y}{H}) \, dy .$$

Setting $\xi = y/H$, we get

$$\sum_{k=1}^{\infty} B_k \int_0^1 \sin(k\pi\xi) \sin(n\pi\xi) \, d\xi = V \int_0^1 (1 - \xi) \sin(n\pi\xi) \, d\xi .$$

Due to the orthogonality property (6.141), the only nonzero term on the left hand side is that for $k=n$; hence,

$$B_k \frac{1}{2} = V \int_0^1 (1 - \xi) \sin(k\pi\xi) \, d\xi = V \frac{1}{k\pi} \quad \Longrightarrow$$

$$B_k = \frac{2V}{k\pi} . \tag{6.142}$$

Substituting into Eq. (6.139) gives

$$u_x'(y,t) = \frac{2V}{\pi} \sum_{k=1}^{\infty} \frac{1}{k} \sin(\frac{k\pi y}{H}) e^{-\frac{k^2\pi^2}{H^2}\nu t} . \tag{6.143}$$

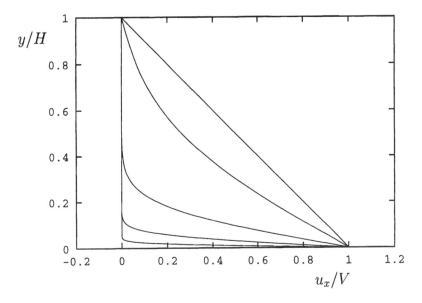

Figure 6.23. *Transient plane Couette flow. Velocity profiles at $\nu t/H^2$=0.0001, 0.001, 0.01, 0.1, and 1.*

Finally, for the original dependent variable $u_x(y,t)$ we get

$$u_x(y,t) = V\left(1 - \frac{y}{H}\right) - \frac{2V}{\pi} \sum_{k=1}^{\infty} \frac{1}{k} \sin(\frac{k\pi y}{H}) e^{-\frac{k^2\pi^2}{H^2}\nu t} . \tag{6.144}$$

The evolution of the solution is illustrated in Fig. 6.23. Initially, the presence of the stationary plate does not affect the development of the flow, and thus the solution is similar to the one of the previous example. This is evident when comparing Figs. 6.21 and 6.23. □

Example 6.6.3. Flow due to an oscillating plate

Consider flow of a semi-infinite Newtonian liquid, set in motion by an oscillating plate of velocity

$$V = V_0 \cos \omega t, \quad t > 0. \tag{6.145}$$

The governing equation, the initial condition, and the boundary condition at $y \to \infty$ are the same as those of Example 6.6.1. At $y=0$, u_x is now equal to $V_0 \cos \omega t$. Hence, we have the following problem:

$$\frac{\partial u_x}{\partial t} = \nu \frac{\partial^2 u_x}{\partial y^2}, \tag{6.146}$$

with

$$\left. \begin{array}{lll} u_x = V_0 \cos \omega t & \text{at} & y = 0, \ t > 0 \\ u_x \to 0 & \text{at} & y \to \infty, \ t \geq 0 \\ u_x = 0 & \text{at} & t = 0, \ 0 \leq y \leq \infty \end{array} \right\}. \tag{6.147}$$

This is known as *Stokes' problem* or *Stokes' second problem*, first studied by Stokes in 1845.

Since the period of the oscillations of the plate introduces a time scale, no similarity solution exists to this problem. By virtue of Eq. (6.145), it may be expected that u_x will also oscillate in time with the same frequency, but possibly with a phase shift relative to the oscillations of the plate. Thus, we separate the two independent variables by representing the velocity as

$$u_x(y,t) = \mathcal{R}e \left[Y(y) \, e^{i\omega t} \right], \tag{6.148}$$

where $\mathcal{R}e$ denotes the real part of the expression within the brackets, i is the imaginary unit, and $Y(y)$ is a complex function. Substituting into the governing equation, we have

$$\frac{d^2 Y}{dy^2} - \frac{i\omega}{\nu} Y = 0. \tag{6.149}$$

The general solution of the above equation is

$$Y(y) = c_1 \exp \left\{ -\sqrt{\frac{\omega}{2\nu}} (1+i) \, y \right\} + c_2 \exp \left\{ \sqrt{\frac{\omega}{2\nu}} (1+i) \, y \right\}.$$

The fact that $u_x=0$ at $y \to \infty$, dictates that c_2 be zero. Then, the boundary condition at $y=0$ requires that $c_1=V_0$. Thus,

$$u_x(y,t) = V_0 \, \mathcal{R}e \left[\exp \left\{ -\sqrt{\frac{\omega}{2\nu}} (1+i) \, y \right\} e^{i\omega t} \right], \tag{6.150}$$

The resulting solution,

$$u_x(y,t) = V_0 \exp \left(-\sqrt{\frac{\omega}{2\nu}} \, y \right) \cos \left(\omega t - \sqrt{\frac{\omega}{2\nu}} \, y \right), \tag{6.151}$$

describes a damped transverse wave of wavelength $2\pi\sqrt{2\nu/\omega}$ propagating in the y-direction with phase velocity $\sqrt{2\nu\omega}$. The amplitude of the oscillations decays exponentially with y. The depth of penetration of vorticity is $\delta \sim \sqrt{2\nu/\omega}$, suggesting that the distance over which the fluid feels the motion of the plate gets smaller as the frequency of the oscillations increases.   □

Example 6.6.4. Transient plane Poiseuille flow

Let us now consider a transient flow which is induced by a suddenly applied constant pressure gradient. A Newtonian liquid of density ρ and viscosity η, is contained between two horizontal plates separated by a distance $2H$ (Fig. 6.24). The liquid is initially at rest; at time $t=0^+$, a constant pressure gradient, $\partial p/\partial x$, is applied, setting the liquid into motion.

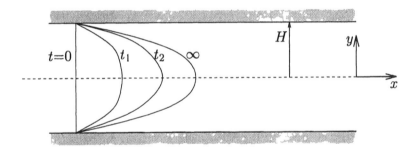

Figure 6.24. *Schematic of the evolution of the velocity in transient plane Poiseuille flow.*

The governing equation for this flow is

$$\rho\frac{\partial u_x}{\partial t} = -\frac{\partial p}{\partial x} + \eta\frac{\partial^2 u_x}{\partial y^2} \, . \tag{6.152}$$

Positioning the x-axis on the symmetry plane of the flow (Fig. 6.24), the boundary and initial conditions become:

$$\left.\begin{array}{lll} u_x = 0 & \text{at} & y = H,\ t \geq 0 \\ \frac{\partial u_x}{\partial y} = 0 & \text{at} & y = 0,\ t \geq 0 \\ u_x = 0 & \text{at} & t = 0,\ 0 \leq y \leq H \end{array}\right\} \, . \tag{6.153}$$

The problem of Eqs. (6.152) and (6.153) is solved using separation of variables. Since the procedure is similar to that used in Example 6.6.2, it is left as an exercise

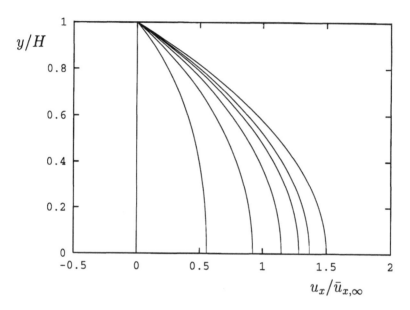

Figure 6.25. *Transient plane Poiseuille flow. Velocity profiles at $\nu t/H^2 = 0.2$, 0.4, 0.6, 0.8, 1, and ∞.*

for the reader (Problem 6.8) to show that

$$
u_x(y,t) = -\frac{1}{2\eta}\frac{\partial p}{\partial x}H^2\left\{1 - \left(\frac{y}{H}\right)^2\right.
$$

$$
\left. -\frac{32}{\pi^3}\sum_{k=1}^{\infty}\frac{(-1)^{k+1}}{(2k-1)^3}\cos\left[\frac{(2k-1)\pi}{2}\frac{y}{H}\right]\exp\left[-\frac{(2k-1)^2\pi^2}{4H^2}\nu t\right]\right\}. \qquad (6.154)
$$

The evolution of the velocity towards the parabolic steady-state profile is shown in Fig. 6.25.   □

Example 6.6.5. Transient axisymmetric Poiseuille flow

Consider a Newtonian liquid of density ρ and viscosity η, initially at rest in an infinitely long horizontal cylindrical tube of radius R. At time $t=0^+$, a constant pressure gradient, $\partial p/\partial z$, is applied, setting the liquid into motion.

This is obviously a transient axisymmetric rectilinear flow. Since gravity is zero, the governing equation is

$$
\rho\frac{\partial u_z}{\partial t} = -\frac{\partial p}{\partial z} + \eta\left(\frac{\partial^2 u_z}{\partial r^2} + \frac{1}{r}\frac{\partial u_z}{\partial r}\right), \qquad (6.155)
$$

subject to the following boundary conditions:

$$
\left.
\begin{array}{lll}
u_z = 0 & \text{at} & r = R,\ t \geq 0 \\
u_z \text{ finite} & \text{at} & r = 0,\ t \geq 0 \\
u_z = 0 & \text{at} & t = 0,\ 0 \leq r \leq R
\end{array}
\right\} .
\tag{6.156}
$$

By decomposing $u_z(r,t)$ into the steady-state Poiseuille flow component (expected to prevail at large times) and a new dependent variable,

$$
u_z(r,t) = -\frac{1}{4\eta}\frac{\partial p}{\partial z}(R^2 - r^2) - u_z'(r,t),
\tag{6.157}
$$

the inhomogeneous pressure-gradient term in Eq. (6.155) is eliminated, and the following homogeneous problem is obtained:

$$
\frac{\partial u_z'}{\partial t} = \nu\left(\frac{\partial^2 u_z'}{\partial r^2} + \frac{1}{r}\frac{\partial u_z'}{\partial r}\right)
\tag{6.158}
$$

with

$$
\left.
\begin{array}{lll}
u_z' = 0 & \text{at} & r = R,\ t \geq 0 \\
u_z' \text{ finite} & \text{at} & r = 0,\ t \geq 0 \\
u_z' = -\frac{1}{4\eta}\frac{\partial p}{\partial z}(R^2 - r^2) & \text{at} & t = 0,\ 0 \leq r \leq R
\end{array}
\right\} .
\tag{6.159}
$$

Using separation of variables, we express $u_z'(r,t)$ in the form

$$
u_z'(r,t) = X(r)\,T(t).
\tag{6.160}
$$

Substituting into Eq. (6.158) and separating the functions X and T, we get

$$
\frac{1}{\nu T}\frac{dT}{dt} = \frac{1}{X}\left(\frac{d^2 X}{dr^2} + \frac{1}{r}\frac{dX}{dr}\right).
$$

Equating both sides of the above expression to $-\alpha^2/R^2$, where α is a dimensionless constant, we obtain two ordinary differential equations:

$$
\frac{dT}{dt} + \frac{\nu\alpha^2}{R^2}T = 0,
\tag{6.161}
$$

$$
\frac{d^2 X}{dr^2} + \frac{1}{r}\frac{dX}{dr} + \frac{\alpha^2}{R^2}X = 0.
\tag{6.162}
$$

The solution to Eq. (6.161) is

$$
T = c_0\,e^{-\frac{\nu\alpha^2}{R^2}t},
\tag{6.163}
$$

where c_0 is an integration constant.

Equation (6.162) is a Bessel's differential equation, the general solution of which is given by

$$X(r) = c_1 J_0(\frac{\alpha r}{R}) + c_2 Y_0(\frac{\alpha r}{R}), \tag{6.164}$$

where J_0 and Y_0 are the zeroth-order Bessel functions of the first and second kind, respectively. From the theory of Bessel functions, we know that $Y_0(x)$ and its first derivative are unbounded at $x=0$. Since u_z' and thus X must be finite at $r=0$, we get $c_2=0$.

Differentiating Eq. (6.164) and noting that

$$\frac{dJ_0}{dx}(x) = -J_1(x),$$

where J_1 is the first-order Bessel function of the first kind, we obtain:

$$\frac{dX}{dr}(r) = -c_1 \frac{\alpha}{R} J_1(\frac{\alpha r}{R}) + c_2 \frac{\alpha}{R} \frac{dY_0}{dr}(\frac{\alpha r}{R}).$$

Given that $J_1(0)=0$, we find again that c_2 must be zero so that $dX/dr=0$ at $r=0$. Thus,

$$X(r) = c_1 J_0(\frac{\alpha r}{R}). \tag{6.165}$$

Applying the boundary condition at $r=R$, we get

$$J_0(\alpha) = 0. \tag{6.166}$$

Note that $J_0(x)$ is an oscillating function with an infinite number of roots,

$$\alpha_k, \quad k = 1, 2, \cdots$$

Therefore, $u_z'(r, t)$ is expressed as an infinite sum of the form

$$u_z'(r, t) = \sum_{k=1}^{\infty} B_k J_0(\frac{\alpha_k r}{R}) e^{-\frac{\nu \alpha_k^2}{R^2} t}, \tag{6.167}$$

where the constants B_k are to be determined from the initial condition. For $t=0$, we have

$$\sum_{k=1}^{\infty} B_k J_0(\frac{\alpha_k r}{R}) = -\frac{1}{4\eta} \frac{\partial p}{\partial z} R^2 \left[1 - \left(\frac{r}{R}\right)^2\right]. \tag{6.168}$$

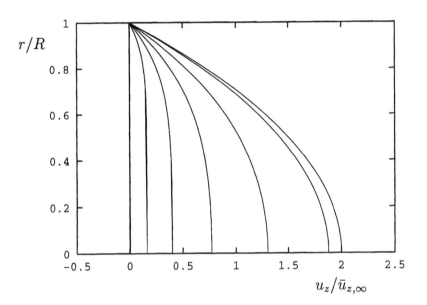

Figure 6.26. *Transient axisymmetric Poiseuille flow.* *Velocity profiles at* $\nu t/R^2 = 0.02,\ 0.05,\ 0.1,\ 0.2,\ 0.5,\ and\ \infty.$

In order to take advantage of the orthogonality property of Bessel functions,

$$\int_0^1 J_0(\alpha_k r)\, J_0(\alpha_n r)\, r\,dr \;=\; \begin{cases} \frac{1}{2}J_1^2(\alpha_k)\,, & k=n \\[2mm] 0\,, & k\neq n \end{cases} \qquad (6.169)$$

where both α_k and α_n are roots of J_0, we multiply both sides of Eq. (6.168) by $J_0(\alpha_n r/R)r\,dr$, and then integrate from 0 to R, to get

$$\sum_{k=1}^{\infty} B_k \int_0^R J_0(\frac{\alpha_k r}{R})\, J_0(\frac{\alpha_n r}{R})\, r\,dr \;=\; -\frac{1}{4\eta}\frac{\partial p}{\partial z} R^2 \int_0^R \left[1-\left(\frac{r}{R}\right)^2\right] J_0(\frac{\alpha_n r}{R})\, r\,dr\,,$$

or

$$\sum_{k=1}^{\infty} B_k \int_0^1 J_0(\alpha_k \xi)\, J_0(\alpha_n \xi)\, \xi\,d\xi \;=\; -\frac{1}{4\eta}\frac{\partial p}{\partial z} R^2 \int_0^1 (1-\xi^2)\, J_0(\alpha_n \xi)\, \xi\,d\xi\,,$$

where $\xi = r/R$. The only nonzero term on the left-hand side corresponds to $k=n$. Hence,

$$B_k \frac{1}{2}J_1^2(\alpha_k) \;=\; -\frac{1}{4\eta}\frac{\partial p}{\partial z} R^2 \int_0^1 (1-\xi^2)\, J_0(\alpha_k \xi)\, \xi\,d\xi\,. \qquad (6.170)$$

Using standard relations for Bessel functions, we find that

$$\int_0^1 (1 - \xi^2)\, J_0(\alpha_k \xi)\, \xi d\xi = \frac{4 J_1(\alpha_k)}{\alpha_k^3}.$$

Therefore,

$$B_k = -\frac{1}{4\eta} \frac{\partial p}{\partial z} \frac{8}{\alpha_k^3 J_1(\alpha_k)},$$

and

$$u_z' = -\frac{1}{4\eta} \frac{\partial p}{\partial z} (8R^2) \sum_{k=1}^{\infty} \frac{J_0\left(\frac{\alpha_k r}{R}\right)}{\alpha_k^3 J_1(\alpha_k)} e^{-\frac{\nu \alpha_k^2}{R^2} t}. \tag{6.171}$$

Substituting into Eq. (6.167) gives

$$u_z(r,t) = -\frac{1}{4\eta} \frac{\partial p}{\partial z} R^2 \left[1 - \left(\frac{r}{R}\right)^2 - 8 \sum_{k=1}^{\infty} \frac{J_0\left(\frac{\alpha_k r}{R}\right)}{\alpha_k^3 J_1(\alpha_k)} e^{-\frac{\nu \alpha_k^2}{R^2} t} \right]. \tag{6.172}$$

The evolution of the velocity is shown in Fig. 6.26.   □

Example 6.6.6. Flow inside a cylinder that is suddenly rotated

A Newtonian liquid of density ρ and viscosity η is initially at rest in a vertical, infinitely long cylinder of radius R. At time $t=0^+$, the cylinder starts rotating about its axis with constant angular velocity Ω, setting the liquid into motion.

This is a transient axisymmetric torsional flow governed by

$$\frac{\partial u_\theta}{\partial t} = \nu \left(\frac{\partial^2 u_\theta}{\partial r^2} + \frac{1}{r} \frac{\partial u_\theta}{\partial r} - \frac{1}{r^2} u_\theta \right), \tag{6.173}$$

subject to the following conditions:

$$\left. \begin{array}{ll} u_\theta = \Omega R & \text{at} \quad r = R,\ t > 0 \\ u_\theta \text{ finite} & \text{at} \quad r = 0,\ t \geq 0 \\ u_\theta = 0 & \text{at} \quad t = 0,\ 0 \leq r \leq R \end{array} \right\}. \tag{6.174}$$

The solution procedure for the problem described by Eqs. (6.173) and (6.174) is the same as in the previous example. The steady-state solution has been obtained in Example 6.3.1. Setting

$$u_\theta(r,t) = \Omega r - u_\theta'(r,t), \tag{6.175}$$

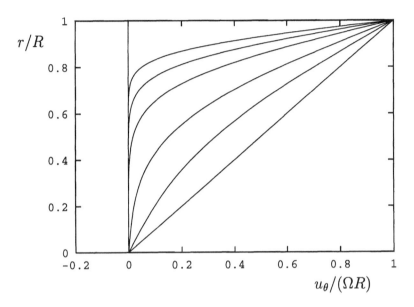

Figure 6.27. *Flow inside a cylinder that is suddenly rotated. Velocity profiles at*
$\nu t/R^2 = 0.005,\ 0.01,\ 0.02,\ 0.05,\ 0.1,\ and\ \infty.$

we obtain the following homogeneous problem

$$\frac{\partial u'_\theta}{\partial t} = \nu \left(\frac{\partial^2 u'_\theta}{\partial r^2} + \frac{1}{r}\frac{\partial u'_\theta}{\partial r} - \frac{1}{r^2}u'_\theta \right) ,\tag{6.176}$$

$$\left.\begin{array}{ll} u'_\theta = 0 & \text{at}\quad r = R,\ t > 0 \\ u'_\theta \text{ finite} & \text{at}\quad r = 0,\ t \geq 0 \\ u'_\theta = \Omega r & \text{at}\quad t = 0,\ 0 \leq r \leq R \end{array}\right\} .\tag{6.177}$$

The independent variables are separated by setting

$$u'_\theta(r,t) = X(r)\, T(t) ,\tag{6.178}$$

which leads to two ordinary differential equations:

$$\frac{dT}{dt} + \frac{\nu \alpha^2}{R^2}\, T = 0 ,\tag{6.179}$$

and

$$\frac{d^2 X}{dr^2} + \frac{1}{r}\frac{dX}{dr} + \left(\frac{\alpha^2}{R^2} - \frac{1}{r^2} \right) X = 0 .\tag{6.180}$$

Equation (6.179) is identical to Eq. (6.161) of the previous example, the general solution of which is

$$T = c_0 \, e^{-\frac{\nu \alpha^2}{R^2} t} \, . \tag{6.181}$$

The general solution of Eq. (6.180) is

$$X(r) = c_1 \, J_1(\frac{\alpha r}{R}) + c_2 \, Y_1(\frac{\alpha r}{R}) \, , \tag{6.182}$$

where J_1 and Y_1 are the first-order Bessel functions of the first and second kind, respectively. Since $Y_1(x)$ is unbounded at $x=0$, c_2 must be zero. Therefore,

$$X(r) = c_1 J_1(\frac{\alpha r}{R}) \, . \tag{6.183}$$

The boundary condition at $r=R$ requires that

$$J_1(\alpha) = 0 \, , \tag{6.184}$$

which has an infinite number of roots. Therefore, $u_\theta'(r,t)$ is expressed as an infinite sum of the form

$$u_\theta'(r,t) = \sum_{k=1}^{\infty} B_k \, J_1(\frac{\alpha_k r}{R}) \, e^{-\frac{\nu \alpha_k^2}{R^2} t} \, , \tag{6.185}$$

where the constants B_k are to be determined from the initial condition. For $t=0$, we have

$$\sum_{k=1}^{\infty} B_k \, J_1(\frac{\alpha_k r}{R}) = \Omega \, r \, . \tag{6.186}$$

The constants B_k are determined by using the orthogonality property of Bessel functions,

$$\int_0^1 J_1(\alpha_k r) \, J_1(\alpha_n r) \, r dr = \begin{cases} \frac{1}{2} J_0^2(\alpha_k) \, , & k = n \\[2mm] 0 \, , & k \neq n \end{cases} \tag{6.187}$$

where both α_k and α_n are roots of J_1. Multiplying both sides of Eq. (6.186) by $J_1(\alpha_n r/R) r dr$, and integrating from 0 to R, we get

$$\sum_{k=1}^{\infty} B_k \int_0^R J_1(\frac{\alpha_k r}{R}) \, J_1(\frac{\alpha_n r}{R}) \, r dr = \Omega \int_0^R J_1(\frac{\alpha_n r}{R}) \, r^2 \, dr \, ,$$

or

$$\sum_{k=1}^{\infty} B_k \int_0^1 J_1(\alpha_k \xi) \, J_1(\alpha_n \xi) \, \xi d\xi = \Omega R \int_0^1 J_1(\alpha_n \xi) \, \xi^2 \, d\xi \, ,$$

where $\xi = r/R$. Invoking Eq. (6.187), we get

$$B_k \frac{1}{2} J_0^2(\alpha_k) = \Omega R \int_0^1 J_1(\alpha_k \xi) \, \xi^2 d\xi = -\Omega R \frac{J_0(\alpha_k)}{\alpha_k} \quad \Longrightarrow$$

$$B_k = -\frac{2\Omega R}{\alpha_k J_0(\alpha_k)} \, .$$

Therefore,

$$u_\theta' = -2\Omega R \sum_{k=1}^{\infty} \frac{J_1\left(\frac{\alpha_k r}{R}\right)}{\alpha_k J_0(\alpha_k)} e^{-\frac{\nu \alpha_k^2}{R^2} t} \tag{6.188}$$

and

$$u_\theta(r,t) = \Omega r + 2\Omega R \sum_{k=1}^{\infty} \frac{J_1\left(\frac{\alpha_k r}{R}\right)}{\alpha_k J_0(\alpha_k)} e^{-\frac{\nu \alpha_k^2}{R^2} t} . \tag{6.189}$$

The evolution of the u_θ is shown in Fig. 6.27. $\qquad\qquad\square$

6.7 Steady, Two-Dimensional Rectilinear Flows

As explained in Section 6.1, in steady, rectilinear flows in the x direction, $u_x = u_x(y, z)$ and the x-momentum equation is reduced to a *Poisson equation*,

$$\frac{\partial^2 u_x}{\partial y^2} + \frac{\partial^2 u_x}{\partial z^2} = \frac{1}{\eta} \frac{\partial p}{\partial x} - \frac{1}{\nu} g_x \, . \tag{6.190}$$

Equation (6.190) is an *elliptic* PDE. Since $\partial p/\partial x$ is a function of x alone and u_x is a function of y and z, Eq. (6.190) can be satisfied only when $\partial p/\partial x$ is constant. Therefore, the right-hand side term of Eq. (6.190) is a constant. This inhomogeneous term can be eliminated by introducing a new dependent variable which satisfies the *Laplace equation*.

Two classes of flows governed by Eq. (6.190) are:

(a) Poiseuille flows in tubes of arbitrary but constant cross section; and

(b) gravity-driven rectilinear film flows.

One-dimensional Poiseuille flows have been encountered in Sections 6.1 and 6.2. The most important of them, i.e., plane, round, and annular Poiseuille flows, are summarized in Fig. 6.28. In the following, we will discuss two-dimensional Poiseuille

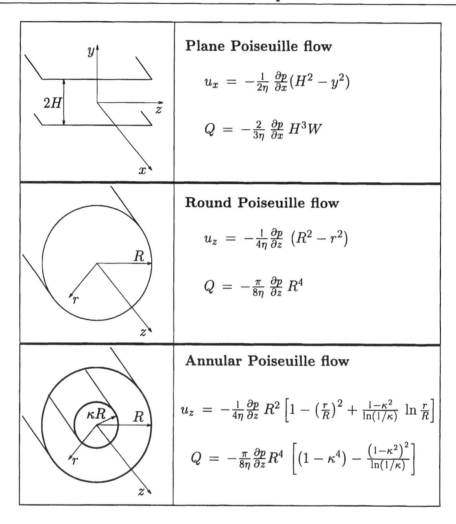

Figure 6.28. *One-dimensional Poiseuille flows.*

flows in tubes of elliptical, rectangular and triangular cross sections, illustrated in Fig. 6.29. In these rather simple geometries, Eq. (6.190) can be solved analytically. Analytical solutions for other cross sectional shapes are given in Refs. [10] and [11].

Example 6.7.1. Poiseuille flow in a tube of elliptical cross section

Consider fully-developed flow of an incompressible Newtonian liquid in an infinitely long tube of elliptical cross section under constant pressure gradient $\partial p/\partial x$. Gravity

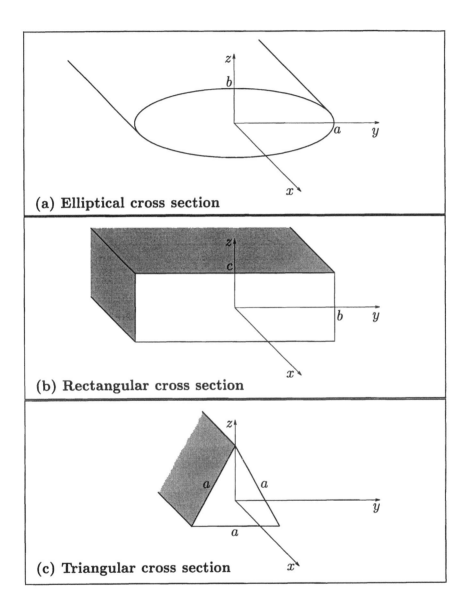

(a) Elliptical cross section

(b) Rectangular cross section

(c) Triangular cross section

Figure 6.29. *Two-dimensional Poiseuille flow in tubes of various cross sections.*

is neglected, and thus Eq. (6.190) becomes

$$\frac{\partial^2 u_x}{\partial y^2} + \frac{\partial^2 u_x}{\partial z^2} = \frac{1}{\eta}\frac{\partial p}{\partial x} \quad \text{in} \quad \frac{y^2}{a^2} + \frac{z^2}{b^2} \leq 1, \tag{6.191}$$

where a and b are the semi-axes of the elliptical cross section, as shown in Fig. 6.29a. The velocity is zero at the wall, and thus the boundary condition is:

$$u_x = 0 \quad \text{on} \quad \frac{y^2}{a^2} + \frac{z^2}{b^2} = 1. \tag{6.192}$$

Let us now introduce a new dependent variable u_x', such that

$$u_x(y,z) = u_x'(y,z) + c_1 y^2 + c_2 z^2, \tag{6.193}$$

where c_1 and c_2 are nonzero constants to be determined so that (a) u_x' satisfies the Laplace equation, and (b) u_x' is constant on the wall. Substituting Eq. (6.193) into Eq. (6.191), we get

$$\frac{\partial^2 u_x'}{\partial y^2} + \frac{\partial^2 u_x'}{\partial z^2} + 2c_1 + 2c_2 = \frac{1}{\eta}\frac{\partial p}{\partial x}. \tag{6.194}$$

Evidently, u_x' satisfies the Laplace equation,

$$\frac{\partial^2 u_x'}{\partial y^2} + \frac{\partial^2 u_x'}{\partial z^2} = 0, \tag{6.195}$$

if

$$2c_1 + 2c_2 = \frac{1}{\eta}\frac{\partial p}{\partial x}. \tag{6.196}$$

From boundary condition (6.192), we have

$$u_x'(y,z) = -c_1 y^2 - c_2 z^2 = -c_1\left[y^2 + \frac{c_2}{c_1}z^2\right] \quad \text{on} \quad \frac{y^2}{a^2} + \frac{z^2}{b^2} = 1.$$

Setting

$$\frac{c_2}{c_1} = \frac{a^2}{b^2}, \tag{6.197}$$

u_x' becomes constant on the boundary,

$$u_x'(y,z) = -c_1 a^2 \quad \text{on} \quad \frac{y^2}{a^2} + \frac{z^2}{b^2} = 1. \tag{6.198}$$

The *maximum principle* for the Laplace equation states that u'_x has both its minimum and maximum values on the boundary of the domain [12]. Therefore, u'_x is constant over the whole domain,

$$u'_x(y, z) = -c_1 a^2 . \tag{6.199}$$

Substituting into Eq. (6.193) and using Eq. (6.197), we get

$$u_x(y, z) = -c_1 a^2 + c_1 y^2 + c_2 z^2 = -c_1 a^2 \left[1 - \frac{y^2}{a^2} - \frac{c_2}{c_1} \frac{z^2}{a^2} \right] \implies$$

$$u_x(y, z) = -c_1 a^2 \left[1 - \frac{y^2}{a^2} - \frac{z^2}{b^2} \right] . \tag{6.200}$$

The constant c_1 is determined from Eqs. (6.196) and (6.197),

$$c_1 = \frac{1}{2\eta} \frac{\partial p}{\partial x} \frac{b^2}{a^2 + b^2} ; \tag{6.201}$$

consequently,

$$u_x(y, z) = -\frac{1}{2\eta} \frac{\partial p}{\partial x} \frac{a^2 b^2}{a^2 + b^2} \left[1 - \frac{y^2}{a^2} - \frac{z^2}{b^2} \right] . \tag{6.202}$$

Obviously, the maximum velocity occurs at the origin. Integration of the velocity profile (6.202) over the elliptical cross section yields the volumetric flow rate

$$Q = -\frac{\pi}{4\eta} \frac{\partial p}{\partial x} \frac{a^3 b^3}{a^2 + b^2} . \tag{6.203}$$

Equation (6.202) degenerates to the circular Poiseuille flow velocity profile when $a = b = R$,

$$u_x(y, z) = -\frac{1}{4\eta} \frac{\partial p}{\partial x} R^2 \left[1 - \frac{y^2 + z^2}{R^2} \right] .$$

Setting $r^2 = y^2 + z^2$, and switching to cylindrical coordinates, we get

$$u_z(r) = -\frac{1}{4\eta} \frac{\partial p}{\partial z} (R^2 - r^2) . \tag{6.204}$$

If now $a = H$ and $b \gg H$, Eq. (6.202) yields the plane Poiseuille flow velocity profile,

$$u_x(y) = -\frac{1}{2\eta} \frac{\partial p}{\partial x} (H^2 - y^2) . \tag{6.205}$$

Note that, due to symmetry, the shear stress is zero along symmetry planes. The zero shear stress condition along such a plane applies also in gravity-driven flow of a film of semielliptical cross section. Therefore, the velocity profile for the latter flow can be obtained by replacing $-\partial p/\partial x$ by ρg_x. Similarly, Eqs. (6.204) and (6.205) can be modified to describe the gravity-driven flow of semicircular and planar films, respectively.   □

Example 6.7.2. Poiseuille flow in a tube of rectangular cross section

Consider steady, pressure-driven flow of an incompressible Newtonian liquid in an infinitely long tube of rectangular cross section of width $2b$ and height $2c$, as shown in Fig. 6.29b. The flow is governed by the Poisson equation

$$\frac{\partial^2 u_x}{\partial y^2} + \frac{\partial^2 u_x}{\partial z^2} = \frac{1}{\eta}\frac{\partial p}{\partial x}. \tag{6.206}$$

Taking into account the symmetry with respect to the planes $y=0$ and $z=0$, the flow can be studied only in the first quadrant (Fig. 6.30). The boundary conditions can then be written as follows:

$$\left.\begin{array}{ll} \dfrac{\partial u_x}{\partial y} = 0 & \text{on } y = 0 \\[2mm] u_x = 0 & \text{on } y = b \\[2mm] \dfrac{\partial u_x}{\partial z} = 0 & \text{on } z = 0 \\[2mm] u_x = 0 & \text{on } z = c \end{array}\right\}. \tag{6.207}$$

Equation (6.206) can be transformed into the Laplace equation by setting

$$u_x(y,z) = -\frac{1}{2\eta}\frac{\partial p}{\partial x}(c^2 - z^2) + u'_x(y,z). \tag{6.208}$$

Note that the first term in the right-hand side of Eq. (6.208) is just the Poiseuille flow profile between two infinite plates placed at $z=\pm c$. Substituting Eq. (6.208) into Eqs. (6.206) and (6.207), we get

$$\frac{\partial^2 u'_x}{\partial y^2} + \frac{\partial^2 u'_x}{\partial z^2} = 0, \tag{6.209}$$

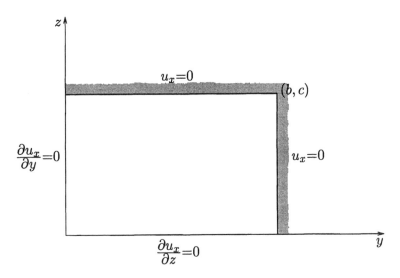

Figure 6.30. *Boundary conditions for the flow in a tube of rectangular cross section.*

subject to

$$
\left.
\begin{array}{lll}
\dfrac{\partial u'_x}{\partial y} = 0 & \text{on} & y = 0 \\[2mm]
u'_x = \dfrac{1}{2\eta}\dfrac{\partial p}{\partial x}\left(c^2 - z^2\right) & \text{on} & y = b \\[2mm]
\dfrac{\partial u'_x}{\partial z} = 0 & \text{on} & z = 0 \\[2mm]
u'_x = 0 & \text{on} & z = c
\end{array}
\right\} .
\tag{6.210}
$$

The above problem can be solved using separation of variables (see Problem 6.13). The solution is

$$
u_x(y, z) = -\frac{1}{2\eta}\frac{\partial p}{\partial x}c^2\left[1 - \left(\frac{z}{c}\right)^2 + 4\sum_{k=1}^{\infty}\frac{(-1)^k}{\alpha_k^3}\frac{\cosh\left(\frac{\alpha_k y}{c}\right)}{\cosh\left(\frac{\alpha_k b}{c}\right)}\cos\left(\frac{\alpha_k z}{c}\right)\right]
\tag{6.211}
$$

where

$$
\alpha_k = (2k - 1)\frac{\pi}{2}, \quad k = 1, 2, \cdots
\tag{6.212}
$$

In Fig. 6.31, we show the velocity contours predicted by Eq. (6.211) for different values of the width-to-height ratio. It is observed that, as this ratio increases,

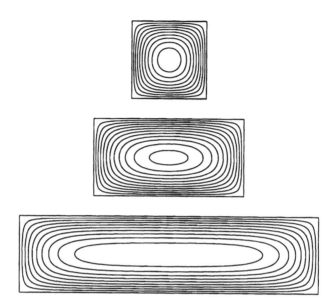

Figure 6.31. *Velocity contours for steady, unidirectional flow in tubes of rectangular cross section with width-to-height ratio equal to 1, 2, and 4.*

the velocity contours become horizontal away from the two vertical walls. This indicates that the flow away from the two walls is approximately one-dimensional (the dependence of u_x on y is weak).

The volumetric flow rate is given by

$$Q = -\frac{4}{3\eta} \frac{\partial p}{\partial x} bc^3 \left[1 - \frac{6c}{b} \sum_{k=1}^{\infty} \frac{\tanh\left(\frac{\alpha_k b}{c}\right)}{\alpha_k^5} \right] . \tag{6.213}$$

□

Example 6.7.3. Poiseuille flow in a tube of triangular cross section

Consider steady, pressure-driven flow of a Newtonian liquid in an infinitely long tube, the cross section of which is an equilateral triangle of side a, as shown in Fig. 6.29c. Once again, the flow is governed by the Poisson equation

$$\frac{\partial^2 u_x}{\partial y^2} + \frac{\partial^2 u_x}{\partial z^2} = \frac{1}{\eta} \frac{\partial p}{\partial x} . \tag{6.214}$$

If the origin is set at the centroid of the cross section, as in Fig. 6.32, the three sides of the triangle lie on the lines

$$2\sqrt{3}z + a = 0, \quad \sqrt{3}z + 3y - a = 0 \quad \text{and} \quad \sqrt{3}z - 3y - a = 0.$$

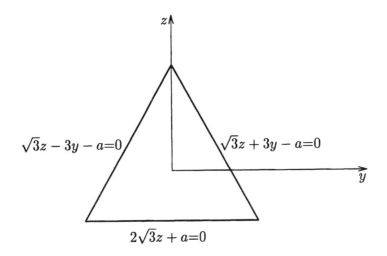

Figure 6.32. *Equations of the sides of an equilateral triangle of side a when the origin is set at the centroid.*

Since the velocity $u_x(y, z)$ is zero on the wall, the following solution form is prompted

$$u_x(y, z) = A \left(2\sqrt{3}z + a\right) \left(\sqrt{3}z + 3y - a\right) \left(\sqrt{3}z - 3y - a\right), \tag{6.215}$$

where A is a constant to be determined so that the governing Eq. (6.214) is satisfied. Differentiation of Eq. (6.215) gives

$$\frac{\partial^2 u_x}{\partial y^2} = -18A \left(2\sqrt{3}z + a\right) \quad \text{and} \quad \frac{\partial^2 u_x}{\partial z^2} = 18A \left(2\sqrt{3}z - a\right).$$

It turns out that Eq. (6.214) is satisfied provided that

$$A = -\frac{1}{36\eta} \frac{\partial p}{\partial x} \frac{1}{a}. \tag{6.216}$$

Thus, the velocity profile is given by

$$u_x(y, z) = -\frac{1}{36\eta} \frac{\partial p}{\partial x} \frac{1}{a} \left(2\sqrt{3}z + a\right) \left(\sqrt{3}z + 3y - a\right) \left(\sqrt{3}z - 3y - a\right). \tag{6.217}$$

The volumetric flow rate is

$$Q = -\frac{\sqrt{3}}{320\eta} \frac{\partial p}{\partial x} a^4 .$$ (6.218)

□

The unidirectional flows examined in this chapter are good approximations to many important industrial and processing flows. Channel, pipe, and annulus flows are good prototypes of liquid transferring systems. The solutions to these flows provide the means to estimate the power required to overcome friction and force the liquid through, and the residence or traveling time. Analytical solutions are extremely important to the design and operation of viscometers [13]. In fact, the best known viscometers were named after the utilized flow: Couette viscometer, capillary or pressure viscometer, and parallel plate viscometer [14].

The majority of the flows studied in this chapter are easily extended to *nearly unidirectional* flows in nonparallel channels or pipes and annuli, and to nonuniform films under the action of surface tension, by means of the *lubrication approximation* [15] examined in detail in Chapter 9. Transient flows that involve vorticity generation and diffusion are dynamically similar to steady flows overtaking submerged bodies, giving rise to *boundary layers* [9], which are studied in Chapter 8.

6.8   Problems

6.1. Consider flow of a thin, uniform film of an incompressible Newtonian liquid on an infinite, inclined plate that moves upwards with constant speed V, as shown in Fig. 6.33. The ambient air is assumed to be stationary, and the surface tension is negligible.
(a) Calculate the velocity $u_x(y)$ of the film in terms of V, δ, ρ, η, g, and θ.
(b) Calculate the speed V of the plate at which the net volumetric flow rate is zero.

6.2. A thin Newtonian film of uniform thickness δ is formed on the external surface of a vertical, infinitely long cylinder which rotates at angular speed Ω, as illustrated in Fig. 6.34. Assume that the flow is steady, the surface tension is zero, and the ambient air is stationary.
(a) Calculate the two nonzero velocity components.
(b) Sketch the streamlines of the flow.
(c) Calculate the volumetric flow rate Q.
(d) What must be the external pressure distribution, $p(z)$, so that uniform thickness is preserved?

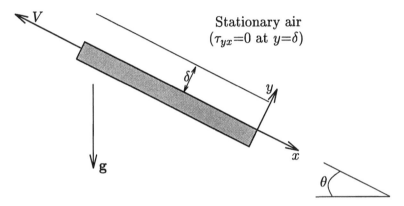

Figure 6.33. *Film flow down a moving inclined plate.*

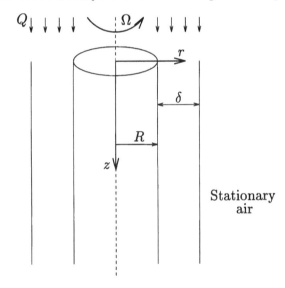

Figure 6.34. *Thin film flow down a vertical rotating cylinder.*

6.3. A spherical bubble of radius R_A and of constant mass m_0 grows radially at a rate

$$\frac{dR_A}{dt} = k,$$

within a spherical incompressible liquid droplet of density ρ_1, viscosity η_1, and volume V_1. The droplet itself is contained in a bath of another Newtonian liquid of density ρ_2 and viscosity η_2, as shown in Fig. 6.35. The surface tension of the inner liquid is σ_1, and its interfacial tension with the surrounding liquid is σ_2.

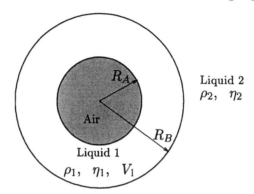

Figure 6.35. *Liquid film growing around a gas bubble.*

(a) What is the growth rate of the droplet?
(b) Calculate the velocity distribution in the two liquids.
(c) What is the pressure distribution within the bubble and the two liquids?
(d) When does the continuity of the thin film of liquid around the bubble break down?

6.4. The equations

$$\frac{\partial u_x}{\partial x} + \frac{\partial u_y}{\partial y} = 0$$

and

$$\rho \left(u_y \frac{\partial u_x}{\partial y} + u_x \frac{\partial u_x}{\partial x} \right) = \eta \frac{\partial^2 u_x}{\partial y^2}$$

govern the (bidirectional) *boundary layer* flow near a horizontal plate of infinite dimensions coinciding with the xz-plane. The boundary conditions for $u_x(x, y)$ and $u_y(x, y)$ are

$$u_x = u_y = 0 \quad \text{at} \quad y=0$$
$$u_x = V, \quad u_y = 0 \quad \text{at} \quad y=\infty$$

Does this problem admit a similarity solution? What is the similarity variable?

6.5. Consider a semi-infinite, incompressible Newtonian liquid of viscosity η and density ρ, bounded below by a plate at $y=0$, as illustrated in Fig. 6.36. Both the plate and liquid are initially at rest. Suddenly, at time $t=0^+$, a constant shear stress τ is applied along the plate.

(a) Specify the governing equation, and the boundary and initial conditions for this flow problem.

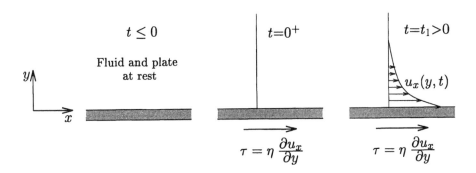

Figure 6.36. *Flow near a plate along which a constant shear stress is suddenly applied.*

(b) Assuming that the velocity u_x is of the form

$$u_x = \frac{\tau}{\eta} \sqrt{\nu t}\, f(\xi)\,, \tag{6.219}$$

where

$$\xi = \frac{y}{\sqrt{\nu t}}\,, \tag{6.220}$$

show that

$$f(\xi) - \xi f'(\xi) = 2\, f''(\xi)\,. \tag{6.221}$$

(The primes denote differentiation with respect to ξ.)
(c) What are the boundary conditions for $f(\xi)$?
(d) Show that

$$u_x = \frac{\tau}{\eta} \sqrt{\nu t} \left\{ \frac{2}{\sqrt{\pi}} e^{-\xi^2/4} - \xi \left[1 - \mathrm{erf}\left(\frac{\xi}{2}\right) \right] \right\}\,. \tag{6.222}$$

6.6. A Newtonian liquid is contained between two horizontal, infinitely long and wide plates separated by a distance $2H$, as illustrated in Fig. 6.37. The liquid is initially at rest; at time $t=0^+$, both plates start moving with constant speed V.
(a) Identify the governing equation, and the boundary and initial conditions for this transient flow.
(b) What is the solution for $t \leq 0$?
(c) What is the solution for $t \to \infty$?
(d) Find the time-dependent solution $u_x(y,t)$ using separation of variables.
(e) Sketch the velocity profiles at $t=0$, 0^+, $t_1 >0$, and ∞.

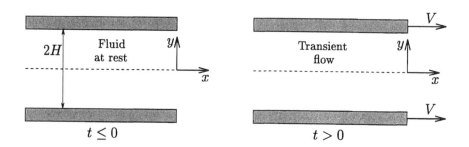

Figure 6.37. *Transient Couette flow (Problem 6.6).*

6.7. A Newtonian liquid is contained between two horizontal, infinitely long and wide plates separated by a distance H, as illustrated in Fig. 6.38. Initially, the liquid flows steadily, driven by the motion of the upper plate which moves with constant speed V while the lower plate is held stationary. Suddenly, at time $t=0^+$, the speed of the upper plate changes to $2V$, resulting in transient flow.

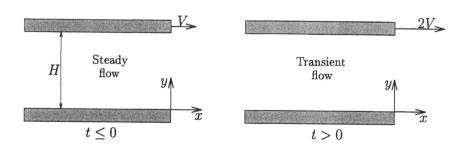

Figure 6.38. *Transient Couette flow (Problem 6.7).*

(a) Identify the governing equation, and the boundary and initial conditions for this transient flow.
(b) What is the solution for $t \leq 0$?
(c) What is the solution for $t \to \infty$?
(d) Find the time-dependent solution $u_x(y, t)$.
(e) Sketch the velocity profiles at $t=0$, 0^+, $t_1 > 0$, and ∞.

6.8. Using separation of variables, show that Eq. (6.154) is indeed the solution of the transient plane Poiseuille flow described in Example 6.6.4.

6.9. A Newtonian liquid contained between two concentric, infinitely long, vertical cylinders of radii R_1 and R_2, where $R_2 > R_1$, is initially at rest. At time $t=0^+$, the inner cylinder starts rotating about its axis with constant angular velocity Ω_1.
(a) Specify the governing equation for this transient flow.
(b) Specify the boundary and the initial conditions.
(c) Calculate the velocity $u_\theta(r,t)$.

6.10. An infinitely long, vertical rod of radius R is initially held fixed in an infinite pool of Newtonian liquid. At time $t=0^+$, the rod starts rotating about its axis with constant angular velocity Ω.
(a) Specify the governing equation for this transient flow.
(b) Specify the boundary and the initial conditions.
(c) Calculate the velocity $u_\theta(r,t)$.

6.11. Consider a Newtonian liquid contained between two concentric, infinitely long, horizontal cylinders of radii κR and R, where $\kappa < 1$. Assume that the liquid is initially at rest. At time $t=0^+$, the outer cylinder starts translating parallel to its axis with constant speed V. The geometry of the flow is shown in Fig. 6.13.
(a) Specify the governing equation for this transient flow.
(b) Specify the boundary and the initial conditions.
(c) Calculate the velocity $u_z(r,t)$.

6.12. A Newtonian liquid is initially at rest in a vertical, infinitely long cylinder of radius R. At time $t=0^+$, the cylinder starts both translating parallel to itself with constant speed V and rotating about its axis with constant angular velocity Ω.
(a) Calculate the corresponding steady-state solution.
(b) Specify the governing equation for the transient flow.
(c) Specify the boundary and the initial conditions.
(d) Examine whether the superposition principle holds for this transient *bidirectional* flow.
(e) Show that the time-dependent velocity and pressure profiles evolve to the steady-state solution as $t \to \infty$.

6.13. Using separation of variables, show that Eq. (6.211) is the solution of steady Newtonian Poiseuille flow in a tube of rectangular cross section described in Example 6.7.2.

6.14. Consider steady Newtonian Poiseuille flow in a horizontal tube of square cross section of side $2b$. Find the velocity distribution in the following cases:
(a) The liquid does not slip on any wall.
(b) The liquid slips on only two opposing walls with constant slip velocity u_w.
(c) The liquid slips on all walls with constant slip velocity u_w.

(d) The liquid slips on only two opposing walls according to the slip law

$$\tau_w = \beta\, u_w\,, \tag{6.223}$$

where τ_w is the shear stress and β is a material slip parameter. (Note that, in this case, the slip velocity u_w is not constant.)

6.15. Integrate $u_x(y, z)$ over the corresponding cross sections to calculate the volumetric flow rates of the Poiseuille flows discussed in the three examples of Section 6.7.

6.16. Consider steady, unidirectional, gravity-driven flow of a Newtonian liquid in an inclined, infinitely long tube of rectangular cross section of width $2b$ and height $2c$, illustrated in Fig. 6.39.

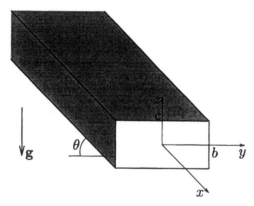

Figure 6.39. *Gravity-driven flow in an inclined tube of rectangular cross section.*

(a) Simplify the three components of the Navier–Stokes equation for this two-dimensional, unidirectional flow.
(b) Calculate the pressure distribution $p(z)$.
(c) Specify the boundary conditions on the first quadrant.
(d) Calculate the velocity $u_x(y, z)$. How is this related to Eq. (6.211)?

6.17. Consider steady, gravity-driven flow of a Newtonian rectangular film in an inclined, infinitely long channel of width $2b$, illustrated in Fig. 6.40. The film is assumed to be of uniform thickness H, the surface tension is negligible, and the air above the free surface is considered stationary.
(a) Taking into account possible symmetries, specify the governing equation and the boundary conditions for this two-dimensional, unidirectional flow.
(b) Is the present flow related to that of the previous problem?
(c) Calculate the velocity $u_x(y, z)$.

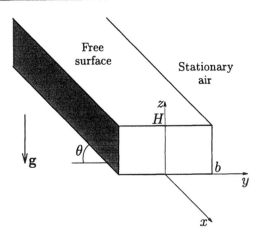

Figure 6.40. *Gravity-driven film flow in an inclined channel.*

6.9   References

1. C. Pozrikidis, *Introduction to Theoretical and Computational Fluid Dynamics*, Oxford University Press, New York, 1997.

2. H.F. Weinberger, *Partial Differential Equations*, Blaidsdell Publishing Company, Massachusetts, 1965.

3. A.G. Hansen, *Similarity Analysis of Boundary Value Problems in Engineering*, Prentice-Hall, Englewood Cliffs, New Jersey, 1964.

4. D.J. Acheson, *Elementary Fluid Dynamics*, Clarendon Press, Oxford, 1995.

5. R.B. Bird, R.C. Armstrong, and O. Hassager, *Dynamics of Polymeric Liquids: Fluid Mechanics*, John Wiley & Sons, New York, 1987.

6. J.D. Logan, *Applied Mathematics*, John Wiley & Sons, New York, 1987.

7. R.S. Brodkey, *The Phenomena of Fluid Motions*, Addison-Wesley Series in Chemical Engineering, 1967.

8. Lord Rayleigh, *Scientific Papers*, Dover, 1964.

9. H. Schlichting, *Boundary Layer Theory*, McGraw-Hill, New York, 1968.

10. R. Berger, "Intégration des équations du mouvement d'un fluide visqueux incompressible," in *Handbuch der Physik* **8**(2), Springer, Berlin, 1-384 (1963).

11. R.K. Shah, and A.L. London, *Laminar Flow Forced Convection in Ducts*, Academic Press, New York, 1978.

12. M.H. Protter and H.F. Weinberger, *Maximum Principles in Differential Equations*, Prentice-Hall, Englewood Cliffs, New Jersey, 1967.

13. H.A. Barnes, J.F. Hutton and K. Walters, *An Introduction to Rheology*, Elsevier, Amsterdam, 1989.

14. J.M. Dealy, *Rheometers for Molten Plastics*, Van Nostrand Rheinhold, New York, 1982.

15. O. Reynolds, "Papers on Mechanical and Physical Aspects," *Phil. Trans. Roy. Soc.* **177**, 157 (1886).

Chapter 7

APPROXIMATE METHODS

In Chapter 6, we studied incompressible, unidirectional flows in which the equations of motion can be solved analytically. In *bidirectional* or *tridirectional* flows, the governing equations can rarely be solved analytically. One can, of course, solve such problems numerically by using finite elements or finite differences or other methods. Another alternative, however, is to use *approximate methods*. The most widely used approximate techniques are the so-called *perturbation methods*. These are based on *order of magnitude analyses* of the governing equations. Individual terms are first made comparable by *dimensional analysis*, and *relatively small* terms are then eliminated. This simplifies the governing equations and leads to either analytic solutions to the truncated form of the governing equations, or to the construction of approximate solutions.

Dimensional analysis is important on its own merit, too. It is useful in scaling-up lab experiments; in producing *dimensionless numbers* that govern the behavior of the solution without solving the governing equations; in defining regions of *stability* and *instability*, e.g., the transition from *laminar* to *turbulent* flow; and in uncovering the competing forces or driving gradients. A flow depends on relevant dimensionless numbers rather than separately on individual geometrical dimensions (such as width and length), physical properties of the fluid (such as, density, viscosity, and surface tension), or processing variables (such as flow and heat rate).

In Section 7.1, we introduce the basic concepts of the dimensional analysis and discuss the nondimensionalization of the equations of motion. In Section 7.2, we provide a brief introduction to *perturbation methods* (for further study, References [1]-[5] are recommended). Finally, in Section 7.3, we discuss the use of perturbation methods in fluid mechanics.

7.1 Dimensional Analysis

Consider the plane Couette–Poiseuille flow, shown in Fig. 7.1. This flow was solved in Example 6.1.4. The governing equation and the boundary conditions for the dependent variable, $u_x = u_x(y)$, are:

$$\left.\begin{array}{c} \dfrac{\Delta p}{\Delta L} + \eta \dfrac{d^2 u_x}{dy^2} = 0 \\[2mm] u_x(y=0) = 0 \\[2mm] u_x(y=a) = V \end{array}\right\} , \tag{7.1}$$

where $\Delta p / \Delta L = -\partial p / \partial x$ is the constant pressure gradient.

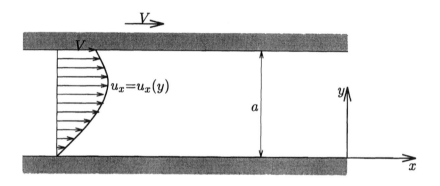

Figure 7.1. *Plane Poiseuille flow with the upper plate moving with constant speed.*

The solution to problem (7.1),

$$u_x = \frac{1}{2\eta} \frac{\Delta p}{\Delta L} (ay - y^2) + \frac{V}{a} y , \tag{7.2}$$

can be rearranged as

$$\frac{u_x}{V} = \frac{a^2}{2\eta} \frac{\Delta p}{V \Delta L} \left[\frac{y}{a} - \left(\frac{y}{a}\right)^2 \right] + \frac{y}{a} . \tag{7.3}$$

By defining the *dimensionless variables*

$$u_x^* = \frac{u_x}{V} \quad \text{and} \quad y^* = \frac{y}{a} , \tag{7.4}$$

Eq. (7.3) takes the *dimensionless form*

$$u_x^* = \frac{\Pi}{2} (y^* - y^{*2}) + y^* , \tag{7.5}$$

where

$$\Pi \equiv \frac{a^2}{\eta \, V} \frac{\Delta p}{\Delta L} \tag{7.6}$$

is referred to as a *dimensionless number*. Π can be interpreted as the ratio of the driving pressure gradient to the driving velocity of the upper plate or, equivalently, as the ratio of viscous to drag forces.

The dimensionless Eq. (7.5) is by far more useful than its dimensional counterpart, Eq. (7.2). Given that $0 \leq y^* \leq 1$, some conclusions about the behavior of the solution and the competing driving forces can be deduced from the numerical value of Π alone:

(i) If $\Pi \ll 1$, then

$$u_x^* \simeq y^* \, ,$$

which corresponds to plane Couette flow.

(ii) If $\Pi = O(1)$, then

$$u_x^* = \frac{\Pi}{2} \left(y^* - y^{*2} \right) + y^* \, ,$$

which corresponds to plane Couette–Poiseuille flow.

(iii) If $\Pi \gg 1$, then

$$u_x^* = \frac{\Pi}{2} \left(y^* - y^{*2} \right) \, ,$$

which corresponds to plane Poiseuille flow.

Under the constraints imposed by the boundary conditions (which come from nature, not from mathematics), the exact form of the solution can be deduced using elementary calculus of maxima and minima.

The plane Couette–Poiseuille flows of different liquids under different flow conditions are said to be *dynamically similar* if they correspond to the same value of Π, i.e., if

$$\frac{a_1^2}{\eta_1 \, V_1} \left(\frac{\Delta p}{\Delta L} \right)_1 = \frac{a_2^2}{\eta_2 \, V_2} \left(\frac{\Delta p}{\Delta L} \right)_2 = \Pi \, ,$$

where the indices 1 and 2 denote the corresponding quantities in the two flows. In such a case,

$$(u_x^*)_1 = (u_x^*)_2 \quad \Longrightarrow \quad (u_x)_2 = (u_x)_1 \, \frac{V_2}{V_1} \, .$$

The ratio V_2/V_1 is the *scaling factor*, say from a small-scale lab experiment of velocity $(u_x)_1$ to a real large scale application of velocity $(u_x)_2$.

If the exact form of the solution were unknown, an *a priori* knowledge of the fact that u^* is a function of y^* and Π alone, i.e., knowing simply that

$$u^* = f(y^*, \Pi) \,, \tag{7.7}$$

would guide a systematic experimental procedure for uncovering the unknown function f. Equation (7.7) indicates that experiments need to be carried out only for different values of Π, and not for different values of each of $\eta, V, \Delta p/\Delta L$, and a. Hence, an additional advantage of the dimensional analysis is the minimization of the number of experiments required for the complete study of a certain flow. The procedure for obtaining the functional form (7.7) from the governing equation and the boundary conditions is described in the following subsection.

7.1.1 Nondimensionalization of the Governing Equations

The discussion on the advantages of the dimensionless variables and solutions is meaningful under a vital precondition: the conclusions drawn by inspecting and rearranging the already known solution (7.2) ought to be possible independently of the existence or knowledge of the solution itself. Otherwise, Eq. (7.2) is perfectly adequate to fully describe the flow. What happens in flow situations in two- and three-dimensions where analytic solutions cannot be constructed? Does the dimensional analysis stop short of addressing these complicated problems? Fortunately, the answer is no, as dimensional analysis can be carried out by means of the governing differential equations. Indeed, the functional form (7.7) for the plane Couette–Poiseuille flow can be arrived at without invoking the known solution.

Pretend that the solution to Eqs. (7.1) is unknown and define the dimensionless variables,

$$u_x^* = \frac{u_x}{V} \quad \text{and} \quad y^* = \frac{y}{a} \,, \tag{7.8}$$

to reduce the magnitude of the dependent and independent variables to order one. Substituting into the governing equation (7.1), we get

$$\frac{\Delta p}{\Delta L} + \eta \frac{d^2(u_x^* V)}{d(y^* a)^2} = 0 \implies \frac{\Delta p}{\Delta L} + \frac{\eta V}{a^2} \frac{d^2 u_x^*}{dy^{*2}} = 0 \implies \frac{a^2}{\eta V} \frac{\Delta p}{\Delta L} + \frac{d^2 u_x^*}{dy^{*2}} = 0 \implies$$

$$\Pi + \frac{d^2 u_x^*}{dy^{*2}} = 0 \,.$$

The above equation is the dimensionless form of the governing equation with the scalings defined in Eq. (7.8). The boundary conditions are easily nondimensionalized to arrive at the following dimensionless problem:

$$
\left.
\begin{array}{r}
\Pi + \dfrac{d^2 u_x^*}{dy^{*2}} = 0 \\[2mm]
u_x^*(y^* = 0) = 0 \\[2mm]
u_x^*(y^* = 1) = 1
\end{array}
\right\} . \tag{7.9}
$$

From the above equations, it is obvious that $u_x^* = f(y^*, \Pi)$. We have thus reached the functional form (7.7) by nondimensionalizing the governing equations and boundary conditions. The function f may be found experimentally by measuring $u_x^* = u_x/V$ at several locations $y^* = y/a$ for several values of Π.

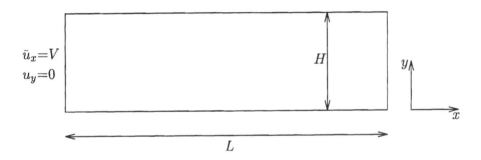

Figure 7.2. *Geometry of a two-dimensional, bidirectional flow.*

Let us now consider a more complicated example of a two-dimensional, bidirectional flow in Cartesian coordinates. Suppose that the flow domain is a rectangle of length L and width H, and that the fluid moves from left to right (Fig. 7.2). Assuming that the flow is incompressible and that $\mathbf{g} = g\mathbf{i}$, the continuity equation and the two relevant components of the momentum equation are

$$
\frac{\partial u_x}{\partial x} + \frac{\partial u_y}{\partial y} = 0 , \tag{7.10}
$$

$$
\rho \left(\frac{\partial u_x}{\partial t} + u_x \frac{\partial u_x}{\partial x} + u_y \frac{\partial u_x}{\partial y} \right) = -\frac{\partial p}{\partial x} + \eta \left(\frac{\partial^2 u_x}{\partial x^2} + \frac{\partial^2 u_x}{\partial y^2} \right) + \rho g , \tag{7.11}
$$

$$
\rho \left(\frac{\partial u_y}{\partial t} + u_x \frac{\partial u_y}{\partial x} + u_y \frac{\partial u_y}{\partial y} \right) = -\frac{\partial p}{\partial y} + \eta \left(\frac{\partial^2 u_y}{\partial x^2} + \frac{\partial^2 u_y}{\partial y^2} \right) . \tag{7.12}
$$

In the above dimensional equations, we have:
(i) three independent variables: x, y, and t;

(ii) three dependent variables: u_x, u_y, and p; and

(iii) five parameters: H, L, ρ, η, and g.

Other parameters are introduced from the boundary and the initial conditions. Here, we assume that the average velocity at the inlet of the domain is equal to V (Fig. 7.2).

The terms of Eqs. (7.10)-(7.12) can be brought to comparable order of magnitude by dividing each of the dimensional independent and dependent variables by appropriate scales, i.e., by *characteristic units* of measure. If L and H are of the same order of magnitude, one of them can be used for scaling both x and y. Assuming that this is the case, we set:

$$x^* = \frac{x}{L} \quad \text{and} \quad y^* = \frac{y}{L}. \tag{7.13}$$

The velocity components can be scaled by a characteristic velocity of the flow, such as V:

$$u_x^* = \frac{u_x}{V} \quad \text{and} \quad u_y^* = \frac{u_y}{V}. \tag{7.14}$$

Substituting Eqs. (7.13) and (7.14) into Eq. (7.10), we get the dimensionless continuity equation:

$$\frac{\partial(u_x^* V)}{\partial(x^* L)} + \frac{\partial(u_y^* V)}{\partial(y^* L)} = 0 \quad \Longrightarrow$$

$$\frac{\partial u_x^*}{\partial x^*} + \frac{\partial u_y^*}{\partial y^*} = 0. \tag{7.15}$$

In order to nondimensionalize the two components of the momentum equation, we still need to find characteristic units for the remaining dimensional variables, i.e., for t and p. Such scales can be found by *blending existing characteristic units and/or physical parameters*. For example, the expressions

$$\frac{L}{V} \quad \text{and} \quad \frac{L^2}{\nu}$$

are both characteristic time units. Similarly, the expressions

$$\frac{\eta V}{L} \quad \text{and} \quad \rho V^2$$

are characteristic pressure units. The choice among different possible characteristic units is usually guided by the physics of the flow. In convection-dominated flows, L/V is the obvious characteristic time. In flows dominated by diffusion of vorticity, L^2/ν is more relevant. In confined viscous flows, $\eta V/L$ is the obvious choice for pressure. In fact, in these flows the motion is mostly due to competing pressure and shear-stress gradients according to

$$\frac{\partial p}{\partial x} \approx \frac{\partial \tau_{yx}}{\partial y} = \eta \frac{\partial^2 u_x}{\partial y^2} .$$

Therefore, the pressure can be seen as equivalent to a viscous stress, which is measured in units of $\eta V/L$. In open inviscid flows, the effects of viscosity are minimized and thus the viscous pressure unit, $\eta V/L$, is not appropriate. These flows are driven by pressure gradients and/or inertia according to Euler's equation,

$$\rho \left(\frac{\partial \mathbf{u}}{\partial t} + \mathbf{u} \cdot \nabla \mathbf{u} \right) = -\nabla p .$$

Thus, the pressure can be viewed as equivalent to kinetic energy, which is measured in units of ρV^2. Any choice among possible units is fundamentally admissible. However, the *advantages of the dimensional analysis are obtained only by appropriate choice of units.*

To proceed with our example, let us nondimensionalize t and p as follows:

$$t^* = \frac{t}{L/V} \quad \text{and} \quad p^* = \frac{p}{\eta V/L} . \tag{7.16}$$

Substituting Eqs. (7.13), (7.14) and (7.16) into the two components of the Navier–Stokes equation yields the following dimensionless equations

$$\frac{\rho V L}{\eta} \left(\frac{\partial u_x^*}{\partial t^*} + u_x^* \frac{\partial u_x^*}{\partial x^*} + u_y^* \frac{\partial u_x^*}{\partial y^*} \right) = -\frac{\partial p^*}{\partial x^*} + \left(\frac{\partial^2 u_x^*}{\partial x^{*2}} + \frac{\partial^2 u_x^*}{\partial y^{*2}} \right) + \frac{\rho g L^2}{\eta V}$$

and

$$\frac{\rho V L}{\eta} \left(\frac{\partial u_y^*}{\partial t^*} + u_x^* \frac{\partial u_y^*}{\partial x^*} + u_y^* \frac{\partial u_y^*}{\partial y^*} \right) = -\frac{\partial p^*}{\partial y^*} + \left(\frac{\partial^2 u_y^*}{\partial x^{*2}} + \frac{\partial^2 u_y^*}{\partial y^{*2}} \right) .$$

Since all terms are dimensionless, the two groups of parameters that appear in the above equations are dimensionless. The first dimensionless group is the *Reynolds number,*

$$Re \equiv \frac{\rho V L}{\eta} , \tag{7.17}$$

which is the ratio of inertia forces to viscous forces. The second dimensionless group is the *Stokes number,*

$$St \equiv \frac{\rho g L^2}{\eta V} , \tag{7.18}$$

which represents the ratio of gravity forces to viscous forces. Therefore,

$$
Re\left(\frac{\partial u_x^*}{\partial t^*} + u_x^*\frac{\partial u_x^*}{\partial x^*} + u_y^*\frac{\partial u_x^*}{\partial y^*}\right) = -\frac{\partial p^*}{\partial x^*} + \left(\frac{\partial^2 u_x^*}{\partial x^{*2}} + \frac{\partial^2 u_x^*}{\partial y^{*2}}\right) + St \qquad (7.19)
$$

and

$$
Re\left(\frac{\partial u_y^*}{\partial t^*} + u_x^*\frac{\partial u_y^*}{\partial x^*} + u_y^*\frac{\partial u_y^*}{\partial y^*}\right) = -\frac{\partial p^*}{\partial y^*} + \left(\frac{\partial^2 u_y^*}{\partial x^{*2}} + \frac{\partial^2 u_y^*}{\partial y^{*2}}\right). \qquad (7.20)
$$

The *aspect ratio*,

$$
\epsilon \equiv \frac{H}{L}, \qquad (7.21)
$$

is an additional dimensionless number which depends solely on the geometry. This ratio provides the constant scale factor required to model a full-scale flow in the lab. The model and the full-scale flows must be *geometrically similar*, i.e., the value of ϵ must be the same in both flows. In case H and L are not of the same order of magnitude, then the two spatial coordinates are scaled as follows:

$$
x^* = \frac{x}{L} \quad \text{and} \quad y^* = \frac{y}{H}.
$$

Such scaling is preferable in problems involving different length scales, as in the *lubrication approximation* (see Chapter 9), since it provides a natural perturbation parameter.

Assuming that no other dimensionless numbers are introduced via the boundary and the initial conditions, the dimensionless governing equations (7.15), (7.19), and (7.20) dictate the following functional forms for the dimensionless dependent variables:

$$
p^* = p^*(x^*, y^*, t^*, Re, St, \epsilon) \quad \text{and} \quad \mathbf{u}^* = \mathbf{u}^*(x^*, y^*, t^*, Re, St, \epsilon).
$$

Some other dimensionless numbers of significance in fluid mechanics are discussed below.

The *capillary number*,

$$
Ca \equiv \frac{\eta V}{\sigma}, \qquad (7.22)
$$

is the ratio of viscous forces to surface tension or capillary forces. It arises in flows involving free surfaces or interfaces between immiscible fluids.

The *Weber number* is the ratio of inertia forces to surface tension forces and is defined by

$$
We \equiv \frac{\rho V^2 L}{\sigma}. \qquad (7.23)
$$

Note that

$$We = Re\, Ca\,. \tag{7.24}$$

The Weber number arises naturally in place of the capillary number when the inviscid pressure scale ρV^2 is used (instead of the viscous scale $\eta V/L$).

The *Froude number*,

$$Fr \equiv \frac{V^2}{gL}\,, \tag{7.25}$$

is the ratio of inertia forces to gravity forces. It arises in gravity-driven flows and in open channel flows. Note that

$$Fr = \frac{Re}{St}\,. \tag{7.26}$$

The *Euler number Eu*, defined by

$$Eu \equiv \frac{\Delta p}{\frac{1}{2}\,\rho V^2}\,, \tag{7.27}$$

is the ratio of pressure forces to viscous forces. It appears mostly in inviscid flows and is useful in aerodynamics.

7.2   Perturbation Methods

Asymptotic analysis is the study of a problem under the assumption that an involved parameter is vanishingly small or infinitely large. Consider the following initial value problem,

$$(1+\epsilon)\,\frac{du}{dx} + u = 0\,, \quad u(0) = a\,, \tag{7.28}$$

which involves one dimensionless parameter, ϵ. The exact solution to problem (7.28) is

$$u = a\,e^{-x/(1+\epsilon)}\,. \tag{7.29}$$

The behavior of the solution is illustrated in Fig. 7.3 for various values of ϵ.

Asymptotic analysis can be carried out at different levels of approximation and accuracy. One approach is to simplify the governing equation by eliminating some terms based on an *order of magnitude analysis* and solve the resulting truncated equation. Coming back to our example, in case $\epsilon \ll 1$, problem (7.28) can be simplified by assuming that $\epsilon=0$. This leads to the truncated problem

$$\frac{du}{dx} + u = 0\,, \quad u(0) = a\,, \tag{7.30}$$

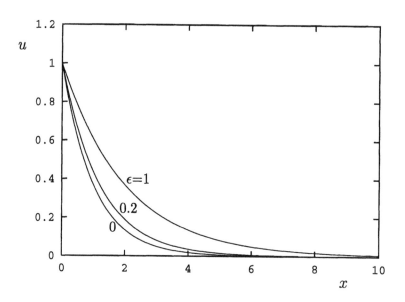

Figure 7.3. *Behavior of the solution to problem (7.28) for a=1 and various values of ϵ.*

the solution of which is

$$u = a\,e^{-x}.\qquad(7.31)$$

This solution is identical to the limit of the exact solution, Eq. (7.29), for $\epsilon \to 0$:

$$\lim_{\epsilon \to 0}\left[a\,e^{-x/(1+\epsilon)}\right] = a\,e^{-x}.$$

It is thus called the *asymptotic solution* to problem (7.28). The asymptotic solution (7.31) is *regular* in the sense that it satisfies the initial condition and is valid uniformly for all $x \geq 0$. This is due to the fact that the truncated differential equation retains the first derivative of x which allows the satisfaction of the boundary condition. In case ϵ is very small, the asymptotic solution might be used as an approximation to the exact solution (see Fig. 7.3). This approximation is said to be of *zeroth order*. Higher-order approximations may be obtained by means of straightforward parameter expansion as discussed in the two subsections below.

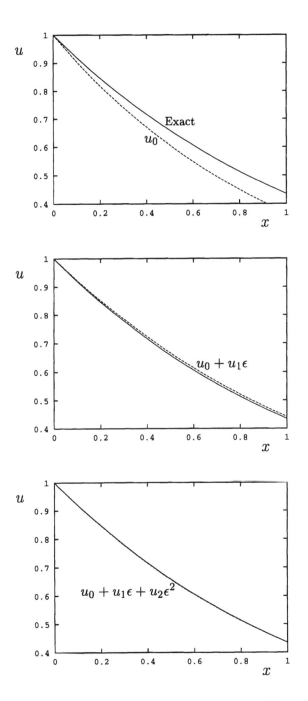

Figure 7.4. *Approximation of the solution to problem (7.28) in the case of $\epsilon=0.2$ and $a=1$, using regular perturbations of zeroth, first, and second order.*

7.2.1   Regular Perturbations

Consider again problem (7.28) and assume that the parameter ϵ is *very small* but not zero. Let us seek a series solution of the form

$$u(x,\epsilon) = u_0(x) + u_1(x)\epsilon + u_2(x)\epsilon^2 + O(\epsilon^3), \qquad (7.32)$$

where $u_0, u_1, \cdots$ are unknown functions of x. Equation (7.32) is called *asymptotic expansion* or *perturbation* of the solution in terms of the parameter ϵ. By substituting Eq. (7.32) into the differential equation (7.28) and collecting powers of ϵ, we obtain the following *perturbation equation*:

$$\left(\frac{du_0}{dx} + u_0\right) + \left(\frac{du_0}{dx} + \frac{du_1}{dx} + u_1\right)\epsilon + \left(\frac{du_1}{dx} + \frac{du_2}{dx} + u_2\right)\epsilon^2 + O(\epsilon^3) = 0. \quad (7.33)$$

This equation must be satisfied for any value of ϵ, as $\epsilon \to 0$. Therefore, all the coefficients in Eq. (7.33) must vanish, which yields:

$$\left.\begin{array}{ll} Order\ \epsilon^0 &: \quad \dfrac{du_0}{dx} + u_0 = 0 \\[2mm] & u_0(0) = a \end{array}\right\} \implies u_0(x) = a\,e^{-x}$$

$$\left.\begin{array}{ll} Order\ \epsilon^1 &: \quad \dfrac{d}{dx}(a\,e^{-x}) + \dfrac{du_1}{dx} + u_1 = 0 \\[2mm] & u_1(0) = 0 \end{array}\right\} \implies u_1(x) = a\,xe^{-x}$$

$$\left.\begin{array}{ll} Order\ \epsilon^2 &: \quad \dfrac{d}{dx}(a\,xe^{-x}) + \dfrac{du_2}{dx} + u_2 = 0 \\[2mm] & u_2(0) = 0 \end{array}\right\} \implies u_2(x) = a\left(\dfrac{x^2}{2} - x\right)e^{-x}$$

The resulting series solution is

$$u(x,\epsilon) = a\,e^{-x} + (a\,xe^{-x})\,\epsilon + \left[a\left(\frac{x^2}{2} - x\right)e^{-x}\right]\epsilon^2 + O(\epsilon^3). \qquad (7.34)$$

The first few terms of Eq. (7.34) form an approximate solution to problem (7.28) which is called *perturbation solution*. Note that the zeroth-order perturbation solution, $u_0(x)$, is identical to the exact solution for $\epsilon=0$. The expansion (7.34) holds uniformly for all values of x and any value of ϵ near zero. Note that there is no indefinite term, neither in the series expansion nor in the exact solution, for either $\epsilon \to 0$

or $x \to 0$ and, therefore, the initial condition $u(0)=a$ is satisfied exactly. Since the expansion (7.34) is valid uniformly over the domain of interest, the perturbation is said to be *regular*. Figure 7.4 shows the exact solution and the asymptotic solutions of zeroth, first, and second orders for $\epsilon=0.2$ and $a=1$. We observe that the asymptotic solution converges to the exact solution as the order of the approximation is increased.

Perturbation analyses can be carried out also in case the parameter ϵ appears in the initial (or boundary) conditions and not in the governing differential equation. If two parameters ϵ_1 and ϵ_2 are involved, a double perturbation expansion can be used, i.e,[1]

$$
\begin{aligned}
u(x, \epsilon_1, \epsilon_2) =\ & \left[u_0^{(1)}(x) + u_1^{(1)}(x)\epsilon_1 + u_2^{(1)}(x)\epsilon_1^2 + O(\epsilon_1^3) \right] \epsilon_2^0 \\
& + \left[u_0^{(2)}(x) + u_1^{(2)}(x)\epsilon_1 + u_2^{(2)}(x)\epsilon_1^{(2)} + O(\epsilon_1^3) \right] \epsilon_2^1 \\
& + \left[u_0^{(3)}(x) + u_1^{(3)}(x)\epsilon_1 + u_2^{(3)}(x)\epsilon_1^2 + O(\epsilon_1^3) \right] \epsilon_2^2 + O(\epsilon_2^3) \,.
\end{aligned}
$$

Example 7.2.1. Regular perturbation

Consider the problem

$$
\left. \begin{aligned}
\frac{d^2u}{dx^2} + \epsilon \frac{du}{dx} &= 0 \\[2mm]
u(0) = 1, \quad u(1) &= 0
\end{aligned} \right\} \,,
$$

where $\epsilon \ll 1$. Approximating the solution as

$$
u(x, \epsilon) \approx u_0(x) + u_1(x)\epsilon + u_2(x)\epsilon^2 \,,
$$

we get the following perturbation equation

$$
\frac{d^2u_0}{dx^2} + \left(\frac{d^2u_1}{dx^2} + \frac{du_0}{dx} \right) \epsilon + \left(\frac{d^2u_2}{dx^2} + \frac{du_1}{dx} \right) \epsilon^2 = 0 \,.
$$

By collecting the appropriate terms, we get:

$$
\left. \begin{aligned}
Order\ \epsilon^0 \quad : \quad \frac{d^2u_0}{dx^2} &= 0 \\[2mm]
u_0(0) = 1, \quad u_0(1) &= 0
\end{aligned} \right\} \implies u_0(x) = 1 - x
$$

---

[1] Note that the asymptotically correct form depends on the relative orders of ϵ_1 and ϵ_2.

$$Order\ \epsilon^1\ :\ \left.\begin{array}{l} \dfrac{d^2u_1}{dx^2} - 1 = 0 \\[2ex] u_1(0) = 0, \quad u_1(1) = 0 \end{array}\right\} \implies u_1(x) = -\dfrac{x}{2}\left(1 - x\right)$$

$$Order\ \epsilon^2\ :\ \left.\begin{array}{l} \dfrac{d^2u_2}{dx^2} - \dfrac{1}{2} + x = 0 \\[2ex] u_2(0) = 0, \quad u_2(1) = 0 \end{array}\right\} \implies u_2(x) = \dfrac{x}{6}\left(x - \dfrac{1}{2}\right)\left(1 - x\right)$$

The resulting approximate solution is

$$u(x,\epsilon) \approx (1 - x)\left[1 - \frac{x}{2}\epsilon + \frac{x}{6}\left(x - \frac{1}{2}\right)\epsilon^2\right].$$

□

7.2.2   Singular Perturbations

In many problems involving a perturbation parameter ϵ, an expansion of the form

$$u(x,\epsilon) = u_0(x) + u_1(x)\epsilon + u_2(x)\epsilon^2 + O(\epsilon^3)$$

may not be *uniformly valid* over the entire interval of interest. Problems leading to nonuniform expansions are called *singular perturbation* or *boundary layer* problems. These are problems that have multiple length or time scales.

A typical singular perturbation problem involves two different length scales. A perturbation expansion in the original independent variable is, in general, good over a large interval corresponding to one length scale, but breaks down in a boundary layer, i.e., in a layer near a boundary where the other length scale is relevant and the dependent variable changes rapidly. This expansion is called *outer approximation* to the problem, and the region over which it is valid is called *outer region*. By properly rescaling the independent variable in the boundary layer, it is often possible to obtain an *inner approximation* to the solution which is valid in the boundary layer and breaks down in the outer region. A *composite approximation* uniformly valid over the entire domain can then be constructed by *matching* the inner and outer approximations. Due to the matching procedure, singular perturbation methods are also called *matched asymptotic expansions*. The most characteristic application of singular perturbation in fluid mechanics is the matching of potential solutions (outer approximation) to the boundary layer solutions (inner approximation) of the Navier–Stokes equation, with the inverse of the Reynolds number serving as the perturbation parameter.

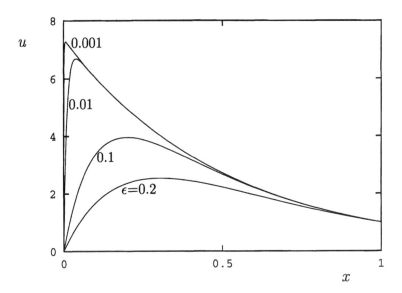

Figure 7.5. *Behavior of the solution (7.36) for various values of ϵ.*

In the following, we describe briefly the method of matched asymptotic expansions and introduce the main ideas behind the construction and matching of inner and outer expansions. For additional reading, we refer the reader to References [1]-[5].

Consider the following boundary value problem

$$\left.\begin{array}{c} \epsilon\dfrac{d^2u}{dx^2} + (1 + 2\epsilon)\,\dfrac{du}{dx} + 2u = 0 \\[2mm] u(0) = 0, \quad u(1) = 1 \end{array}\right\} , \tag{7.35}$$

where $\epsilon \ll 1$. Note that the perturbation parameter multiplies the highest derivative in Eq. (7.35). Setting ϵ equal to zero completely *changes the type* of the problem by reducing it to a first-order differential equation. The exact solution is given by

$$u = \frac{e^{-2x} - e^{-x/\epsilon}}{e^{-2} - e^{-1/\epsilon}} . \tag{7.36}$$

In Fig. 7.5, the behavior of u for various values of ϵ is shown. When ϵ is small, the solution changes dramatically in a small interval near the origin and then appears to approach asymptotically a certain curve. Note also that the curves for $\epsilon=0.001$ and 0.01 appear to coincide outside the vicinity of $x=0$.

For the zeroth-order approximation of u, we get

$$\frac{du_0}{dx} + 2u_0 = 0 \quad \Longrightarrow$$

$$u_0(x) = c_1 e^{-2x} . \tag{7.37}$$

Our perturbation leads to a first-order differential equation. As a consequence, only one of the two boundary conditions,

$$u_0(0) = 0 \quad \text{and} \quad u_0(1) = 1 ,$$

can be satisfied. Application of the first condition leads to $u_0(x)=0$, which violates the condition at $x=1$. Application of the boundary condition $u_0(1)=1$ gives $c_1=e^2$ and

$$u_0(x) = e^{2(1-t)} . \tag{7.38}$$

This function does not satisfy the boundary condition at $x=0$.

In general, there is no rule as to which boundary condition must be satisfied by the outer solution; one may have to follow a trial and error procedure in order to determine that condition. In many cases, however, a choice is hinted by the physics or the mathematics of the problem [3,6]. In the present case, it turns out that the proper choice is to satisfy the boundary condition at $x=1$. Note that the approximation (7.38) can be obtained directly from Eq. (7.35) by setting ϵ equal to zero and then solving the resulting reduced differential equation together with the boundary condition at $x=1$. This solution is the *outer expansion* of problem (7.35) and is denoted by u^o; hence,

$$u^o = e^{2(1-x)} . \tag{7.39}$$

The outer approximation (7.39) is plotted in Fig. 7.6 together with the exact solution for $\epsilon=0.05$. We observe that u^o does an excellent job far from $x=0$ but breaks down in the boundary layer where the exact solution changes very rapidly, complying with the boundary condition $u(0)=0$. This failure is due to the fact that d^2u/dx^2 attains large values within the boundary layer; although ϵ is very small, the term $\epsilon d^2u/dx^2$ is not, and should not have been omitted. In other words, the length scale in the boundary layer is different from that of the outer region. This can be corrected by rescaling the differential equation in the boundary layer region. The independent variable x is stretched to a new variable ξ by means of a *stretching transformation* of the general form

$$\xi = \frac{x}{\delta(\epsilon)} , \tag{7.40}$$

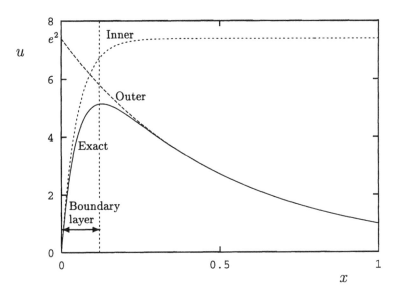

Figure 7.6. *Outer and inner solutions to problem (7.35) for ϵ=0.05.*

where $\delta(\epsilon)$ is a function representing the length scale of the boundary layer. This must be chosen carefully so that the transformed differential equation reflects the physics in the boundary layer and the second derivative term is retained. For an interesting discussion on the selection of the stretching transformation, see Reference [3].

In the present example, the proper choice is $\delta(\epsilon)=\epsilon$; hence,

$$\xi = \frac{x}{\epsilon} . \tag{7.41}$$

By means of the above transformation, Eq. (7.35) is transformed to

$$\frac{d^2 u}{d\xi^2} + (1 + 2\epsilon) \frac{du}{d\xi} + 2\epsilon\, u = 0 . \tag{7.42}$$

Equation (7.42) is amenable to a regular perturbation analysis. However, we will focus on the leading-term behavior setting $\epsilon=0$; this leads to

$$\frac{d^2 u}{d\xi^2} + \frac{du}{d\xi} = 0 . \tag{7.43}$$

The above equation has general solution

$$u(\xi) = A_1 + A_2\, e^{-\xi}, \tag{7.44}$$

where A_1 and A_2 are constants. By applying the boundary condition $u(0){=}0$, we get $A_2{=}{-}A_1$, and so

$$u(\xi) = A_1\,(1 - e^{-\xi}). \tag{7.45}$$

We denote this solution by u^i and call it the *inner expansion*,

$$u^i = A_1\,(1 - e^{-\xi}) = A_1\,(1 - e^{-x/\epsilon}). \tag{7.46}$$

The constant A_1 is determined by *matching* the inner and outer solutions. Using Prandtl's matching rule, we require that

$$\lim_{x \to 0} u^o(x) = \lim_{\xi \to \infty} u^i(\xi),$$

which is equivalent to

$$\lim_{\substack{\epsilon \to 0 \\ \xi\ fixed}} u^o(\xi, \epsilon) = \lim_{\substack{\epsilon \to 0 \\ x\ fixed}} u^i(x, \epsilon). \tag{7.47}$$

The left-hand limit is called *inner limit* of the outer solution and is denoted by $(u^o)^i$; the right-hand limit is called *outer limit* of the inner solution and is denoted by $(u^i)^o$. In words, the inner limit of the outer solution is equal to the outer limit of the inner solution. Thus, the above matching principle can be written as

$$(u^o)^i = (u^i)^o. \tag{7.48}$$

Since

$$(u^o)^i = \lim_{x \to 0} e^{2(1-x)} = e^2 \quad \text{and} \quad (u^i)^o = \lim_{\xi \to \infty} A_1\,(1 - e^{-\xi}) = A_1,$$

we get

$$A_1 = e^2.$$

Therefore, the inner expansion is

$$u^i = e^2\,(1 - e^{-\xi}) = e^2\,(1 - e^{-x/\epsilon}). \tag{7.49}$$

As shown in Fig. 7.6, u^i provides a reasonable approximation in the boundary layer.

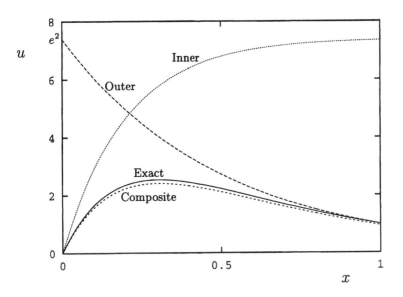

Figure 7.7. *Outer, inner, and composite solutions to problem (7.35) for ε=0.2.*

The next step is to combine the outer and inner solutions to a *composite expansion*, u^c, that is uniformly valid throughout the interval $[0,1]$. This can be achieved by adding u^o and u^i and subtracting the common limit (7.48),

$$u^c = u^o + u^i - (u^o)^i. \tag{7.50}$$

Substitution into the above equation gives

$$u^c = e^{2(1-x)} + e^2 (1 - e^{-x/\epsilon}) - e^2 \quad \Longrightarrow$$

$$u^c = e^{2(1-x)} - e^2 e^{-x/\epsilon}. \tag{7.51}$$

The composite solution u^c provides a uniform approximate solution throughout the interval $[0,1]$. It is easily verified that Eq. (7.51) satisfies the differential equation (7.35). Moreover,

$$u^c(0) = 0 \quad \text{and} \quad u^c(1) = 1 - e^{2-1/\epsilon} ;$$

the boundary condition at $x=0$ is satisfied exactly while the condition at $x=1$ is satisfied asymptotically. In Fig. 7.7, we compare the outer, the inner, and the composite solutions with the exact solution for $\epsilon=0.2$. The value of ϵ was intentionally chosen to be "large" so that the differences between u^c and the exact solution can be observed. As ϵ gets smaller, the composite solution gets closer to the corresponding exact solution and the two curves become indistinguishable.

In addition to Prandtl's matching rule, a number of other, more refined matching methods have been proposed [2]-[4]. The most widely matching principle is the one introduced by Van Dyke [2]. This requires that the "m-term inner expansion of the n-term outer expansion be equal to the n-term outer expansion of the m-term inner expansion," where the integers m and n are not necessarily equal.

7.3   Perturbation Methods in Fluid Mechanics

One of the most important perturbation parameters in fluid mechanics is the Reynolds number,

$$Re \equiv \frac{\rho V L}{\eta}. \tag{7.52}$$

In confined viscous flows, the pressure is scaled by $\eta V/L$. As shown in Section 7.1, this scaling leads to the following dimensionless form of the steady-state Navier–Stokes equation

$$Re\,(\mathbf{u}\cdot\nabla\mathbf{u}) = -\nabla p + \nabla^2\mathbf{u}, \tag{7.53}$$

where gravity has been assumed negligible. If the Reynolds number is vanishingly small, $Re \ll 1$, we get the the so-called *creeping* or *Stokes flow*, governed by

$$-\nabla p + \nabla^2\mathbf{u} = 0. \tag{7.54}$$

Since the Reynolds number multiplies only lower-derivative terms, Eq. (7.53) is amenable to regular perturbation analysis. It is also clear that the Stokes flow solution is the zeroth-order approximation to the solution. This is usually obtained in terms of the *stream function* (see Chapter 10). Corrections to the Stokes solution may then be obtained by regular perturbations.

The appropriate pressure scale for open, almost inviscid *laminar* flows is ρV^2. In this case, the nondimensionalized steady-state Navier–Stokes equation under negligible gravitational forces is given by

$$\mathbf{u}\cdot\nabla\mathbf{u} = -\nabla p + \frac{1}{Re}\,\nabla^2\mathbf{u}. \tag{7.55}$$

If the Reynolds number is infinitely large, Eq. (7.55) is reduced to the *Euler* or *potential flow* equation

$$\mathbf{u}\cdot\nabla\mathbf{u} = -\nabla p. \tag{7.56}$$

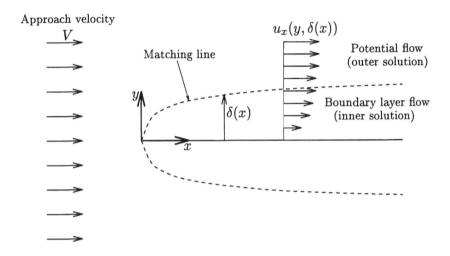

Figure 7.8. *Boundary layer and potential flow when an approaching stream over-takes a thin plate at high Reynolds numbers.*

At large values of the Reynolds number, Eq. (7.55) is amenable to perturbation analysis in terms of $1/Re$. Given that $1/Re$ multiplies the higher-order derivatives of **u**, the perturbation is singular. The potential flow solution is the outer solution at high values of the Reynolds number. For flow in the x-direction, the stretching variable

$$\xi = x\sqrt{Re} \tag{7.57}$$

leads to the *boundary layer* equation

$$\mathbf{u} \cdot \nabla_\xi \mathbf{u} = -\nabla p + \nabla_\xi^2 \mathbf{u}, \tag{7.58}$$

the solution of which is the inner solution. The outer (potential) and the inner (boundary layer) solutions are matched at the *boundary layer thickness*, as shown in Fig. 7.8.

Geometrical parameters, such as aspect ratios and inclinations, are often used as perturbation parameters. *Domain perturbation* is possible when the domain variation can be expressed in terms of these parameters [7,8]. *Lubrication* and *stretching flows* are two important classes of almost rectilinear flows amenable to domain perturbation. These are examined in detail in Chapter 9.

A typical example of lubrication flow is the two-dimensional flow in a converging channel of inclination α, as shown in Fig. 7.9. The two relevant nondimensionalized

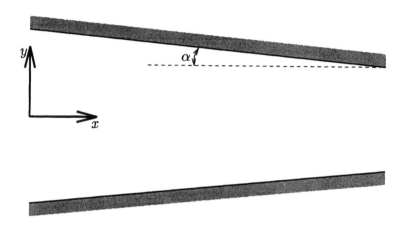

Figure 7.9. *Flow in a slightly converging channel.*

components of the Navier–Stokes equation are

$$\alpha Re\,\frac{Du_x}{Dt} = -\frac{\partial p}{\partial x} + \alpha^2\frac{\partial^2 u_x}{\partial x^2} + \frac{\partial^2 u_x}{\partial y^2} \tag{7.59}$$

and

$$\alpha^3 Re\,\frac{Du_y}{Dt} = -\frac{\partial p}{\partial y} + \alpha^3\frac{\partial^2 u_y}{\partial x^2} + \alpha\frac{\partial^2 u_x}{\partial y^2}\,. \tag{7.60}$$

Lubrication flow corresponds to $\alpha Re \ll 1$ and $\alpha \ll 1$; hence, it can be solved by regular perturbation. The zeroth-order approximation is the rectilinear flow solution corresponding to $\alpha=0$ [9].

Stretching flows are good prototypes of important processing flows such as spinning of fibers and casting of films [10]. A properly defined aspect ratio (e.g., thickness to length) may be used as the perturbation parameter. Consider, for example, the fiber spinning flow depicted in Fig. 7.10. As shown in Chapter 9, the nondimensionalized z-component of the Navier–Stokes equation becomes

$$Re\,u_z\frac{du_z}{dz} = -\frac{dp}{dz} + \frac{d^2 u_z}{dz^2} + St\,g_z\,. \tag{7.61}$$

Averaging Eq. (7.61) over the cross sectional area of the fiber and invoking the continuity equation and the free surface boundary condition result in the following

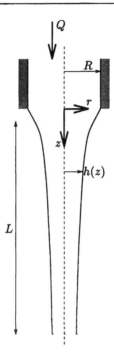

Figure 7.10. *Fiber-spinning flow.*

ordinary differential equation:

$$\frac{d}{dz}\left(\frac{3}{u_z}\frac{du_z}{dz}\right) + \frac{1}{\epsilon\,Ca}\frac{d}{dz}\sqrt{\frac{1}{u_z}} - Re\,\frac{du_z}{dz} + St\,\frac{1}{u_z} = 0\,, \qquad (7.62)$$

where ϵ is the ratio of the initial radius of the fiber to its length. Equation (7.62) can be solved for limiting values of the involved parameters (ϵ, Re, Ca, and St) by regular or singular perturbation.

We conclude this chapter by noting that perturbation analysis applies only to limiting values of the involved parameters and does not provide solutions for the entire spectrum of their values. Such solutions can then be obtained by numerical modeling and/or experimentation.

7.4   Problems

7.1. In almost inviscid laminar flows, the pressure is customarily scaled by ρV^2, where V is a characteristic velocity of the flow of interest. Show that this scaling leads to the following nondimensionalized form of the Navier–Stokes equation

$$\frac{D\mathbf{u}}{Dt} = -\nabla p + \frac{1}{Re}\nabla^2\mathbf{u} + \frac{1}{Fr}\hat{\mathbf{g}}, \tag{7.63}$$

where $\hat{\mathbf{g}}$ is the unit vector in the direction of the gravitational acceleration.

7.2. Consider a spherical gas bubble of radius R growing at a rate $\dot{R}$ in a liquid bath of density ρ, viscosity η, and surface tension σ. Nondimensionalize the equation governing the stress jump across the interface, and identify the resulting dimensionless numbers and their physical significance.

7.3. Construct an approximate solution to the problem

$$\frac{du}{dx} + a\left(\frac{du}{dx} + u\right) = 0, \quad u(0) = 0,$$

where $a \gg 1$. Plot and compare the asymptotic and the exact solutions for various values of a.

7.4. The problem

$$\left.\begin{array}{c} \epsilon\dfrac{d^2u}{dx^2} + \dfrac{du}{dx} = -\frac{3}{2}(1-3\epsilon)\,e^{-3x} \\[2mm] u(0) = 0 \quad \text{and} \quad u(\infty) = 1 \end{array}\right\},$$

where $\epsilon \ll 1$, is amenable to singular perturbation analysis. Construct the composite solution by taking the outer solution to satisfy (a) $u(\infty)=1$, and (b) $u(0)=0$. In each case, plot the inner, the outer, and the composite solutions for $\epsilon=0.2$ and compare them with the exact solution.

7.5. Find the asymptotic solution to the problem

$$\left.\begin{array}{c} a\dfrac{d^2u}{dx^2} + \dfrac{du}{dx} = 0 \\[2mm] u(0) = 1 \quad \text{and} \quad u(1) = 0 \end{array}\right\},$$

where $a \gg 1$. Choose a physical model that can be described by the above equations and bring out the physical significance of your findings.

7.6. A one-dimensional, steady, convection-reaction-diffusion problem is modeled by the following nondimensionalized equations

$$\left.\begin{array}{c} \dfrac{dc}{dx} = N_1\dfrac{d^2c}{dx^2} + N_2\,c^2 \\[2mm] u(0) = 0 \quad \text{and} \quad u(1) = 1 \end{array}\right\},$$

where c is the dimensionless concentration, and N_1 and N_2 are dimensionless numbers. Construct the asymptotic solution when

(a) $N_1 \ll 1$;   (b) $N_1 \gg 1$;   (c) $N_2 \ll 1$;   (d) $N_2 \gg 1$.

What is the physical significance of your solutions? Calculate the exact solution and validate your results by constructing representative plots.

7.5   References

1. L.G. Leal, *Laminar Flow and Convective Transport Processes*, Butterworth-Heinemann, Boston, 1992.

2. M. Van Dyke, *Perturbation Methods in Fluid Mechanics*, Academic Press, New York, 1964.

3. A.H. Nayfeh, *Perturbation Methods*, Wiley-Interscience, New York, 1973.

4. M.H. Holmes, *Introduction to Perturbation Methods*, Springer, New York, 1995.

5. J.D. Logan, *Applied Mathematics*, John Wiley & Sons, New York, 1987.

6. G.M. Bender and S.A. Orszag, *Advanced Mathematical Methods for Scientists and Engineers*, McGraw-Hill, New York, 1978.

7. D.D. Joseph, "Domain perturbations: the higher order theory of infinitesimal water waves," *Arch. Rat. Mech. Anal.* **51**, 295-303 (1973).

8. J. Tsamopoulos and R.A. Brown, "Nonlinear oscillations of inviscid drops and bubbles," *J. Fluid Mech.* **127**, 519-537 (1983).

9. T.C. Papanastasiou, "Lubrication flows," *Chem. Eng. Educ.* **24**, 50 (1989).

10. J.R.A. Pearson, *Mechanics of Polymer Processing*, Elsevier, London, 1985.

Chapter 8

LAMINAR BOUNDARY LAYER FLOWS

8.1 Boundary Layer Flow

In this chapter, we consider flows near solid surfaces known as *boundary layer flows*. One way of describing these flows is in terms of vorticity dynamics, i.e., generation, diffusion, convection, and intensification of vorticity. The presence of vorticity distinguishes boundary layer flows from potential flows which are free of vorticity.

In two-dimensional flow along the xy-plane, the vorticity is given by

$$\boldsymbol{\omega} \equiv \nabla \times \mathbf{u} = \left(\frac{\partial u_y}{\partial x} - \frac{\partial u_x}{\partial y} \right) \mathbf{e}_k , \qquad (8.1)$$

and is a measure of rotation in the fluid. As discussed in Chapter 6, vorticity is generated at solid boundaries. For example, if $u_x = u_x(x, y)$ and the plane $y=0$ corresponds to an impermeable wall, then along this wall $u_y=0$ and $\partial u_y/\partial x=0$. Due to the no-slip boundary condition, $\partial u_x/\partial y$ is nonzero and, thus, vorticity is generated according to

$$\boldsymbol{\omega} = -\frac{\partial u_x}{\partial y} \mathbf{e}_k . \qquad (8.2)$$

Vorticity diffuses away from the generator wall at a rate of $(\nu \nabla^2 \omega)$, and competes with convection at a rate of $(\mathbf{u} \cdot \nabla \omega)$ (Fig. 8.1). Due to the effects of convection, the vorticity is confined within a parabolic-like envelope which is commonly known as *boundary layer*. Therefore, the area away from the solid wall remains free of vorticity.

The line separating boundary-layer and potential flows, i.e., the line where the velocity changes from a parabolic to a flat profile, is defined by the orbit of vorticity "particles" generated at a solid surface and diffused away to a *penetration* or *boundary layer thickness*, $\delta(x)$. As already discussed in Section 7.3, along the edge

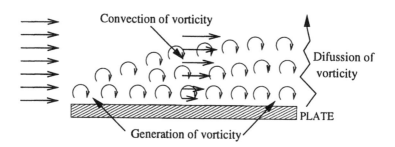

Figure 8.1. *Generation, diffusion, and convection of vorticity in the vicinity of a solid wall.*

of the boundary layer, convection and diffusion of vorticity are of the same order of magnitude, i.e.,

$$V\frac{\partial \omega}{\partial x} \simeq k^2 \nu \frac{\partial^2 \omega}{\partial y^2}, \tag{8.3}$$

where k is a constant. Consequently,

$$\left[\frac{V}{x}\right] \simeq k^2 \left[\frac{\nu}{\delta^2(x)}\right], \tag{8.4}$$

where x is the distance from the leading edge of the plate. Therefore, the expression

$$\delta(x) = k\sqrt{\frac{\nu x}{V}}, \tag{8.5}$$

provides an order of magnitude estimate for the boundary layer thickness.

Consider now the flow past a submerged body as shown in Fig. 8.2. Across the boundary layer, the velocity increases from zero – due to the no-slip boundary condition – to the finite value of the free stream flow. The thickness of the boundary layer, $\delta(x)$, is a function of the distance from the leading edge of the body and depends on the local Reynolds number, $Re \equiv \rho V x/\eta$; $\delta(x)$ can be infinitesimal, finite, or practically infinite. When $Re \ll 1$ (which leads to creeping flow), the distance $\delta(x)$ is practically infinite. In this case, the solution to the Navier–Stokes equations for creeping flow (discussed in Chapter 10) holds uniformly over the entire flow area. For $1 \ll Re < 10^4$, $\delta(x)$ is small but finite, i.e., $\delta(x)/L \ll 1$. For higher Reynolds numbers, the flow becomes turbulent leading to a turbulent boundary layer. Under certain flow conditions, the boundary layer flow detaches from the

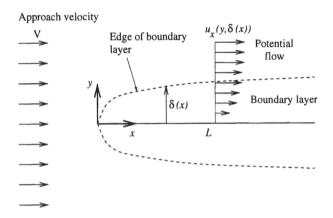

Figure 8.2. *Boundary layer and potential flow regions around a plate.*

solid surface, resulting in shedding of vorticity that eventually accumulates into periodically spaced, traveling vortices that constitute the wake.

From the physical point of view, the boundary layer thickness $\delta(x)$ defines the region where the effect of diffusion of vorticity away from the generating solid surface competes with convection from bulk motion. A rough estimate of the thickness $\delta(x)$ is provided by Eq. (8.5). The presence of vorticity along and across the boundary layer is indicated in the schematic of Fig. 8.1. As discussed in Chapter 7, from a mathematical point of view, the solution within the boundary layer is an inner solution to the Navier-Stokes equations which satisfies the no-slip boundary condition, but not the potential velocity profile away from the body.

8.2   Boundary Layer Equations

Boundary layer flow of Newtonian fluids can be studied by means of the Navier-Stokes equations. However, the characteristics of the flow suggest the use of simplified governing equations. Indeed, using order of magnitude analysis, a more simplified set of equations known as the *boundary layer equations* [1] can be developed. In reference to Fig. 8.2, the Navier-Stokes equations are made dimensionless by means of characteristic quantities that bring the involved terms to comparable orders of magnitude:

$$x^* = \frac{x}{L}, \quad y^* = \frac{y}{L}Re^{1/2},$$

$$u_x^* = \frac{u_x}{V} , \qquad u_y^* = \frac{u_y}{V} Re^{1/2} ,$$

$$\tau_{ij}^* = \frac{\tau_{ij}}{\eta \frac{V}{\delta}} , \qquad p^* = \frac{p}{\rho V^2} ,$$

where $Re \equiv VL/\nu$ is the Reynolds number. For steady flow, the resulting dimensionless equations are:

$$\frac{\partial u_x^*}{\partial x^*} + \frac{\partial u_y^*}{\partial y^*} = 0 ; \tag{8.6}$$

$$\left(u_x^* \frac{\partial u_x^*}{\partial x^*} + u_y^* \frac{\partial u_x^*}{\partial y^*} \right) = -\frac{\partial p^*}{\partial x^*} + \frac{\partial^2 u_x^*}{\partial y^{*2}} + \frac{1}{Re} \frac{\partial^2 u_x^*}{\partial x^{*2}} ; \tag{8.7}$$

$$\frac{1}{Re} \left(u_x^* \frac{\partial u_y^*}{\partial x^*} + u_y^* \frac{\partial u_y^*}{\partial y^*} \right) = -\frac{\partial p^*}{\partial y^*} + \frac{1}{Re} + \frac{\partial^2 u_y^*}{\partial y^{*2}} + \frac{1}{Re} \frac{\partial^2 u_y^*}{\partial x^{*2}} . \tag{8.8}$$

If $Re \gg 1$, these equations reduce to

$$\frac{\partial u_x^*}{\partial x^*} + \frac{\partial u_y^*}{\partial y^*} = 0 , \tag{8.9}$$

$$u_x^* \frac{\partial u_x^*}{\partial x^*} + u_y^* \frac{\partial u_x^*}{\partial y^*} = -\frac{\partial p^*}{\partial x^*} + \frac{\partial^2 u_x^*}{\partial y^{*2}} , \tag{8.10}$$

and

$$p^* = p^*(x^*) . \tag{8.11}$$

The appropriate, dimensionless boundary conditions to Eqs. (8.9) to (8.11) are:

at $y^* = 0$, $u_x^* = u_y^* = 0$ (no-slip);

at $y^* = 1$, $u_x^* = 1$, $\dfrac{\partial u_x^*}{\partial y^*} = 0$ (continuity of velocity and stress);

at $x^* = 0$, $u_x^* = u_y^* = 0$ (stagnation point).

The pressure gradient, dp^*/dx^*, is identical to that of the outer, potential flow, $(dp^*/dx^*)_p$,

$$\frac{dp^*}{dx^*} = \left(\frac{dp^*}{dx^*} \right)_p = \left(-u_x^* \frac{du_x^*}{dx^*} \right)_p = \begin{cases} 0, & \text{slender body} \\ \\ \text{known,} & \text{nonslender body} \end{cases} \tag{8.12}$$

Thus, the only unknowns in Eqs. (8.9) and (8.10) are the two velocity components, u_x^* and u_y^*. The latter is eliminated by means of the continuity equation,

$$u_y^* = -\int_0^{y^*} \frac{\partial u_x^*}{\partial x^*} \, dy^* , \tag{8.13}$$

leading to a single equation,

$$u_x^* \frac{\partial u_x^*}{\partial x^*} + \frac{\partial u_x^*}{\partial y^*} \left(-\int_0^{y^*} \frac{\partial u_x^*}{\partial x^*} dy^* \right) = -\frac{dp^*}{dx^*} + \frac{\partial^2 u_x^*}{\partial y^{*2}} .$$

(8.14)

The corresponding dimensional forms of the boundary layer equations for laminar flow are

$$\frac{\partial u_x}{\partial x} + \frac{\partial u_y}{\partial y} = 0 ,$$

(8.15)

$$u_x \frac{\partial u_x}{\partial x} + u_y \frac{\partial u_x}{\partial y} = -\frac{1}{\rho} \frac{\partial p}{\partial x} + \nu \frac{\partial^2 u_x}{\partial y} ,$$

(8.16)

and

$$p = p(x) .$$

(8.17)

These constitute a set of nonlinear parabolic equations for which an exact, closed-form solution is not possible. Blasius [2] developed an approximate similarity solution for flow past a flat plate, i.e., for $dp^*/dx^*=0$. He introduced a dimensionless stream function which, of course, satisfies the dimensionless continuity Eq. (8.9), of the form

$$\psi^* \equiv \frac{\psi}{\sqrt{\nu V x}} = f(\xi) ,$$

(8.18)

where ξ is a similarity coordinate variable, defined as

$$\xi = y \sqrt{\frac{V}{\nu x}} .$$

By recognizing that an estimate of the boundary layer thickness is

$$\delta(x) \approx \sqrt{\frac{\nu x}{V}} ,$$

the variable ξ scales appropriately the coordinate across the thickness of the boundary layer,

$$\xi = \frac{y}{\delta(x)} .$$

(8.19)

From the definition of the stream function (Chapter 2),

$$u_x = \frac{\partial \psi}{\partial y} = V f' ,$$

(8.20)

and

$$u_y = -\frac{\partial \psi}{\partial x} = \frac{1}{2}\sqrt{\frac{\nu V}{x}} \left(\xi f' - f \right) . \tag{8.21}$$

Substitution of Eqs. (8.20) and (8.21) in Eq. (8.16) leads to the *Blasius equation*,

$$2\frac{d^3 f}{d\xi^3} + f\frac{d^2 f}{d\xi^2} = 0 . \tag{8.22}$$

This is a nonlinear, ordinary differential equation subject to the boundary conditions

$$f(0) = f'(0) = 0 \tag{8.23}$$

and

$$f'(\xi \to \infty) = 1 . \tag{8.24}$$

In order to avoid the infinite domain of the two-point boundary value problem, Blasius solved Eq. (8.22) by transforming it to an equivalent forward numerical integration scheme of ordinary differential equations from $\xi=0$ up to a point ξ_∞ where the outer-edge boundary condition (8.24) is satisfied. These numerical results are tabulated in Table 8.1. In Fig. 8.3, the dimensionless velocity u_x/V is plotted versus the similarity variable $\xi=y/\delta(x)=y/(\nu x/V)^{1/2}$. Several other useful quantities can be calculated by means of the results given in Table 8.1, such as:

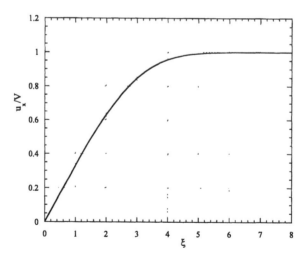

Figure 8.3. *Solution of the Blasius equation.*

ξ	f	f'	f''
0.0000000E+00	0.0000000E+00	0.0000000E+00	0.3320600
0.2500000	1.0376875E-02	8.3006032E-02	0.3319165
0.5000000	4.1494049E-02	0.1658866	0.3309136
0.7500000	9.3284436E-02	0.2483208	0.3282084
1.000000	0.1655748	0.3297825	0.3230098
1.250000	0.2580366	0.4095603	0.3146356
1.500000	0.3701439	0.4867927	0.3025829
1.750000	0.5011419	0.5605230	0.2866015
2.000000	0.6500322	0.6297698	0.2667536
2.250000	0.8155764	0.6936100	0.2434452
2.500000	0.9963216	0.7512640	0.2174131
2.750000	1.190646	0.8021722	0.1896628
3.000000	1.396821	0.8460487	0.1613615
3.250000	1.613085	0.8829061	0.1337038
3.500000	1.837715	0.9130443	0.1077739
3.750000	2.069094	0.9370083	8.4430709E-02
4.000000	2.305766	0.9555216	6.4236179E-02
4.250000	2.546470	0.9694083	4.7435224E-02
4.500000	2.790157	0.9795170	3.3984236E-02
4.750000	3.035983	0.9866555	2.3614302E-02
5.000000	3.283299	0.9915444	1.5911143E-02
5.250000	3.531620	0.9947910	1.0394325E-02
5.500000	3.780600	0.9968815	6.5830108E-03
5.750000	4.029997	0.9981864	4.0417481E-03
6.000000	4.279651	0.9989761	2.4056220E-03
6.250000	4.529459	0.9994394	1.3880555E-03
6.500000	4.779355	0.9997029	7.7646959E-04
6.750000	5.029300	0.9998482	4.2111927E-04
7.000000	5.279273	0.9999259	2.2145297E-04
7.250000	5.529260	0.9999662	1.1292604E-04
7.500000	5.779254	0.9999865	5.5846038E-05
7.750000	6.029253	0.9999964	2.6787688E-05
8.000000	6.279252	1.000001	1.2465006E-05

Table 8.1. *Solution of the Blasius equation.*

(a) Boundary layer thickness, $\delta(x)$

This is defined as the location where $u_x/V = u_x^* = 0.99$. From Fig. 8.3 (or Table 8.1), this occurs at $\xi \approx 5$. Hence,

$$\delta(x) = y(u_x^* = 0.99) = 5\sqrt{\frac{\nu x}{V}}. \tag{8.25}$$

(b) Wall shear stress, τ_w, and drag force, F_D

The shear stress at the plate is

$$\tau_w = \eta \frac{\partial u_x}{\partial y}\bigg|_{y=0} = \eta \sqrt{\frac{V^3}{\nu x}} f''(0) = 0.332 \, (\rho V^2) \sqrt{\frac{\nu}{Vx}}. \tag{8.26}$$

Therefore, the *drag force* on the plate per unit width is

$$F_D = \int_0^L \tau_w \, dx = 0.664 \sqrt{V^3 \rho \eta L}. \tag{8.27}$$

The net normal force is zero. For the *drag coefficient*, C_D, defined by

$$C_D \equiv \frac{F_D}{\frac{\rho V^2}{2} L},$$

we get

$$C_D = \frac{1.328}{\sqrt{\frac{VL}{\nu}}}. \tag{8.28}$$

Notice that the local shear stress breaks down at $x = 0$ where τ_w is singular, i.e., $\tau_w \to \infty$ as $x \to 0$. Actually, the formula does not apply there because the potential flow approximation is not valid near $x=0$. Nevertheless, the singularity is *integrable*, i.e., the drag F_D is finite.

(c) The small normal velocity component, u_y

At $\xi = 5$,

$$u_y = -\frac{\partial \psi}{\partial x} = \frac{1}{2}\sqrt{\frac{V\nu}{x}} (\xi f' - f) = 0.837 \sqrt{\frac{V\nu}{x}}. \tag{8.29}$$

(d) Transition to turbulent boundary layer

Transition to turbulent boundary layer occurs at $Re=Vx/\nu \simeq 112,000$, or at distance x downstream from the leading edge, given by

$$x = \frac{112,000 \, \nu}{V}. \tag{8.30}$$

The dimensionless boundary layer equations and their solutions are independent of the Reynolds number. Thus, *all boundary layer flows in similar geometries are dynamically similar. The Reynolds number is only a scaling factor of the boundary layer thickness and the associated variables.*

8.3   Approximate Momentum Integral Theory

Reasonably accurate solutions to boundary layer flows can be obtained from macroscopic mass and momentum balances through the use of finite control volumes. This method was first introduced by von Karman [3] and is highlighted below for boundary layer over a flat plate (Fig. 8.4).

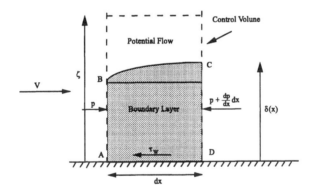

Figure 8.4. *Derivation of von Karman's approximate equations for boundary layer flow over a flat plate.*

In von Karman's method, the flow is integrated across the thickness of the layer. To account for the development of the boundary layer, and to ensure that its thickness is confined within the limits of integration, the control volume is selected so that its size, ζ, is larger than the expected thickness of the layer. Integrating from 0 to ζ gives

$$\int_0^\zeta \left(u_x \frac{\partial u_x}{\partial x} + u_y \frac{\partial u_x}{\partial y} \right) dy = \int_0^\zeta -\frac{1}{\rho} \frac{\partial p}{\partial x} \, dy + \int_0^\zeta \frac{1}{\rho} \frac{\partial \tau_{xy}}{\partial y} \, dy .$$

From the continuity equation,

$$u_y = -\int_0^y \frac{\partial u_x}{\partial x} dy .$$

For steady flow, the pressure gradient is

$$\frac{1}{\rho}\frac{dp}{dx} = -V\frac{dV}{dx} .$$

By substitution, we get

$$\int_0^\zeta u_x \frac{\partial u_x}{\partial x} dy - \int_0^\zeta \left(\int_0^y \frac{\partial u_x}{\partial x} dy \right) \frac{\partial u_x}{\partial y} dy = \int_0^\zeta V\frac{\partial V}{\partial x} dy + \int_0^\zeta \frac{1}{\rho}\frac{\partial \tau_{xy}}{\partial y} dy .$$

Defining now the new variables A and B such that

$$dB = \frac{\partial u_x}{\partial y} dy \quad \Longrightarrow \quad B = u_x ,$$

and

$$A = \int_0^y \frac{\partial u_x}{\partial x} dy \quad \Longrightarrow \quad dA = \frac{\partial u_x}{\partial x} dy ,$$

the second term in the momentum equation can be written as

$$\int_0^\zeta A \, dB = AB|_0^\zeta - \int_0^\zeta B dA = V\int_0^\zeta \frac{\partial u_x}{\partial x} dy - \int_o^\zeta u_x \frac{\partial u_x}{\partial x} dy .$$

Finally, since ζ is independent of x, by integrating and rearranging, the integral form of the momentum equation becomes

$$\frac{\partial}{\partial x} V^2 \int_0^\zeta \left[\frac{u_x}{V}\left(1 - \frac{u_x}{V}\right) \right] dy + V\frac{\partial V}{\partial x} \int_0^\zeta \left(1 - \frac{u_x}{V}\right) dy = \frac{\tau_w}{\rho} , \tag{8.31}$$

where τ_w is the shear stress at the wall.

The above integral equation can be simplified further to

$$\frac{\partial}{\partial x} \left[V^2 \delta_2(x) \right] + \delta_1(x) \, V\frac{\partial V}{\partial x} = \frac{\tau_w}{\rho} , \tag{8.32}$$

where $\delta_1(x)$ is the *displacement thickness* and $\delta_2(x)$ is the *momentum thickness*. Equation (8.32) is known as the *momentum-integral equation* or *von Karman's integral equation*.

The displacement thickness, $\delta_1(x)$, is associated with the reduction in the mass flow rate per unit depth in the boundary layer, as a result of the velocity slowdown $(V - u_x)$, i.e.,

$$\dot{m}_r = \rho \int_0^\delta (V - u_x) \, dy .$$

When the rate of "mass loss", $\dot{m}_r$, is expressed in terms of an equivalent thickness δ_1 of uniform flow with the same mass flow rate, i.e., $\dot{m}_r = \rho\, V\, \delta_1$, we have

$$\dot{m} = \rho\, \delta_1(x)\, V = \rho \int_0^\zeta (V - u_x)\, dy\;,$$

which leads to the definition of $\delta_1(x)$ as

$$\delta_1(x) = \int_0^\zeta \left(1 - \frac{u_x}{V}\right) dy\;. \tag{8.33}$$

The *momentum thickness*, $\delta_2(x)$, is related to the deficiency in the momentum per unit depth, $\dot{J}_D$, associated with the slowdown of the velocity within the boundary layer,

$$\dot{J}_D = \rho \int_0^\delta u\,(V - u)\; dy\;.$$

This momentum deficiency produces a net force along the direction of flow. Thus, $\delta_2(x)$ is defined as the thickness of uniform flow that carries the same momentum, i.e., $\dot{J}_D = \rho\, \delta_2\, V^2$. Therefore,

$$\rho\, \delta_2(x)\, V^2 = \rho \int_0^\zeta u_x\,(V - u_x)\, dy$$

or

$$\delta_2(x) = \int_0^\zeta \left[\frac{u_x}{V}\left(1 - \frac{u_x}{V}\right)\right] dy\;. \tag{8.34}$$

Since the boundary layer thickness increases in the direction of flow, both δ_1 and δ_2 are functions of x. The approximate integral momentum equation (8.32) is derived without any assumption concerning the nature of the flow. Therefore, it applies to both laminar and turbulent flows.

Example 8.3.1. Von Karman's boundary layer solution

Consider the velocity profile

$$\frac{u_x}{V} = u_x^* = a\xi^3 + b\xi^2 + c\xi + d \quad \text{with} \quad \xi = \frac{y}{\delta}\;. \tag{8.35}$$

At $\quad \xi = 0\,,\quad u_x^* = 0\,;\quad$ therefore, $d = 0\,.$

At $\quad \xi = 1\,,\quad \dfrac{\partial u_x^*}{\partial \xi} = 0\,;\quad$ therefore, $3a + 2b + c = 0\,.$

At $\quad \xi = 1\,,\quad u_x^* = 1\,;\quad$ therefore, $a + b + c = 1\,.$

An additional boundary condition is obtained at $y=0$ by satisfying the momentum equation there

$$\left[u_x^* \frac{\partial u_x^*}{\partial x^*} + u_y^* \frac{\partial u_x^*}{\partial y^*}\right]_{y^*=0} = -\left.\frac{dp^*}{dx^*}\right|_{y^*=0} + \left.\frac{\partial^2 u_x^*}{\partial y^{*2}}\right|_{y^*=0}. \tag{8.36}$$

Since $u_x^*=u_y^*=0$ and $dp^*/dx^*=0$ (for flow past a flat plate),

$$\left.\frac{\partial^2 u_x^*}{\partial y^{*2}}\right|_{y^*=0} = 0. \tag{8.37}$$

Therefore,

$$\text{At} \quad \xi = 0, \quad \frac{\partial^2 u_x^*}{\partial \xi^2} = 0 \quad \text{and} \quad b = 0. \tag{8.38}$$

The four boundary conditions determine the unknown coefficients as

$$b = d = 0, \quad c = 3/2, \quad \text{and} \quad a = -1/2.$$

Therefore, the admissible velocity profile is

$$u_x^* = \frac{u_x}{V} = \frac{3}{2}\xi - \frac{1}{2}\xi^3. \tag{8.39}$$

Equation (8.39) is substituted in von Karman's equation, Eq. (8.32), to yield

$$\tau_w = \rho V^2 \frac{d\delta}{dx} \int_0^1 \left(1 - \frac{u_x}{V}\right) \frac{u_x}{V} \, d\xi = 0.139 \, \rho V^2 \frac{d\delta}{dx}. \tag{8.40}$$

Also,

$$\tau_w = \eta \frac{du_x}{dy} = \eta \frac{V}{\delta} \left.\frac{du_x^*}{d\xi}\right|_{\xi=0} = \frac{3}{2}\eta \frac{V}{\delta}. \tag{8.41}$$

Equation (8.40) then becomes

$$\delta \, d\delta = 8.791 \frac{\eta}{\rho V} dx \quad \text{with} \quad \delta(x = 0) = 0, \tag{8.42}$$

which is solved for the boundary layer thickness,

$$\delta(x) = 4.646 \sqrt{\frac{\nu x}{V}}. \tag{8.43}$$

The exact value given by Eq. (8.25) is, therefore, represented reasonably well by the approximate expression shown in Eq. (8.43) with a difference of less than 10%. A similar analysis can be repeated by means of exponential or other similar functional forms for the velocity distribution. The error between the exact and the approximate solutions depends on the approximating velocity profile.   □

Example 8.3.2

Another approximate qualitative expression for the thickness $\delta(x)$ can be derived by approximating the velocity profile throughout the flow as:

$$u_x(x,y) = V\left(1 - e^{-ay}\right),\tag{8.44}$$

where a is a function of x and accounts for the growth of the boundary layer. In this case, the match with the potential flow $u_x = V$ occurs at a large distance away from the boundary layer, i.e., as $y \to \infty$. For this velocity profile,

$$\delta_2(x) = \int_0^\infty \left[\frac{u_x}{V}\left(1 - \frac{u_x}{V}\right)\right] dy = \int_0^\infty \exp^{-ay}\left(1 - \exp^{-ay}\right) dy = \frac{1}{2a}.\tag{8.45}$$

The wall shear stress is given by

$$\tau_w = \eta \left.\frac{\partial u_x}{\partial x}\right|_{y=0} = \eta a V.\tag{8.46}$$

Substitution of Eqs. (8.45) and (8.46) into Eq. (8.32) for $dp/dx = 0$ yields

$$\frac{da}{a^3} = -\frac{2\eta}{\rho V}\,dx,\tag{8.47}$$

which is integrated to

$$a = \left(\frac{V}{4\nu x}\right)^{1/2}.\tag{8.48}$$

Therefore, the velocity profile becomes

$$u_x(x,y) = V\left[1 - \exp\left(-y\sqrt{\frac{V}{4\nu x}}\right)\right].\tag{8.49}$$

In practice, the boundary layer thickness is taken as the distance where $u_x = 0.99V$. Therefore,

$$\delta(x) = \ln 100\,\sqrt{\frac{\nu x}{V}} = 4.54\sqrt{\frac{4\nu x}{V}}.\tag{8.50}$$

The velocity profile can also be written as

$$u_x(x,y) = V \left[1 - \exp\left(-4.54\frac{y}{\delta} \right) \right] . \tag{8.51}$$

The vorticity distribution is

$$w(x,y) = \frac{\partial u_x}{\partial y} = 4.54 \frac{V}{\delta} \exp\left(-4.54 \frac{y}{\delta} \right) . \tag{8.52}$$

At the edge of the boundary layer, i.e., at $y=\delta$,

$$w(x,\delta) = 4.54 \frac{V}{\delta} e^{-4.54} = 0.0454 \frac{V}{\delta} . \tag{8.53}$$

Therefore, the vorticity decays from its boundary value,

$$w(x,0) = 4.54 \frac{V}{\delta} , \tag{8.54}$$

to the minimum vorticity of the boundary layer,

$$w(x,\delta) = 0.0454 \frac{V}{\delta} , \tag{8.55}$$

and, further, quite sharply to

$$w(x,2\delta) \simeq 10^{-3} \frac{V}{\delta} \tag{8.56}$$

at a distance just twice the boundary layer thickness. □

Example 8.3.3. Blasius' and Sakiades' boundary layers

Consider the two flow configurations shown in Fig. 8.5, studied by Blasius [2] and Sakiades [4], respectively. In light of the previous discussions, the boundary conditions for the two flows are:

Blasius flow	Sakiades flow
At $y=0$, $u_x = u_y = 0$	At $y=0$, $u_x = V$, $u_y = 0$
As $y \to \infty$, $u_x \to V$	As $y \to \infty$, $u_x \to 0$
As $\left.\begin{array}{l} x \to 0 \\ y \neq 0 \end{array}\right\} u_x \to V$	As $\left.\begin{array}{l} x \to 0 \\ y \neq 0 \end{array}\right\} u_x \to 0$

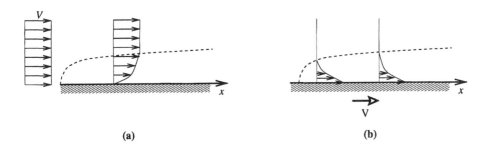

Figure 8.5. *(a) Blasius' boundary layers arise when liquid streams overtake stationary bodies; (b) Sakiades' boundary layers arise when bodies travel in stationary liquids.*

Although the boundary conditions in the two flows are different, the boundary layer equations with no streamwise pressure gradient apply to both cases:

$$u_x \frac{\partial u_x}{\partial x} + u_y \frac{\partial u_x}{\partial y} = \nu \frac{\partial^2 u_x}{\partial y^2} \,, \tag{8.57}$$

and

$$\frac{\partial u_x}{\partial x} + \frac{\partial u_y}{\partial y} = 0 \,. \tag{8.58}$$

By means of the stream function, ψ,

$$u_x = \frac{\partial \psi}{\partial y} \quad \text{and} \quad u_y = -\frac{\partial \psi}{\partial x} \,, \tag{8.59}$$

and the system of Eqs. (8.57) and (8.58) reduces to

$$\frac{\partial \psi}{\partial y} \frac{\partial^2 \psi}{\partial x \partial y} - \frac{\partial \psi}{\partial x} \frac{\partial^2 \psi}{\partial y^2} = \nu \frac{\partial^2 \psi}{\partial y^3} \,. \tag{8.60}$$

In terms of the stream function, the boundary conditions are:

Blasius flow

At $y = 0$, $\dfrac{\partial \psi}{\partial y} = \dfrac{\partial \psi}{\partial x} = 0$

As $y \to \infty$, $\psi \to V y$

As $\left. \begin{array}{c} x \to 0 \\ y \neq 0 \end{array} \right\}$ $\psi \to V y$

Sakiades flow

At $y = 0$, $\psi = V y$, $\dfrac{\partial \psi}{\partial x} = 0$

As $y \to \infty$, $\psi \to$ constant $(0, \text{say})$

As $\left. \begin{array}{c} x \to 0 \\ y \neq 0 \end{array} \right\}$ $\psi \to$ constant $(0, \text{say})$

Since there is no length scale in this problem, the solution must be independent of the unit chosen for length. By dimensional analysis, it follows that

$$\frac{\psi}{\sqrt{2\nu V x}} = f\left(y\sqrt{\frac{V}{2\nu x}}\right).$$

(8.61)

Note that a factor of 2 is included for convenience. The most important aspects of the two boundary layer solutions are:

Velocity and boundary layer thickness

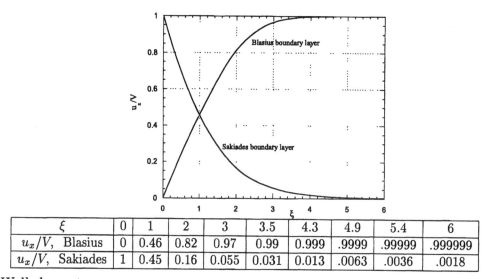

ξ	0	1	2	3	3.5	4.3	4.9	5.4	6
u_x/V, Blasius	0	0.46	0.82	0.97	0.99	0.999	.9999	.99999	.999999
u_x/V, Sakiades	1	0.45	0.16	0.055	0.031	0.013	.0063	.0036	.0018

Wall shear stress
By means of

$$\tau_w = \eta \left.\frac{\partial u_x}{\partial y}\right|_{y=0} = \eta\sqrt{\frac{V^3}{2\nu x}} f''(0),$$

we find that

$$\tau_w^B = 0.332(\rho V^2)\sqrt{\frac{\nu}{Vx}} \qquad \text{and} \qquad \tau_w^S = -0.444\rho V^2\sqrt{\frac{\nu}{Vx}},$$

(8.62)

where superscripts B and S denote the Blasius and the Sakiades solutions, respectively.

Drag on a finite length
The drag on a finite length, ℓ, per unit width is given by

$$F_D = \int_0^\ell \tau_w \, dx,$$

which gives

$$F_D^B = 0.664 \sqrt{V^3 \rho \eta \ell} \qquad \text{and} \qquad F_D^S = -0.888 \sqrt{V^3 \rho \eta \ell} \, . \qquad (8.63)$$

Traverse velocity
This is given by

$$u_y = \sqrt{\frac{\nu V}{2x}} \left[\xi f'(\xi) - f(\xi) \right] ,$$

which yields

$$\frac{u_y^B}{V} \to 0.837 \sqrt{\frac{\nu}{V x}} \qquad \text{and} \qquad \frac{u_y^S}{V} = -0.808 \sqrt{\frac{\nu}{V x}} \, . \qquad (8.64)$$

Therefore, the laminar flow drag force on a flat surface moving through still fluid with velocity V is about 34% greater than the drag force on the same surface due to fluid flowing past it with velocity V. In the former case there is a drift velocity *towards* the plate because the fluid is *accelerated*, while in the latter case this situation is reversed. □

8.4 Boundary Layers within Accelerating Potential Flow

The boundary layer thickness is defined by the competition between convection, which tends to confine vorticity close to its generating source, and diffusion that drives to vorticity uniformity away from the solid surface. Besides these effects, due to the dominant velocity component, vorticity penetration is enhanced by a small vertical velocity component away from the boundary, as in the case of Blasius' boundary layer, and is inhibited by a small vertical velocity component towards the boundary, as in the case of Sakiades' boundary layer. At the outer part of the boundary layer, the vertical velocity component is related to the potential velocity profile, $V(x)$, by the continuity equation,

$$u_y = - \int_0^y \frac{\partial u_x}{\partial x} dy \simeq - \int_0^y \frac{\partial V(x)}{\partial x} dy \, . \qquad (8.65)$$

Equation (8.65) suggests that an accelerating potential flow of $\partial V / \partial x > 0$ induces a small vertical velocity component towards the boundary, which in turn confines vorticity from penetrating away and, therefore, reduces the thickness of the boundary layer. The opposite is true for a decelerating potential flow, of $\partial V / \partial x < 0$, that induces a small vertical velocity component away from the boundary, and therefore increases the boundary layer thickness.

The Bernoulli equation along a potential streamline of arc length s takes the form

$$V\frac{\partial V}{\partial s} = -\frac{1}{\rho}\frac{\partial p}{\partial s}. \tag{8.66}$$

Thus, an accelerating potential velocity results in a negative pressure gradient which, according to the boundary layer momentum equation, tends to diminish the variation of the velocity across the boundary layer; therefore, it tends to decrease the boundary layer thickness. The opposite is true for a decelerating potential velocity.

Falkner and Skan [5] extended Blasius' boundary layer analysis to cases with an external potential velocity field of the type

$$V = c\,x^m, \tag{8.67}$$

where c and m are positive constants. The stream function,

$$\psi = (\nu V x)^{1/2} f(\eta), \tag{8.68}$$

where η is a similarity variable defined by

$$\eta = (V/\nu x)^{1/2} y, \tag{8.69}$$

transforms the momentum equation to the ordinary differential equation,

$$m f'^2 - \frac{1}{2}(m+1)f f'' = m + f'''. \tag{8.70}$$

Note that the above equation reduces to Eq. (8.22) in the limit of $m=0$. In stagnation flow, where a jet impinges on a vertical wall ($m=1$), Eq. (8.70) becomes

$$f'^2 - f f'' = 1 + f'''. \tag{8.71}$$

The corresponding boundary conditions are

$$f(0) = f'(0) = 0, \tag{8.72}$$

and

$$f'(\eta \to \infty) = 1. \tag{8.73}$$

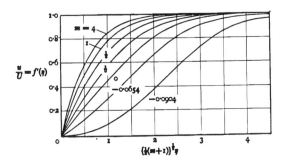

Figure 8.6.   *Similarity distributions of velocity across the boundary layer in an external potential field of velocity, $V = cx^m$. [Taken from Reference 6, by permission.]*

The numerical solution of Eqs. (8.71) to (8.73) is shown in Fig. 8.6, where the velocity distributions at different values of $\eta(x, y) = (V/\nu x)^{1/2} y$ are given. The boundary layer thickness is defined so that

$$\frac{u_x}{V}\left(\eta = \delta^{(m)}\sqrt{\frac{V}{\nu x}}\right) = 0.99 \, . \tag{8.74}$$

Other useful quantities such as the boundary shear stress,

$$\tau_w^{(m)} = \eta \left.\frac{\partial u_x^{(m)}}{\partial y}\right|_{y=0} = \rho \left(\frac{\nu V^3}{x}\right)^{1/2} \left(f_0^{(m)}\right)'' \, , \tag{8.75}$$

the vertical velocity component, and the total drag force are computed accordingly.

Example 8.4.1. Stagnation flow boundary layer

Consider the stagnation point flow shown in Fig. 8.7. The free stream velocity described by the stream function

$$\psi_p = k \, xy \, , \tag{8.76}$$

with velocity components

$$U = kx \quad V = -ky \, , \tag{8.77}$$

impinges normal to the plate and forms a boundary layer of thickness $\delta(x)$. Since the potential velocity component, V, depends only on y and the other component, U, depends on x, the following form of the stream function is suggested (within the boundary layer)

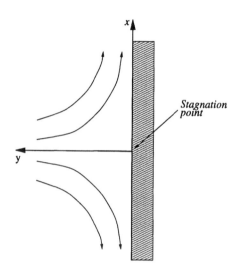

Figure 8.7. *Stagnation point flow.*

$$\psi = x f(y) . \tag{8.78}$$

The individual velocity components are

$$u_x = x f'(y) \quad \text{and} \quad u_y = -f(y) , \tag{8.79}$$

whereas the vorticity is given by

$$\omega = \frac{\partial u_y}{\partial x} - \frac{\partial u_x}{\partial y} = -x f''(y) . \tag{8.80}$$

Substituting these expressions in the boundary layer vorticity equation, we get

$$u_x \frac{\partial \omega}{\partial x} + u_y \frac{\partial \omega}{\partial y} = \nu \left(\frac{\partial^2 \omega}{\partial x^2} + \frac{\partial^2 \omega}{\partial y^2} \right) , \tag{8.81}$$

which leads to the ordinary differential equation,

$$-f'f'' + f f''' + \nu f'''' = 0 , \tag{8.82}$$

with boundary conditions,

$$f(y = 0) = f'(y = 0) = 0 , \tag{8.83}$$

and

$$f(y \to \infty) = k\, y . \tag{8.84}$$

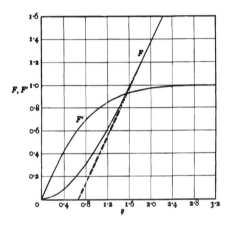

Figure 8.8. *Boundary layer results according to Eqs. (8.79) to (8.88). [Taken from Reference 6, by permission.]*

By means of

$$y = \left(\frac{\nu}{k}\right)^{1/2} \eta \quad \text{and} \quad f(y) = (\nu k)^{1/2} F(\eta) , \tag{8.85}$$

Eqs. (8.82) to (8.84) are made dimensionless, as follows:

$$F'^2 - FF'' - F''' = 1 , \tag{8.86}$$
$$F(\eta = 0) = F'(\eta = 0) = 0 , \tag{8.87}$$
$$F(\eta \to \infty) = \eta . \tag{8.88}$$

Hiemenz solved these equations numerically [7]. The most important of his calculations are shown in Fig. 8.8. The resulting boundary layer thickness is

$$\delta(x) = 2.4\sqrt{\frac{\nu}{k}} , \tag{8.89}$$

which is independent of x. This is a consequence of the fact that the controlling velocity component of the potential flow, V, is uniform over x whereas the x-component varies linearly with x; hence, vorticity is convected downstream parallel to the wall within the boundary layer.

The corresponding axisymmetric case, that may occur when bodies of revolution move parallel to their axes of symmetry and form a stagnation point at the leading edge, was treated by Homann [8]. Both the two-dimensional and the axisymmetric flows are special cases of Howarth's analysis of the general stagnation flow [9].   □

8.5   Flow over Nonslender Planar Bodies

In flow past nonslender bodies, Eqs. (8.15) to (8.17) still apply. However, Blasius' exact solution does not apply because

$$\frac{dp}{dx} = V(x)\frac{dV(x)}{dx} \neq 0 \,, \tag{8.90}$$

due to the fact that, outside the boundary layer, the velocity $V(x)$ of the potential flow is deflected by the two-dimensional body and made different from the approaching stream velocity, V. In this case, the solution is found iteratively; initially dp/dx is assumed zero, u_x is defined, and δ is calculated from the corresponding von Karman equation which, for nonslender bodies, takes the form

$$\frac{d}{dx}\left[V^2(x)\int_0^\delta \frac{u_x}{V(x)}\left(1 - \frac{u_x}{V(x)}\right)dy\right] + V(x)\frac{dV(x)}{dx}\int_0^\delta \left(1 - \frac{u_x}{V(x)}\right)dy = \frac{\tau_w}{\rho}\,, \tag{8.91}$$

where

$$\frac{dp}{dx} = \rho\,V(x)\,\frac{dV(x)}{dx}\,. \tag{8.92}$$

Then dp/dx is calculated and the iteration is repeated until consecutive values of dp/dx do not differ much. In cases where the potential velocity is known (e.g., flow around a wedge-like, two-dimensional body), dp/dx is substituted directly in Eq. (8.91) and the iterative procedure is avoided.

8.6   Rotational Boundary Layers

A solid body moving with relative velocity V with respect to its surrounding liquid that cannot slip generates an average vorticity, ω, of the order of $\omega \simeq V/\delta(x)$

that penetrates to a distance $\delta(x)$ under the combined action of convection and diffusion. Similarly, a disk in relative rotation Ω with respect to its surrounding liquid generates vorticity ω equal to twice its angular speed Ω that spreads in the vertical direction. Since the vorticity gradient in the azimuthal direction is zero, vorticity also spreads in the radial directions by convection and diffusion. At very low rotational speeds or Reynolds numbers, where centrifugal effects are negligible and Coriolis effects are dominant, a strictly azimuthal motion near the disk may be conserved. However, at high rotational speeds and Reynolds numbers, where the ratio of centrifugal/Coriolis effects is reversed, a strictly circular motion cannot be maintained and the fluid near the disk spirals outwards. Under these high Reynolds numbers, an axial motion towards the disk is induced by virtue of mass conservation. This vertical velocity component confines the vorticity at the disk's surface within a finite distance and defines a rotational boundary layer. Briefly, the rotating disk operates as a centrifugal fan that receives fluid vertically and delivers it nearly radially.

For this class of problems, von Karman was the first to suggest that a solution of the form, $u_r = rf(z)$, $u_\theta = rg(z)$ and $u_z = h(z)$ is possible [2]. The boundary conditions are

$$u_r(z = 0) = u_z(z = 0) = 0 \, ; \tag{8.93}$$

$$u_\theta(z = 0) = \Omega r \, ; \tag{8.94}$$

$$u_r(z \to \infty) = u_z(z \to \infty) = 0 \, . \tag{8.95}$$

The z-momentum equation becomes

$$\frac{\partial p}{\partial z} = \eta \frac{du_z}{dz} - \frac{\rho}{2} u_z^2 + c \, , \tag{8.96}$$

which suggests that

$$\frac{\partial p}{\partial r} = \frac{\partial p}{\partial \theta} = 0 \, . \tag{8.97}$$

The remaining momentum equations, in the r- and θ-directions, become

$$\frac{u_r^2}{r^2} + u_z \frac{d}{dz}\left(\frac{u_r}{r}\right) - \frac{u_\theta^2}{r^2} = \nu \frac{d^2}{dz^2}\left(\frac{u_r}{r}\right) \, , \tag{8.98}$$

and

$$\frac{2u_r u_\theta}{r^2} + u_z \frac{d}{dz}\left(\frac{u_\theta}{r}\right) = \nu \frac{d^2}{dz^2}\left(\frac{u_\theta}{r}\right) \, , \tag{8.99}$$

respectively. Finally, the continuity equation simplifies to

$$\frac{2u_r}{r} + \frac{du_z}{dz} = 0 . \tag{8.100}$$

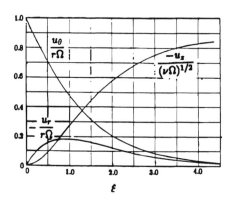

Figure 8.9. *Velocity components, according to Eqs. (8.102) and (8.107), over a fast rotating disk. [Taken from Reference 10, by permission.]*

By introducing the transformation

$$z = \left(\frac{\nu}{\Omega}\right)^{1/2} \xi , \quad \frac{u_\theta}{r} = \Omega \, g(\xi) \quad \text{and} \quad u_z = (\nu\Omega)^{1/2} \, h(\xi) , \tag{8.101}$$

the continuity equation is transformed to

$$\frac{u_r}{r} = \frac{\Omega}{2} \, h'(\xi) . \tag{8.102}$$

Equations (8.98) and (8.99) become

$$\frac{1}{4}h'^2 - \frac{1}{2}hh'' - g^2 = -\frac{1}{2}h''' \tag{8.103}$$

and

$$-gh' + g'h = g'' , \tag{8.104}$$

subject to

$$h(\xi = 0) = h'(\xi) = 0 ; \tag{8.105}$$

$$g(\xi = 0) = 1 ; \tag{8.106}$$

$$h'(\xi \to \infty) = g(\xi \to \infty) = 0 . \tag{8.107}$$

Cochran solved the above equations numerically [9]. His main results are shown in Fig. 8.9. By defining the thickness of the boundary layer as the distance ξ where $u_\theta = 0.01\Omega r$, it turns out that the thickness is uniform,

$$\delta = 5.4 \left(\frac{\nu}{\Omega}\right)^{1/2} = 5.4 \left(\frac{\nu r}{V}\right)^{1/2}, \tag{8.108}$$

which is similar to a stagnation-type boundary layer thickness.

8.7 Problems

8.1. Water approaches an infinitely long and thin plate with uniform velocity. (a) Determine the velocity distribution u_x in the boundary layer given that

$$u_x(x,y) = a(x)y^2 + b(x)y + c(x).$$

(b) What is the flux of mass (per unit length of plate) across the boundary layer?
(c) Calculate the magnitude and the direction of the force needed to keep the plate in place.

8.2. In applying von Karman's momentum balance method to boundary layer flows, one is not restricted to piecewise differentiable approximations of the form

$$u_x^* = \begin{cases} f(\eta), & 0 \leq \eta \leq 1 \\ V_\infty^*, & \eta > 1 \end{cases}$$

for the velocity distribution. The exact solution for the case of flow near a wall suddenly set in motion (derived in Chapter 6) suggests using the continuously differentiable velocity distribution

$$u_x^* = erf(\eta), \quad \eta \geq 0$$

for the boundary layer flow past a flat plate. This distribution satisfies the following conditions:

$$u_x^* = 0 \quad \text{at} \quad \eta = 0;$$
$$u_x^* \to 1 \quad \text{as} \quad \eta \to \infty;$$
$$\left(\frac{\partial^2 u_x^*}{\partial y^{*2}}\right)_{\eta=0} = \frac{dp^*}{dx^*} = 0.$$

Derive the boundary layer thickness, the displacement thickness, the momentum thickness, the shear stress at the wall, and the drag force over a length L. Compare with the results based on the traditional piecewise differentiable distributions.

8.3. Derive a formula for estimating how far downstream of a smooth, rounded inlet the parabolic velocity profile of plane Poiseuille flow becomes fully developed. Assume that the core flow in the entry region is rectilinear and irrotational, and that the velocity distribution in the boundary layers is self-similar:

$$\frac{u_x(x, y)}{U(x)} = f\left(\frac{y}{\delta(x)}\right).$$

Use von Karman's integral momentum equation. Find the corresponding equation for pressure drop in the entry length.

8.4. Consider the *solid jet* flow induced by a continuous solid sheet emerging at constant velocity from a slit into fluid at rest.

(a) Justify the boundary-layer approximation for the flow.

(b) Show that the boundary conditions differ from those for flow past a flat plate, although the Blasius equation for the stream function applies.

(c) Employ von Karman's momentum equation to obtain approximate solutions for the local wall shear stress, the total drag on the two surfaces of the sheet, the displacement thickness, and the momentum thickness, using for the velocity profile

    i. a fourth-degree polynomial in $\eta \equiv y/a \sqrt{V/\nu x}$, and

    ii. the complementary error function, erfc $\eta = 1-$ erf η,

where a is an arbitrary constant. Compare the results with those for flow past a flat plate. (According to the exact solution for the solid jet, the dimensionless local stress is 0.444 as compared with 0.332 for flow past a flat plate.)

8.5. *Laminar, incompressible, two-dimensional jet.*

(a) Justify the boundary-layer approximation for the flow caused by a fine sheet-jet emerging from a slit into fluid at rest.

(b) Noting that the total momentum flux in the x-direction must be independent of distance x from the slit, i.e.,

$$\int_{-\infty}^{\infty} \rho u_x^2 \, dy = J = \text{const.},$$

establish that the velocity can depend on position only through the dimensionless combination

$$\eta = \frac{y}{\alpha} \left(\frac{J}{\rho \nu^2 x^2} \right)^{1/3}$$

where α is an arbitrary constant.

(c) Introduce a dimensionless stream function $f(\eta)$ and choose the constant α in such a way that the boundary-layer equation reduces to the ordinary differential equation

$$f''' + 2ff'' + 2f'^2 = 0.$$

(d) Show that the solution of this equation is $\alpha \tanh \alpha \eta$ and evaluate the constant α.

(e) Calculate u_x and u_y and the total volumentric flow rate, Q, entrained (as a function of x).

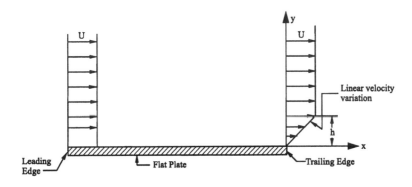

Figure 8.10. *Schematic of the flow in Problem 8.6.*

8.6. Consider a two-dimensional boundary layer flow of water over one side of a flat plate. At the leading edge, the velocity is uniform and equal to U. Downstream at the trailing edge, the velocity profile is as shown in Fig. 8.10.

(a) Find the shear force of the fluid on the plate by using an overall control volume approach.

(b) Sketch the velocity development from the leading to the trailing edge.

(c) Find an approximation for the normal velocity profile.

(d) Split the entire field into a boundary layer and a potential flow region (graphically).

(e) Construct qualitative plots of the vorticity as a function of x at three distances from the plate: $y=0$, $y=h/2$, and $y=2h$.

8.7. For the general case of two-dimensional planar flow, Prandtl's boundary layer equations are [1]

$$u_x \frac{\partial u_x}{\partial x} + u_y \frac{\partial u_x}{\partial y} = U \frac{dU}{dx} + \nu \frac{\partial^2 u_x}{\partial y^2} ,$$

and

$$\frac{\partial u_x}{\partial x} + \frac{\partial u_y}{\partial y} = 0 .$$

Given the geometry of the problem and $U_0(x)$, the solution of these equations yields $u_x(x, y)$ and $u_y(x, y)$ inside a boundary layer. Consider the flow normal to the axis of a cylinder of arbitrary cross section.

(a) Define x and y for this geometry. (Hint: consider flow along a flat plate.)

(b) What does $U(x)$ represent?

(c) What does $U \, dU/dx$ represent physically?

(d) What are the appropriate boundary conditions?

(e) Does this solution correctly predict the drag force on the cylinder? Explain your answer as quantitatively as possible.

8.8. *Boundary layer over wedge.* Derive the governing equations and the von Karman's-type approximation for boundary layer over a 30° wedge of a liquid stream of density ρ and viscosity η, approaching at velocity V.

8.9. *Boundary layer along conical body.* Repeat Problem 8.8 for the corresponding axisymmetric case when a liquid stream approaches at velocity V and overtakes a cone of angle $\phi=30^o$ placed with its leading sharp end facing the stream.

8.8   References

1. H. Schlichting, *Boundary Layer Theory*, McGraw-Hill, New York, 1950.

2. H. Blasius, "Grenzschlichten in Flüssigkeiten mit Kleiner Reibung," *Z. Math. Phys.* **56**, 1 (1908).

3. T. von Karman, "Über laminare und turbulente Reibung," *Z. Angew. Math. Mech.* **1**, 233 (1921).

4. B.C. Sakiades, "Fluid particle mechanics," in *Perry's Chemical Engineers' Handbook*, McGraw-Hill, New York, 1984.

5. V.M. Falkner and S.W. Skan, *Aero. Res. Coun.*, Rep. and Mem. **1314**, 1 (1930).

6. G.K. Batchelor, *An Introduction to Fluid Dynamics*, Cambridge University Press, Cambridge, 1979.

7. K. Hiemenz, *Die grenzschicht an einem in den gleichformigen flüssigkeitsstrom eingetauchten geraden krieszylinder*, Ph.D. Thesis, University of Göttingen, 1911.

8. F. Homann, "Der einfluss grösser zähigkeit bei der strömung um den Zylinder und um die kugel," *Z. Angew. Math. Mech.* **16**, 153 (1936).

9. L. Howarth, "On the calculation of steady flow in the boundary layer near the surface of a cylinder in a stream," *Aeron. Res. Council-Britain*, 1962.

10. W.G. Cochran, "The flow due to a rotating disc," *Proc. Cambr. Phil. Soc.* **30**, 365 (1934).

Chapter 9

ALMOST UNIDIRECTIONAL FLOWS

In this chapter, two different classes of flows are examined in the limit of almost rectilinear flow domains by using perturbation analysis of the full Navier–Stokes equations. These are:

(a) *Lubrication flows:* these are confined or free surface flows with parabolic velocity profiles under almost rectilinear boundaries or free surfaces. Typical examples are flow in converging and diverging channels, flow in pipes, and flow of thin films on substrates.

(b) *Stretching flows:* these are free surface flows of plug-like velocity profile under almost rectilinear free surfaces, such as jet flows.

Prototypes of these flows, such as flows in nonrectilinear domains, development of wet films under surface tension, and spinning/casting/blowing of polymeric fibers/sheets/films, are depicted in Fig. 9.1.

9.1 Lubrication Flows

Lubrication flows are most applicable to the processing of materials in liquid form, such as polymers, metals, composites, and others. One-dimensional approximations can be derived from simplified mass and momentum balances by means of control volume principles, or by simplifying the general equations of change. This leads to the celebrated *Reynolds equation* [1],

$$F(h, p, St, Ca) = 0 , \tag{9.1}$$

where $h(x)$ is the thickness of the narrow channel or of the thin film; p is the pressure; St is the Stokes number, $St \equiv \rho g L^2 / \eta V$; and Ca is the capillary number,

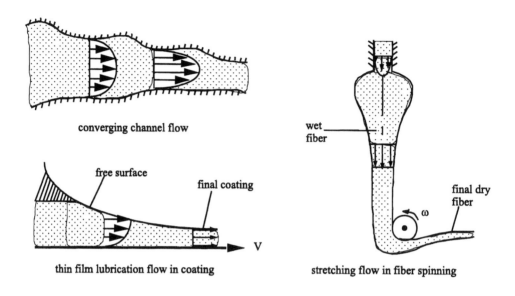

converging channel flow

free surface

final coating

wet fiber

final dry fiber

ω

thin film lubrication flow in coating

stretching flow in fiber spinning

Figure 9.1. *Confined and film lubrication flows and stretching flows.*

$Ca \equiv \eta V/\sigma$, that appears due to surface tension along an interface. Equation (9.1) can be solved:

(a) for the pressure distribution and other relevant quantities, such as load capacity, friction, cavitation, etc., when the thickness $h(x)$ is known. Typical applications are lubrication of solid surfaces in relative motion, such as journal-bearing, piston cylinder, and piston rings of engines [2].

(b) for the thickness $h(x)$, when the pressure is known. Typical examples are formation of thin films and coating applications [3].

9.1.1 Lubrication vs. Rectilinear Flow

The lubrication approximation for flows in nearly rectilinear channels or pipes with nearly parallel walls can be derived intuitively from the complete set of flow equations. Mass conservation requires constant flow rate:

$$\frac{\partial u_x}{\partial x} = 0, \quad u_z = 0, \quad u_x = f(z). \tag{9.2}$$

Conservation of linear momentum in the flow direction requires pressure and viscous force balance in the same direction:

$$\frac{\partial p}{\partial x} = \eta \frac{\partial u_x^2}{\partial z^2}.$$

(9.3)

The pressure gradient, $\partial p/\partial x$, is usually imposed mechanically. For rectilinear channels and steady motion, $\partial p/\partial x$ is constant along the channel, equal to $\Delta p/\Delta L$, where Δp is the pressure difference over a distance ΔL, Fig. 9.2. For constant pressure gradient, the momentum equation predicts linear shear stress and parabolic velocity profile. In these problems, the mechanism of fluid motion is simple; material flows from regions of high pressure to regions of low pressure (Poiseuille-type flow).

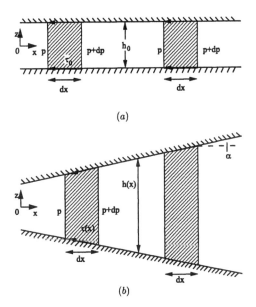

(a)

(b)

Figure 9.2. *Force balance in (a) rectilinear flow, $h_0 dp = 2\tau dx$, and (b) lubrication flow, $h(x)dp(x) = 2\tau(x)dx$.*

When one or both walls are at a slight inclination α relative to each other, the same governing equations are expected to hold. Now, however, they may locally be weak functions of x of order α. Take, for instance, the pressure gradient in lubrication applications where the flow may be accelerating or decelerating in converging or diverging channels, respectively. In such cases, $\partial p/\partial x$ is not constant along the channel. This can be seen in Fig. 9.2(b) where the pressure force needed to move two cones of liquid of the same width, dx, at two different positions along the channel is different. Consequently, both $\partial p/\partial x$ and the velocity are functions of x. Therefore, we have

$$\frac{\partial u_x}{\partial x} + \frac{\partial u_z}{\partial z} = 0 \,, \tag{9.4}$$

and

$$\frac{\partial p(x)}{\partial x} = \eta \frac{\partial^2 u_x}{\partial z^2} \,. \tag{9.5}$$

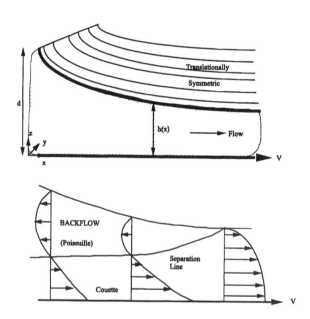

Figure 9.3. *Geometry of one-dimensional lubrication flow. The velocity profiles along the channel are a mixture of Couette and Poiseuille flow.*

Equations (9.3) and (9.5) express conservation of linear momentum for a control volume. They both indicate that, due to negligible convection, there is no accumulation of momentum. Consequently, the forces capable of producing momentum are in equilibrium.

As shown in Fig. 9.2, the forces on a control volume of width dx are the net pressure force $(dp/dx)A(x)$ and the shear stress force $2\tau_{xy}dx$. However, the underlying mechanism in lubrication flows may be more complex than in Poiseuille flow. Consider, for instance, the schematic in Fig. 9.3. Through the action of viscous shear forces, the moving wall on one side sweeps fluid into a narrowing passage. This gives rise to a local velocity profile of Couette-type, $u_x=Vy/h$, with flow rate, $Q=Vh/2$.

Since Q is constant, in order to conserve mass, $h(x)$ is decreasing. The flow then sets up a pressure gradient in order to supply a Poiseuille-type flow component that redistributes the fluid and maintains a constant flow rate.

9.1.2   Derivation of Lubrication Equations

The lubrication equations can be alternatively derived by dimensionless analysis and by order of magnitude comparisons with the full Navier–Stokes equation:

$$\frac{\partial u_x}{\partial x} + \frac{\partial u_z}{\partial z} = 0 \, , \tag{9.6}$$

$$\rho \left(\frac{\partial u_x}{\partial t} + u_x \frac{\partial u_x}{\partial x} + u_z \frac{\partial u_x}{\partial z} \right) = -\frac{\partial p}{\partial x} + \eta \left[\frac{\partial u_x^2}{\partial x^2} + \frac{\partial u_x^2}{\partial z^2} \right] \, , \tag{9.7}$$

$$\rho \left(\frac{\partial u_z}{\partial t} + u_x \frac{\partial u_z}{\partial x} + u_z \frac{\partial u_z}{\partial z} \right) = -\frac{\partial p}{\partial z} + \eta \left[\frac{\partial u_z^2}{\partial x^2} + \frac{\partial u_z^2}{\partial z^2} \right] \, . \tag{9.8}$$

Equations (9.6) to (9.8) are made dimensionless using the following scaling

$$x^* = \frac{x}{L} \, , \quad z^* = \frac{z}{\alpha L} \, , \quad t^* = \frac{tV}{L} \, , \quad h^* = \frac{h}{\alpha L} \, ,$$

$$u_x^* = \frac{u_x}{V} \, , \quad u_z^* = \frac{u_z}{\alpha V} \quad \text{and} \quad p^* = \frac{p}{\eta \frac{V}{\alpha^2 L}} \, ,$$

where α is a small parameter of the same order as the channel slope. The lubrication equation holds in geometries where $a \ll 1$. Upon substitution, the momentum equations yield (with asterisks suppressed hereafter) [4]:

$$\alpha \, Re \left(\frac{\partial u_x}{\partial t} + u_x \frac{\partial u_x}{\partial x} + u_z \frac{\partial u_x}{\partial z} \right) = -\frac{\partial p}{\partial x} + \alpha^2 \frac{\partial^2 u_x}{\partial x^2} + \frac{\partial^2 u_x}{\partial z^2} \, , \tag{9.9}$$

$$\alpha^3 \, Re \left(\frac{\partial u_z}{\partial t} + u_x \frac{\partial u_z}{\partial x} + u_z \frac{\partial u_z}{\partial z} \right) = -\frac{\partial p}{\partial z} + \alpha^3 \frac{\partial^2 u_z}{\partial x^2} + \alpha \frac{\partial^2 u_z}{\partial z^2} \, . \tag{9.10}$$

Since all dimensionless derivative terms in these two equations are of comparable order, the resulting dimensionless lubrication equations, in the limit of $a \approx 0$ or $aRe \approx 0$, are

$$-\frac{\partial p}{\partial x} + \frac{\partial u_x^2}{\partial z^2} = 0 \, , \tag{9.11}$$

and

$$-\frac{\partial p}{\partial z} = 0 \, . \tag{9.12}$$

These equations are similar to those derived intuitively from channel flow, i.e., Eqs. (9.2) and (9.3). Notice that high Reynolds numbers are allowed as far as the product αRe is vanishingly small and the flow remains laminar.

The appropriate boundary conditions are:

(i) at $z=0$, $u_x=V$ (no-slip boundary condition);

(ii) at $z=h$, $u_x=0$ (slit flow, no slip boundary condition), or

(iii) at $z=h$, $\tau_{zx}=0$ (thin film, zero shear stress at free surface).

Under these conditions, the solution to Eq. (9.5) is

$$u_x = -\frac{1}{2\eta}\frac{dp}{dx}(zh - z^2) + V\left(1 - \frac{z}{h}\right) \qquad \text{(slit flow)} , \qquad (9.13)$$

or

$$u_x = -\frac{1}{2\eta}\frac{dp}{dx}(2zh - z^2) + V . \qquad \text{(film flow)} . \qquad (9.14)$$

The volume flux and the pressure distribution in the lubricant layer can be calculated when the total flow rate, Q, and the inclination, α, are known. A lubrication layer will generate a positive pressure and, hence, load capacity normal to the layer only when the layer is arranged so that the relative motion of the two surfaces tends to drag fluid by viscous stresses from the wider to the narrower end of the layer. The load, W, supported by the pressure in slit flow is [5]

$$W = \int_0^L (p - p_0)\, dx = \frac{6\eta V}{\alpha^2}\left[\ln\left(\frac{d}{d - \alpha L}\right) - 2\left(\frac{\alpha L}{2d - \alpha L}\right)\right] , \qquad (9.15)$$

where d is the height of the wide side of the converging channel, and L is the length of the channel. By decelerating the flow and by transmitting momentum, and thus load capacity to the boundary, the slope α is ultimately responsible for the pressure built up.

9.1.3   Reynolds Equation for Lubrication

Mass conservation on an infinitesimal volume yields

$$Q_x - Q_{x+dx} = dx\,\frac{dh}{dt} , \qquad (9.16)$$

which states that the convection of mass in the control volume is used to increase the fluid volume at a rate of $\frac{d}{dt}(dxdh)$, where dx and dh are respectively the width in the flow direction and the height of the volume. By rearranging,

$$-\frac{dQ}{dx} = \frac{dh}{dt} \tag{9.17}$$

which, for confined and film flows, reduces respectively to

$$\frac{d}{dx}\left(-\frac{1}{2\eta}\frac{dp}{dx}\frac{h^3}{6} + \frac{hV}{2}\right) = -\frac{dh}{dt} \tag{9.18}$$

and

$$\frac{d}{dx}\left(-\frac{1}{\eta}\frac{dp}{dx}\frac{h^3}{3} + hV\right) = -\frac{dh}{dt}. \tag{9.19}$$

Equations (9.18) and (9.19) represent the transient lubrication equations. The steady-state form of Eq. (9.18),

$$\frac{d}{dx}\left(-\frac{1}{2\eta}\frac{dp}{dx}\frac{h^3}{6} + \frac{hV}{2}\right) = 0, \tag{9.20}$$

is integrated to

$$-\frac{1}{2\eta}\frac{dp}{dx}\frac{h^3}{6} + \frac{hV}{2} = Q, \tag{9.21}$$

and the pressure is calculated by

$$p(x) = p_0 + 6\eta V \int_0^x \frac{dx}{h^2(x)} - 12\eta Q \int_0^x \frac{dx}{h^3(x)}, \tag{9.22}$$

where

$$Q = \frac{(p_0 - p_L)}{12\eta \int_0^L h^{-3}(x)dx} + \frac{V}{2}\frac{\int_0^L h^{-2}(x)dx}{\int_0^L h^{-3}(x)dx}. \tag{9.23}$$

The load capacity is

$$W = \int_0^L |p_0 - p(x)|\, dx, \tag{9.24}$$

and the shear or friction on the same surface is

$$F = -\int_0^L \tau_{zx}\, dx \, . \tag{9.25}$$

It is easy to show that the load capacity is of order α^{-2} whereas the shear or friction is of order α^{-1}. Thus, the ratio load/friction increases with α^{-1}.

Important applications of the lubrication theory for confined flows are *journal-bearing* [2, 6] and piston ring lubricated systems of engines [7]. Other flows that can be studied by means of the lubrication equations include wire coating [8], roll coating [9], and many polymer applications [10]. Starting from Eq. (9.17), the solution to these problems follows the procedure outlined above. The flow rate is often given by

$$Q = Vh_f \, , \tag{9.26}$$

where V is the speed of production and h_f is the final target thickness. The boundary condition on the pressure at the outlet may vary: $p(L) = 0, dp(L)/dx = 0, p(L) = f_\sigma$, (where f_σ is the force per unit area due to surface tension) and combinations of them [11].

In confined lubrication flows, pressure build up develops due to inclination, α, that may result in backflow of some of the entering liquid. This pressure is used to support loads. In typical thin-film lubrication flows, any pressure build up is primarily due to surface tension. In fact, if surface tension is negligible, then the pressure gradient is zero. For film lubrication flows, the steady-state form of Eq. (9.19),

$$\frac{d}{dx}\left(-\frac{1}{\eta}\frac{dp}{dx}\frac{h^3}{3} + Vh\right) = 0 \, ,$$

is integrated to

$$-\frac{1}{\eta}\frac{dp}{dx}\frac{h^3}{3} + Vh = Q = Vh_f \, , \tag{9.27}$$

where the film thickness, h, is unknown. However, the pressure gradient, dp/dx, can be deduced from surface tension by means of the Young–Laplace equation [12]. By using the lubrication assumption that the slope, dh/dx, must be much smaller than unity, we get

$$-p = \frac{\sigma\dfrac{d^2h}{dx^2}}{[1 + (\dfrac{dh}{dx})^2]^{1/2}} \simeq \sigma\frac{d^2h}{dx^2} \, . \tag{9.28}$$

Here, $h(x)$ is the elevation of the free surface from the x-axis, and σ is the surface tension of the liquid. Then,

$$-\frac{dp}{dx} = \sigma \frac{d^3 h}{dx^3} . \tag{9.29}$$

Substituting Eq. (9.29) in Eq. (9.27), we get

$$\frac{\sigma}{3\eta} h^3 \frac{d^3 h}{dx^3} + hV = Vh_f , \tag{9.30}$$

which is rearranged to

$$h^3 \frac{d^3 h}{dx^3} + 3Ca\,(h - h_f) = 0 . \tag{9.31}$$

Equation (9.31) is nonlinear and cannot be solved analytically.

Some important applications of the thin-film lubrication equations are films falling under surface tension [11], dip and extrusion coating [6], and wetting and liquid spreading [12]. A similar class of problems includes centrifugal spreading, which is common in bell sprayers and in spin coating [3, 12]. A rich collection of lubrication problems from polymer processing can be found in the relevant literature [13, 14], and from recent work on coating [15, 16].

Example 9.1.1. Vertical dip coating [16]

An example of thin lubrication film under gravity, surface tension and viscous drag is found in dip coating, Fig. 9.4. This method of coating is used to cover metals with anticorrosion layers, and to laminate paper and polymer films. The substrate is being withdrawn at speed V from a liquid bath of density ρ, viscosity η, and surface tension σ. The analysis below predicts the final coating thickness as a function of processing conditions (withdrawal speed) and of the physical characteristics of the liquid (ρ, η, and σ).

Solution:
The governing momentum equation, with respect to the Cartesian coordinate system shown in Fig. 9.4, is

$$-\frac{dp}{dz} + \eta \frac{\partial^2 u_z}{\partial y^2} - \rho g = 0 .$$

The boundary conditions are

$$u_z(y = 0) = V \quad \text{and} \quad \tau_{zy}(y = H) = \eta \frac{\partial^2 u_z}{\partial y^2} = 0 .$$

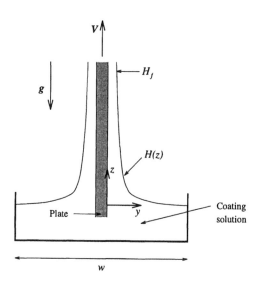

Figure 9.4. *Dip coating: a coated plate is being withdrawn from a coating solution. A final thin film or coating results on the plate under the combined action of gravity, surface tension, and drag by the moving substrate.*

Integration and application of the boundary conditions give the velocity profile

$$u_z = \frac{1}{\eta}\left(\frac{dp}{dz} + \rho g\right)\left(\frac{y^2}{2} - Hy\right) + V \, .$$

The resulting Reynolds equation is

$$-\frac{1}{\eta}\left(\frac{dp}{dz} + \rho g\right)\frac{H^3}{3} + VH = Q = VH_f \, , \tag{9.32}$$

where H_f is the final coating thickness. The pressure gradient,

$$\frac{dp}{dz} = -\sigma\frac{d^3 H}{dz^3} \, ,$$

is substituted in Eq. (9.32) to yield

$$\frac{1}{\eta}\left(\sigma\frac{d^3 H}{dz^3} - \rho g\right)\frac{H^3}{3} + V(H - H_f) = 0 \, ,$$

or

$$\frac{H^3}{3}\frac{d^3H}{dz^3} - \frac{\rho g}{\sigma}\frac{H^3}{3} + \frac{V\eta}{\sigma}(H - H_f) = 0 . \tag{9.33}$$

By identifying the dimensionless capillary and Stokes numbers, defined by

$$Ca \equiv \frac{V\eta}{\sigma} \quad \text{and} \quad St \equiv \frac{\rho g H_f^2}{\eta V} ,$$

respectively, Eq. (9.33) becomes

$$\frac{H^3}{Ca}\frac{d^3H}{dz^3} - St\frac{H^3}{H_f^2} + 3(H - H_f) = 0 . \tag{9.34}$$

The above equation can be solved directly for the following limiting cases:

(i) Negligible surface tension $(Ca \to \infty)$
Equation (9.34) reduces to a third–order algebraic equation,

$$H^3 - \frac{3H_f^2}{St}H + \frac{3H_f^3}{St} = 0 , \tag{9.35}$$

which provides an outer solution to the problem. In the limit of infinite St (i.e., very heavy liquid), we get $H = 0$, i.e., no coating. In the limit of zero St (i.e., horizontal arrangement), $H = H_f$, i.e., plane Couette (plug) flow. For finite values of St, the solution is independent of z and predicts a flat film throughout. The solution to Eq. (9.35) is complemented by the inner solution obtained by the stretching variable $\xi = Ca\, z.$

(ii) Infinitely large surface tension $(Ca \to 0)$
Equation (9.34) reduces to

$$\frac{d^3H}{dz^3} = 0 ,$$

with general solution

$$H(z) = c_1\frac{z^2}{2} + c_2 z + c_3 .$$

Applying the boundary conditions,

$$H(z = 0) = W/2, \quad H(z = L) = H(z = L) = H_f , \quad \text{and} \quad (dH/dz)z = L = 0 ,$$

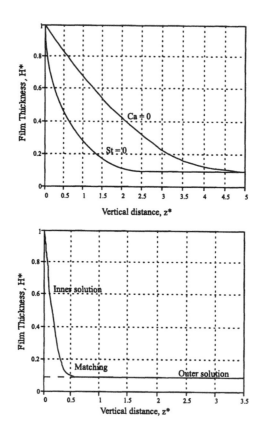

Figure 9.5. *Numerical solution of Eq. (9.34) in the limits of (a) Ca=0 and St=0, and (b) Ca → ∞. The latter case leads to singular perturbation.*

we get a parabolic film thickness,

$$H(z) = \left(\frac{W - 2H_f}{L^2}\right)\left(\frac{z^2}{2} - zL\right) + W/2 \ .$$

(iii) <u>Finite surface tension $(0 < Ca < \infty)$</u>
Equation (9.34) is cast in the form

$$H^3\left(\frac{d^3H}{dz^3} - Ca\,St\,\frac{1}{H_f^2}\right) + 3Ca(H - H_f) = 0 \ , \qquad (9.36)$$

with no apparent analytical solution. For the special case of horizontal coating

$(St = 0)$ with $H_f/W \ll 1$, the transformation

$$H^* = \frac{H}{W} , \quad z^* = \frac{z}{W}$$

reduces Eq. (9.36) to

$$H^{*3}\frac{d^3 H^*}{dz^{*3}} + 3Ca\left(H^* - \frac{H_f}{W}\right) = 0 . \tag{9.37}$$

The above equation predicts that near the inlet, where $H^* \simeq 1$, the film decays at a rate that depends on Ca. Near the other end, the film becomes flat, surface tension becomes unimportant, and the slope is zero. Equation (9.37) can be solved asymptotically by perturbation techniques [17]. Such asymptotic solutions are shown in Fig. 9.5. □

9.1.4   Lubrication Flows in Two Directions

Equations (9.16) to (9.31) are easily generalized to two-dimensional lubrication flows, such as the free-surface and confined flows shown in Figs. 9.6 and 9.7, respectively. The two-dimensional lubrication equations for flow along the xy-plane vertical to the z-direction are

$$-\frac{\partial p}{\partial x} + \eta\frac{\partial^2 u_x}{\partial z^2} = 0 , \tag{9.38}$$

$$-\frac{\partial p}{\partial y} + \eta\frac{\partial^2 u_y}{\partial z^2} = 0 , \tag{9.39}$$

and

$$\frac{\partial p}{\partial z} = 0 . \tag{9.40}$$

The velocity profiles are:

$$u_x = \frac{1}{2\eta}\frac{\partial p}{\partial x}z^2 + b_1 z + c_1 , \tag{9.41}$$

$$u_y = \frac{1}{2\eta}\frac{\partial p}{\partial y}z^2 + b_2 z + c_2 , \tag{9.42}$$

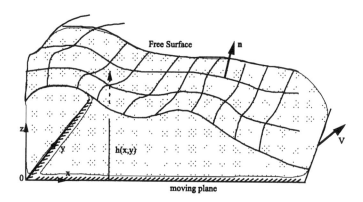

Figure 9.6. *Two-dimensional, thin-film lubrication flow.*

where b_i and c_i are constants depending on the inclination α or the thickness $h(x, y)$. These are determined from appropriate boundary conditions. The resulting dimensionless Reynolds equation is now a partial differential equation,

$$\nabla_{II} \left(-\frac{h^3}{3} \frac{1}{Ca} \nabla_{II} h + h \right) = -\frac{\partial h}{\partial t} , \qquad (9.43)$$

where

$$\nabla_{II}(\cdot) \equiv \mathbf{i} \frac{\partial(\cdot)}{\partial x} + \mathbf{j} \frac{\partial(\cdot)}{\partial y} . \qquad (9.44)$$

For confined flow (Fig. 9.7), we get

$$\nabla_{II} \left(-\frac{h^3}{12} \nabla_{II} p + \frac{h}{2} \right) = -\frac{\partial h}{\partial t} . \qquad (9.45)$$

Two-dimensional confined lubrication flows arise in lubrication of machine parts with different curvature, such as the piston ring systems of internal combustion engines [4]. Two-dimensional, thin-film lubrication flows arise in coating under asymmetric or unstable conditions, and in multilayer extrusion from expanding dies [18].

Example 9.1.2. Lubrication of pistons and piston rings [7]
In this example, the governing equations for lubrication of pistons and piston rings with the necessary boundary conditions are derived. The derivation includes the

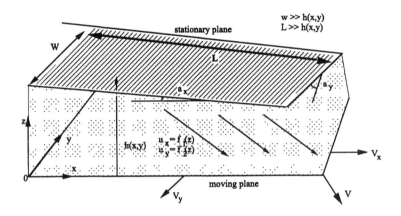

Figure 9.7. *Two-dimensional confined lubrication flow.*

Reynolds equation, which is the main equation in *hydrodynamic lubrication.* In *mixed lubrication* the friction force is calculated using the load on the slider multiplied by a friction coefficient which is taken to be a function of the oil film thickness and *surface roughness.* The actual piston ring arrangement is sketched in Fig. 9.8.

Solution:

Hydrodynamic and Mixed Lubrication

The lubrication of pistons and piston rings can be simulated by the lubrication of a slider which moves over a plane surface and is supported by an oil film. This situation is equivalent to that of a fixed slider and a moving plane surface: Fig. 9.9a is a cross section of the xz-plane through Point A, and Fig. 9.9b is a cross section of the yz-plane through the same point. Let U be the relative velocity of the plane and the slider, and W be the load on the slider which is supported by the oil film pressure.

The oil film thickness at Point A is

$$h = h_m + h_{sx} + h_{sy} , \qquad (9.46)$$

where h_m is the minimum oil film thickness, h_{sx} is the additional oil film thickness due to the slider curvature in the xz-plane, and h_{sy} is the additional oil film thickness due to the slider curvature in the yz-plane.

When the oil film is thick enough so that there is no surface contact between the slider and the plane, the lubrication is hydrodynamic. When there is surface

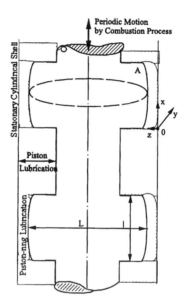

Figure 9.8. *Cross section of piston ring lubrication geometry.[Reproduced from Reference 4, by permission.]*

contact (contact of the asperities of the two surfaces), the lubrication is *mixed*. In this case, part of the load is carried by the oil film and another part is carried by the asperities. The oil film thickness, beyond which hydrodynamic lubrication exists, cannot be determined accurately and depends upon the topography of the involved surfaces and the height of the asperities. The minimum oil film thickness for hydrodynamic lubrication can be taken as a function of surface roughness.

Hydrodynamic Lubrication
The governing equations of the situation of Fig. 9.9 for hydrodynamic lubrication are the Reynolds equation,

$$\frac{\partial}{\partial x}\left(\frac{h^3}{\eta}\frac{\partial p}{\partial x}\right) + \frac{\partial}{\partial y}\left(\frac{h^3}{\eta}\frac{\partial p}{\partial y}\right) = -6U\frac{\partial h}{\partial x} + 12\frac{\partial h}{\partial t} , \qquad (9.47)$$

and the load equation,

$$W = \int_0^l \int_0^L p\,dx dy , \qquad (9.48)$$

where η is the oil viscosity, p is the oil film pressure, and t is time. Equation (9.47) is valid under the assumptions that: (a) body forces are negligible; (b) the pressure

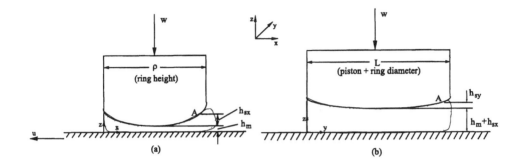

Figure 9.9. *A slider curved in two directions over a plane surface approximates the piston ring lubrication geometry of Fig. 9.8. [Reproduced from Reference 4, by permission.]*

is constant across the thickness of the film; (c) the radius of curvature of surfaces is large compared to the film thickness; (d) slip does not occur at the boundaries; (e) the lubricant is Newtonian; (f) the flow is laminar; and (g) fluid inertia is neglected compared to the viscous forces produced by the high oil viscosity.

To solve the Reynolds equation, four boundary conditions (two in each of the x- and y-directions) and one initial condition are needed. The lubrication problem of piston and piston rings in internal combustion engines is periodic, with period equal to the cycle of the engine. In four-stoke engines, the period is 720 degrees of crank angle (two revolutions). If T is the period, then

$$p(t) = p(t + T) , \qquad (9.49)$$

and

$$h(t) = h(t + T) . \qquad (9.50)$$

The boundary conditions are:

$$\text{at } x = 0, \quad p = p_1 ;$$
$$\text{at } x = \ell, \quad p = p_2 ;$$
$$\text{at } y = 0, \quad \frac{\partial p}{\partial y} = 0 ;$$
$$\text{at } y = L, \quad \frac{\partial p}{\partial y} = 0 .$$

The velocity in the x-direction is

$$u_x = \frac{1}{2\eta} \frac{\partial p}{\partial x} (z^2 - zh) - U \left(\frac{h-z}{h} \right) , \qquad (9.51)$$

whereas, in the y-direction,

$$u_y = -\frac{1}{2\eta} \frac{\partial p}{\partial y} (z^2 - zh) . \qquad (9.52)$$

The friction force per unit length on the xy-plane is

$$f_x = -\frac{h}{2} \frac{\partial p}{\partial x} + \frac{\eta U}{h} \qquad (9.53)$$

in the x-direction, and

$$f_y = \frac{h}{2} \frac{\partial p}{\partial y} \qquad (9.54)$$

in the y-direction.

The total friction force from the slider on the plane surface (Fig. 9.9) in the x-direction (no motion in the y-direction) is

$$F = \int_0^l \int_0^L f_x \, dx \, dy$$

which, in combination with Eq. (9.53), gives

$$F = \int_0^l \int_0^L \left(-\frac{h}{2} \frac{\partial p}{\partial x} + \frac{\eta U}{h} \right) dx \, dy . \qquad (9.55)$$

By means of

$$x^* = \frac{x}{l}, \quad y^* = \frac{y}{L}, \quad h^* = \frac{h}{h_0}, \quad p^* = \frac{p}{p_0}, \quad t^* = \frac{t}{T},$$

the dimensionless equations are

$$\lambda_z \frac{\partial}{\partial x^*} \left(h^{*3} \frac{\partial p^*}{\partial x^*} \right) + \lambda_y \frac{\partial}{\partial y^*} \left(h^{*3} \frac{\partial p^*}{\partial y^*} \right) = -6 \frac{\partial h^*}{\partial x^*} + 12 \frac{l}{T\nu} \frac{\partial h^*}{\partial t^*} , \qquad (9.56)$$

$$\frac{W}{p_0 l L} = \int_0^1 \int_0^1 p^* \, dx^* dy^* , \qquad (9.57)$$

and

$$\frac{F}{p_0 h_0 L} = \int_0^1 \int_0^1 \left(-\frac{h^*}{2} \frac{\partial p^*}{\partial x^*} + \frac{1}{\lambda_x h^*} \right) dx^* dy^* , \qquad (9.58)$$

where

$$\lambda_z = \frac{h_0^2 p_0}{\eta l \nu} , \qquad \lambda_y = \frac{l h_0^2 p_0}{\eta L^2 \nu} , \qquad \text{and} \qquad \lambda_x = \left(\frac{\ell}{L} \right)^2 . \qquad (9.59)$$

The solution of these equations is discussed in Example 9.1.3.   □

Example 9.1.3. Simplified method for ring lubrication [7]

The actual ring profile is shown in Fig. 9.10. When the ring is moving downward, oil pressure builds in the converging region BA. In the diverging region AD, the pressure is zero (more accurately, it is equal to the boundary pressure at the end D of the ring). Therefore, the situation is equivalent to that of Fig. 9.10 where AB has the same curvature as that in Fig. 9.9. Now, since the radius of curvature of the circular profile is large, the curved surface AB can be replaced by the plane surface AB. Finally, the original problem of Fig. 9.9 becomes equivalent to that of Fig. 9.10. When the ring is moving upward, oil pressure builds in the converging region DA and the situation is equivalent to that of Fig. 9.11.

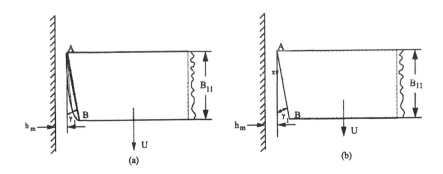

Figure 9.10. *Downward motion of ring (a) and domain approximation (b).*

It is easy to deal with the simplified profiles of Figs. 9.10 and 9.11 where the problem is one-dimensional. Lubrication can be considered to be one-dimensional if both ring and bore are perfectly circular, in which case $h_{sy} = 0$ for all y, and when the ring is large compared to the gap. In this case, the Reynolds equation, (Eq.

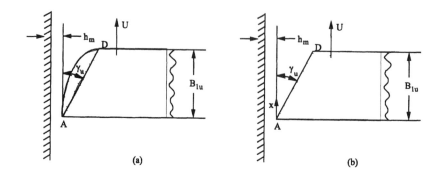

Figure 9.11. *Upward motion of ring (a) and domain approximation (b).*

(9.47)), becomes

$$\frac{\partial}{\partial x}\left(\frac{h^3}{\eta}\frac{\partial p}{\partial x}\right) = -6U\frac{\partial h}{\partial x} + 12\frac{\partial h}{\partial t} ,\tag{9.60}$$

with boundary conditions

$$\text{at} \quad x = 0, \quad p = p_1(t), \quad \text{and}$$
$$\text{at} \quad x = B_1, \quad p = p_2(t)$$

where, depending upon the direction of motion, B_1 is equal to B_{1l} or B_{1u}, and p_1 and p_2 are defined by the engine cycle.

By integrating Eq. (9.60) twice and using the above boundary conditions, the oil film pressure is determined as

$$p = \frac{6(U-\lambda)\eta B}{h_m K}\left[\frac{1}{h} - \frac{h_m h_1}{h^2(h_m+h_1)} - \frac{1}{h_m+h_1}\right] + p_1 + D\frac{h^2-h_m^2}{h^2} ,\tag{9.61}$$

where

$$\lambda = \frac{2B}{h_m K}\left(\frac{\Delta h_m}{\Delta t}\right)$$

and

$$D = (p_2 - p_1)\frac{h_1^2}{h_1^2 - h_m^2} .$$

In the above expressions, h_m is the oil film thickness at $x=0$, h_1 is the oil film thickness at $x=B$, and h is the oil film thickness at arbitrary x; p_1 is the boundary pressure at $x=0$, p_2 is the boundary pressure at $x=B$, Δh_m is the variation of the oil film thickness between the current time step and the previous time step, Δt is the time difference between the two time steps, and K is defined by

$$K = \frac{h_1 - h_m}{h_m} . \tag{9.62}$$

For

$$h = \bar{h} = \frac{2h_m h_1}{h_m + h_1}\left(1 + \frac{p_1 - p_2}{b}\frac{h_m h_1}{h_m - h_1}\right) , \tag{9.63}$$

where

$$b = \frac{6(U - \lambda)\eta B}{h_m K} ,$$

Eq. (9.61) gives the maximum pressure,

$$p_{max} = \frac{6(U - \lambda)\eta B}{h_m K}\left[\frac{1}{\bar{h}} - \frac{h_m h_1}{h^2(h_m + h_1)} - \frac{1}{h_m + h_1}\right] + p_1 + D\frac{\bar{h}^2 - h_m^2}{\bar{h}^2} . \tag{9.64}$$

The load per unit length is

$$\frac{W}{L} = \int_0^B p\,dx ,$$

which gives

$$\frac{W}{LB} = \frac{6(U - \lambda)\eta B}{h_m^2 K^2}\left[ln\frac{h_1}{h_m} - \frac{2(h_1 - h_m)}{h_1 + h_m}\right] + p_1 + (p_2 - p_1)\frac{h_1}{h_1 + h_m} , \tag{9.65}$$

where W/LB is the load per unit projected area.

Finally, the friction force per unit length is

$$\frac{F}{L} = \int_0^B f_x dx ,$$

where f_x is the friction force per unit area, given by

$$f_x = -\frac{h}{2}\frac{\partial p}{\partial x} + \frac{\eta U}{h} .$$

For the case considered, the friction force becomes,

$$\frac{F}{L} = -\frac{1}{2}(p_2 h_1 - p_1 h_m) + \frac{1}{2}\left(\frac{W}{L}\right)\tan\alpha + \frac{\eta U B}{h_m K}ln\frac{h_1}{h_m} , \qquad (9.66)$$

where α is equal to γ_e (Fig. 9.10) or γ_u (Fig. 9.11). Therefore, the ratio of load to friction forces, W/F, is maximized in the limit of $\alpha=0$, where α measures the inclination of the ring, in agreement with Eq. (9.15).   □

9.2   Stretching Flows

Lubrication flows are characterized by dominant velocity gradients in the direction normal to the flow, e.g.,

$$\frac{\partial u_x}{\partial y} \gg \frac{\partial u_x}{\partial x} . \qquad (9.67)$$

These gradients arise from a driving pressure gradient in the flow direction (e.g., flow in nonrectilinear channels and pipes) or by an external velocity gradient due to an inclined boundary moving with respect to a stationary one (e.g., flow in journal-bearing lubrication and coating flow). The pressure and velocity gradients compete to overcome the liquid's adherence to solid boundaries (no-slip boundary condition) in order to initiate and maintain the flow. In the opposite extreme are almost unidirectional extensional flows where the inequality (9.67) is reversed and the dominant velocity gradient is in the direction of flow, i.e.,

$$\frac{\partial u_x}{\partial x} \gg \frac{\partial u_x}{\partial y} . \qquad (9.68)$$

These flows are driven by an external normal *velocity gradient* or a *normal force gradient* in the flow direction. This results in the stretching of material filaments along the streamlines, which is characteristic of extensional flows. Indeed, stretching flows are nearly extensional, irrotational flows with a unique dominant velocity component so that they can be approximated by one-dimensional, cross-section-averaged equations of the *thin-beam approximation* type [13]. An accurate prototype of these flows is the extension of a viscous cylindrical material filament, shown in Fig. 9.12.

Stretching flows are important in a diversity of polymer processes used to produce synthetic fibers by spinning polymer melts, films, and sheets by casting molten metals, bags, and bottles by blowing polymer melts and glass and several three-dimensional polymer articles using extrusion and compression molding. Some of

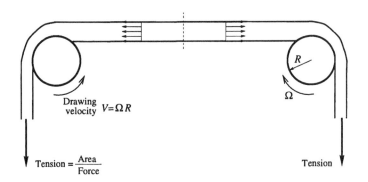

Figure 9.12. *A model of stretching flow.*

these operations are illustrated in Fig. 9.13. Although the involved polymeric materials are often non-Newtonian, the analysis of stretching flows for Newtonian liquid is always instructive and constitutes a useful step towards understanding real situations. Fiber-spinning of Newtonian liquid is a good prototype of these processes and is analyzed below.

9.2.1 Fiber Spinning

Synthetic fibers are produced by drawing melt filaments emerging from a capillary, called *spinneret*, using a rotating drum which directs the fiber to drying and subsequent operations, Fig. 9.13. The ratio of the velocity at the take-up end,

$$u_L = \Omega R \, , \tag{9.69}$$

to the average velocity emerging from the spinneret

$$u_0 = \frac{4Q}{\pi D^2} \, , \tag{9.70}$$

is called the *draw ratio*,

$$D_R = \frac{u_L}{u_0} = \frac{\pi D^2 \omega R}{4Q} > 1 \, . \tag{9.71}$$

The tension required to draw fibers of radius R_L at length L,

$$F = \int_0^{R_L} 2\pi(-p + \tau_{zz})_{z=L} \, r dr \, , \tag{9.72}$$

is called the *drawing force*.

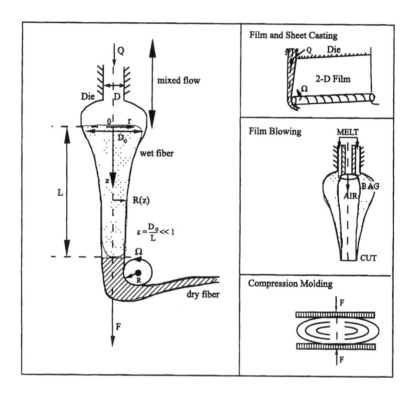

Figure 9.13. *Fiber spinning and other drawing or compressive operations, modeled by stretching flow theory.*

Apart from the flow rearrangement region near the exit of the spinneret, about two diameters downstream, the flow is extensional and the axial velocity increases monotonically. According to the continuity equation, the radius $R(z)$ of the fiber decreases,

$$R^2(z) = \frac{Q}{\pi u(z)} \,. \tag{9.73}$$

Under the preceding assumptions, the dimensionless momentum equation in the flow direction becomes

$$Re\left(\frac{\partial u}{\partial t} + u\frac{\partial u}{\partial z}\right) = -\frac{\partial p}{\partial z} + \frac{\partial \tau_{zz}}{\partial z} + St \,, \tag{9.74}$$

where the velocity is scaled by the characteristic velocity u_0, pressure and stress by $\eta\, u_0/L$, distance by L, and time by L/u_0. With these scalings, $Re \equiv \rho u_0 L/\eta$ and $St \equiv \rho g L^2/(\eta u_0)$.

The pressure is eliminated from Eq. (9.74) by invoking the normal stress boundary condition at the free surface,

$$Ca(-p + \tau_{rr})_{r=R} = -\frac{1}{R(z)} + \frac{\dfrac{d^2R}{dz^2}}{\sqrt{1 + \left(\dfrac{dR}{dz}\right)^2}} \simeq -\frac{1}{R(z)} , \tag{9.75}$$

and by taking into account the fact that both the pressure, p, and the normal stress, τ_{rr}, are virtually constant over the cross section of the fiber. The capillary number is defined here by $Ca \equiv \epsilon \eta u_0 / \sigma$, where $\epsilon \equiv D/L \ll 1$ is the fiber *aspect ratio*. For steady spinning, Eq. (9.74) becomes

$$Re\, u \frac{\partial u}{\partial z} = \frac{\partial}{\partial z}(\tau_{zz} - \tau_{rr}) - \frac{1}{Ca}\frac{\partial}{\partial z}\left(\frac{1}{R}\right) + St , \tag{9.76}$$

which is valid for any fluid. The radius $R(z)$ is eliminated by means of the dimensionless continuity equation, i.e.,

$$R(z) = \sqrt{\frac{Q}{\pi u}} . \tag{9.77}$$

To restrict Eq. (9.76) to Newtonian liquids, we employ the dimensionless form of Newton's law of viscosity

$$\tau_{zz} - \tau_{rr} \simeq 3\frac{\partial u}{\partial z} . \tag{9.78}$$

By substituting Eqs. (9.77) and (9.78) in Eq. (9.76) and integrating across the thickness, the governing equation for the velocity profile along the fiber is

$$\frac{d}{dz}\left(\frac{3}{u}\frac{du}{dz}\right) + \frac{1}{Ca}\left(\frac{1}{\sqrt{u}}\right) - Re\frac{du}{dz} + \frac{St}{u} = 0 , \tag{9.79}$$

which states that the acceleration is due to viscous, capillary, and gravity force gradients.

As for the two required boundary conditions, we define the velocity at the inlet,

$$u(z = 0) = 1 , \tag{9.80}$$

and either the draw ratio at the outlet,

$$D_R = \frac{u_L}{u_0} = \frac{\Omega R_L}{u_0} , \tag{9.81}$$

or the dimensionless drawing force at the take-up end,

$$f = \frac{FL}{Q\eta} \simeq \frac{3}{u}\frac{du}{dz} \ . \tag{9.82}$$

Equations (9.79) to (9.82) can be solved:

(a) numerically, for any value of Re, Ca, St, ϵ, f or D_R;

(b) by perturbation, for limiting values of these parameters;

(c) analytically, when one of $Re, Ca,$ and St is zero.

A reasonable approximation in melt spinning of usually high viscosity and low surface tension is $Re = St = 1/Ca = 0$. The solution is obtained from the truncated form of Eq. (9.79),

$$\frac{3}{u}\frac{du}{dz} = c_1 \ , \tag{9.83}$$

subject to the boundary conditions,

$$u(z = 0) = 1$$

and

$$u(z = 1) = D_R \quad \text{or} \quad \frac{3}{u}\frac{du}{dz}\bigg|_{z=1} = f \ .$$

The general solution is

$$u = \exp(c_1 z/3) \ , \tag{9.84}$$

where the constant c_1 is determined by the boundary condition at $z = 1$. When the draw ratio is specified, the solution is

$$u = e^{z \ln D_R} \tag{9.85}$$

or, in dimensional form,

$$u = u_0 e^{\left(\frac{z}{L} \ln D_R\right)} \ . \tag{9.86}$$

When the drawing force is specified, the solution is

$$u = e^{\frac{fz}{3}} \tag{9.87}$$

or, in dimensional form,

$$u = u_0\, e^{\left(\dfrac{FL}{3Q\eta}\right)\dfrac{z}{}}.$$

(9.88)

The force required to draw fibers of length L at draw ratio D_R is

$$F = \frac{3Q\eta\,\ln D_R}{L}.$$

(9.89)

Similarly, from Eq. (9.72), drawing a fiber of length L using a force F produces a draw ratio of

$$D_R = e^{\dfrac{FL}{3Q\eta}},$$

(9.90)

which corresponds to fibers of final radius

$$R(z = L) = \left[\frac{Q}{\pi} e^{-\left(\dfrac{FL}{3Q\eta}\right)}\right]^{1/2}.$$

(9.91)

9.2.2 Compression Molding

During compression molding under a force F, a polymeric melt is compressed between two hot plates, Fig. 9.14. Due to the high temperature, the hot plates induce slip of the squeezed melt by means of thin layers of low viscosity. This facilitates compression and deformation to the final target shape.

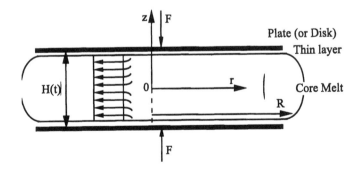

Figure 9.14. *Prototype of compression molding.*

Using the assumptions of perfect slip and negligible viscous effects, the surviving terms of the Navier–Stokes equations are

$$\rho \left(\frac{\partial u_z}{\partial t} + u_z \frac{\partial u_z}{\partial z} + u_r \frac{\partial u_z}{\partial r} \right) = -\frac{\partial p}{\partial z} \tag{9.92}$$

and

$$\rho \left(\frac{\partial u_r}{\partial t} + u_r \frac{\partial u_r}{\partial r} + u_z \frac{\partial u_r}{\partial z} \right) = -\frac{\partial p}{\partial r} . \tag{9.93}$$

The local variation of the velocity with time is eliminated by assuming slow squeezing and slow local variation of velocity. This is known as the *quasi-steady state* assumption. Furthermore, since the viscous terms have been eliminated, the shear stress must vanish as well, which requires that $\partial u_r / \partial z = \partial u_z / \partial r = 0$. Equations (9.92) and (9.93) then become

$$\frac{\partial}{\partial z} \left(\rho \frac{u_z^2}{2} + p \right) = 0 , \tag{9.94}$$

and

$$\frac{\partial}{\partial r} \left(\rho \frac{u_r^2}{2} + p \right) = 0 . \tag{9.95}$$

Combined with the continuity equation,

$$\frac{1}{r} \frac{\partial}{\partial r} (r u_r) + \frac{\partial}{\partial z} (u_z) = 0 , \tag{9.96}$$

and the appropriate boundary conditions,

$$u_z \left(z = \frac{H(t)}{2} \right) = -V = \frac{dH}{dt} \tag{9.97}$$

and

$$p(r = R, z = H/2) = p_0 , \tag{9.98}$$

Eqs. (9.94) and (9.95) yield

$$u_z = -\frac{V}{H} z = \frac{1}{H} \frac{dH}{dt} z = \dot{\epsilon}(t) z \tag{9.99}$$

and

$$u_r = -\dot{\epsilon}(t) \frac{r}{z} , \tag{9.100}$$

where $\dot{\epsilon}(t)$ is the extension rate. The pressure is then given by

$$p(r,z) = \rho \frac{\dot{\epsilon}(t)^2}{2} \left[(H^2 - z^2) + \frac{R^2 - r^2}{4} \right] . \tag{9.101}$$

The required squeezing force is

$$F = 2\pi \int_0^R (-p + \tau_{zz})_{z=H} \, r dr = 2\pi R \left[2\eta - \frac{\dot{\epsilon}\rho R^3}{32} \right] \dot{\epsilon} \simeq 4\pi R \eta \dot{\epsilon} . \tag{9.102}$$

The principles of lubricated squeezing flow are utilized by some commercially available rheometers that measure elongational viscosity, defined by

$$\eta_e \equiv \frac{\tau_{zz} - \tau_{rr}}{\dot{\epsilon}} = \frac{F(t)}{\pi R^2 \dot{\epsilon}(t)} = \frac{F(t)}{\pi R^2 \dfrac{1}{H(t)} \dfrac{dH}{dt}} , \tag{9.103}$$

by recording $F(t)$ and dH/dt in a known geometry [20].

Notice that Eqs. (9.99) to (9.103) provide an exact solution to the full equations of motion. However, these equations are nonlinear and, therefore, the uniqueness of the solution is not guaranteed. Furthermore, the solution is valid only under the extreme assumption of perfect slip. If this assumption is relaxed, the resulting approximate solution for the lubricant-layer and the viscous melt-core velocities and pressures profiles are [21]

$$u_r^L = \frac{V r \eta_L}{4H^3 \left(\dfrac{\delta \eta_V}{H \eta_L} + \dfrac{1}{3} \right) \eta_L} \left[H^2 - z^2 \right] , \tag{9.104}$$

$$u_r^M = \frac{V r}{4H^3 \left(\dfrac{\delta \eta_V}{H \eta_L} + \dfrac{1}{3} \right) \eta_L} \left[2\frac{\eta_V}{\eta_L} H\delta + (H - \delta)^2 - z^2 \right] , \tag{9.105}$$

$$p_L = \frac{2V \eta_V}{4H^3 \left(\dfrac{\delta \eta_V}{H \eta_L} + \dfrac{1}{3} \right)} \left[3\frac{\eta_V}{\eta_L} H\delta + \frac{R^2 - r^2}{2} \right] , \tag{9.106}$$

$$p_M = \frac{2V \eta_V}{4H^3 \left(\dfrac{\delta \eta_V}{H \eta_L} + \dfrac{1}{3} \right)} \left[2\frac{\eta_V}{\eta_L} H\delta + \frac{R^2 - r^2}{2} - (H - \delta)^2 + z^2 \right] , \tag{9.107}$$

where the subscripts L and V refer to the lubricant and viscous liquids of thicknesses δ and $H - \delta$, respectively. The resulting squeezing force is now

$$F = \frac{3\pi V \eta_V R^2}{H} + \int_0^R \frac{16\pi V \eta_L^2 (r^2 - R^4)}{g \eta_V \delta^5} \, r dr , \tag{9.108}$$

which is identical to that under perfect lubrication and slip when

$$\frac{\delta \eta_V}{H \eta_L} \gg 1 \, . \tag{9.109}$$

Figure (9.15) shows the predictions of Eqs. (9.104) to (9.108). By comparing Eqs. (9.102) and (9.108), it is evident that, under certain lubrication conditions, the required squeezing load is reduced since the viscous resistance to flow is lowered by the intervening lubricating layers.

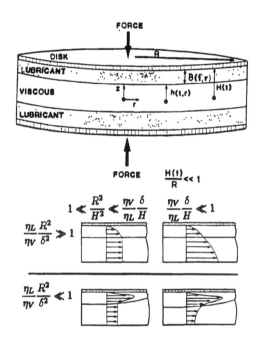

Figure 9.15. *The two limiting flow regimes that are predicted by finite element analysis depend on the dimensionless group $\eta_L R^2 / \eta_V \delta^2$.*

In the case of commercial compression molding, there are no distinct core and lubrication layers of different viscosities. Instead, the viscosity decreases abruptly but continuously from the midplane to the hot plates and, therefore, the preceding analysis provides only a limiting case study. There are cases, however, where two distinct layers exist, such as in lubricated squeezing flow in extensional rheometers, and in transferring highly viscous crude oil by means of lubricated pipes. The latter case, highlighted in Example 9.2.1, generates plug-like flow in the viscous-core liquid which can be viewed as a limiting case of stretching flow.

Example 9.2.1. Lubricated flows [22]

To transfer highly viscous crude oil of viscosity η_V and density ρ_V through commercial pipes of radius R and length L at fixed flow rate

$$Q_o = \frac{\pi R^4}{8\eta_V}\left(\frac{\Delta p_o}{\Delta L}\right),$$

a large pumping power, P, is required to overcome the pressure drop of the Poiseuille flow according to

$$P = Q_o \Delta p_o = Q_o \frac{8\eta_V Q_o}{\pi R^4}(\Delta L). \tag{9.110}$$

To reduce the required power, water of viscosity $\eta_w \ll \eta_V$ and density $\rho_w \simeq \rho_V$ is injected to form a permanent thin lubrication layer between the flowing crude oil and the pipe wall (Fig. 9.16).

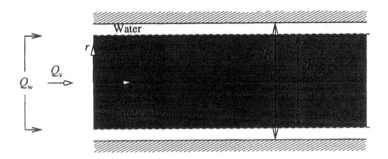

Figure 9.16. *Transferring viscous liquids over large distances by lubricated pipes to reduce pumping power.*

Let us make the following assumptions:

(i) The lubrication is stable and axisymmetric.

(ii) The interfacial tension between water and oil is negligible compared to viscous forces, i.e., $Ca \to \infty$. Therefore, the interface is perfectly cylindrical. Moreover, the velocity and total stress are continuous across the interface, whereas pressure and normal viscous stress are not:

$$\mathbf{u}^w(r = h) \;=\; \mathbf{u}^V(r = h) \quad\Longrightarrow\quad \begin{cases} u_r^w(r = h) = u_r^V(r = h) \\[2mm] u_z^w(r = h) = u_z^V(r = h) \end{cases} \tag{9.111}$$

$$\tau_{rz}^V(r = h) = \tau_{rz}^V(r = h) \,, \tag{9.112}$$

and

$$(-p^V + \tau_{zz}^V)_{r=h} = (-p^V + \tau_{zz}^V)_{r=h} \,. \tag{9.113}$$

(iii) The flow is slow and, hence, inertia effects are negligible.

Under these assumptions, the two velocity profiles are given by

$$u_r^V = \frac{1}{4\eta_V} \frac{\Delta p}{\Delta L} r^2 + c_1^V \ln r + c_2^V \tag{9.114}$$

and

$$u_r^w = \frac{1}{4\eta_w} \frac{\Delta p}{\Delta L} r^2 + c_1^w \ln r + c_2^w \,, \tag{9.115}$$

where the common pressure gradient is different from that of Eq. (9.110). The four constants are evaluated by means of the interface conditions (Eq. (9.111)), the no-slip boundary condition at the pipe wall, and the symmetry condition to get

$$u_r^V = \frac{1}{4\eta_V} \frac{\Delta p}{\Delta L} (r^2 - h^2) + \frac{1}{4\eta_w} \frac{\Delta p}{\Delta L} (h^2 - R^2) \tag{9.116}$$

and

$$u_r^w = \frac{1}{4\eta_w} \frac{\Delta p}{\Delta L} (r^2 - R^2) \,. \tag{9.117}$$

The corresponding volumetric flow rates are

$$Q^V = \frac{\Delta p}{\Delta L} \left[\frac{\pi h^4}{8\eta_V} + \frac{\pi h^2(h^2 - R^2)}{4\eta_w} \right] \tag{9.118}$$

and

$$Q^w = \frac{\Delta p}{\Delta L} \left[\frac{(R^2 - h^2)^2}{8\eta_w} \right] \,. \tag{9.119}$$

By means of Eq. (9.119), the position of the interface is determined as

$$h^2 = R^2 - \left[\frac{8\eta_w Q_w}{(\Delta p/\Delta L)} \right]^{1/2} \,.$$

Since the same amount of crude oil must be transferred ($Q^V = Q_0$), the pressure gradient ratio must be

$$k = \frac{(\Delta p/\Delta L)_0}{(\Delta p/\Delta L)} = \frac{\Delta p_0}{\Delta L} = \frac{16\eta_w Q_w}{2R^4(\Delta p/\Delta L)}$$
$$-2\left(\frac{16\eta_w Q_w}{2R^4\Delta p/\Delta L}\right)^{1/2} + \frac{16\eta_V Q_w}{2R^4(\Delta p/\Delta L)} - \left(\frac{32\eta_v^2 Q_w}{R^4(\Delta p/\Delta L)\eta_w}\right)^{1/2}$$

or, equivalently,

$$k = \frac{(\Delta p/\Delta L)_0}{(\Delta p/\Delta L)} = \left(\frac{\eta_V}{\eta_w}\frac{Q_w}{Q_0}\right)^{1/2}\left[\left(\frac{Q_0}{Q_w}\frac{\eta_V}{\eta_w}\right)^{1/2} + \left(\frac{\eta_w}{\eta_V}\frac{Q_w}{Q_0}\right)^{1/2}\right.$$
$$\left. + 2\left(\frac{\eta_V}{\eta_w}\frac{Q_w}{Q_0}\right)^{1/2} - 2\left(1 + \frac{\eta_V}{\eta_V}\right)\right] \simeq \left(\frac{\eta_V Q_w}{\eta_w Q_0}\right)^{1/2}\left[\left(\frac{Q_0\eta_V}{Q_w\eta_w}\right)^{1/2}\right] = \frac{\eta_V}{\eta_w}.$$

Extreme cases

(i) Perfect lubrication, i.e., $\eta_w/\eta_V \to 0$:

$$k = \left(\frac{\Delta p}{\Delta L}\right)_0 / \left(\frac{\Delta p}{\Delta L}\right) \to \infty \quad \text{and} \quad \frac{\Delta p}{\Delta L} \to 0 .$$

(ii) Lubrication by much more viscous layers, i.e., $\eta_V/\eta_w \to 0$:

$$k \to 0 , \quad \frac{1}{k} \to \infty .$$

(iii) No lubrication, i.e., $\eta_w/\eta_V = 1$:

$$k = 1 .$$

The ratio

$$M = \frac{\eta_V Q_w}{\eta_w Q_v}$$

expresses the ability of one liquid to displace the other and is, therefore, called *mobility ratio*. This parameter is used primarily in flows through porous media where steam is often injected to displace the more viscous oil [12, 23]. □

Intuitively, highly viscous and elastic materials can be stretched the most. This poses significant experimental challenges in producing ideal, extensional flows in order to measure elongational viscosity of viscoelastic liquids of low shear viscosity. This need does not exist in Newtonian liquids, the elongational viscosity of which is exactly three times the shear viscosity by virtue of Newton's law of viscosity. At this stage, fiber-spinning and other related operations (e.g., falling curtains and fibers under gravity) and the recent opposing jet method [24] appear to provide the best means (though not perfect [25]), to measure elongational viscosity. The elongational viscosity is extremely important in industrial polymer processes which may involve any kind of extensional deformation, given that

(a) the common shear viscosity measurements do not provide any indication of the magnitude of the elongational viscosity at even moderate stretching or compression, and

(b) the elongational viscosity may attain values ten-fold or even higher than the shear viscosity, which gives rise to huge normal stresses and, therefore, to excessively high drawing forces and compressive loads required to process highly elastic viscoelastic polymer melts or solutions.

9.3 Problems

9.1. Estimate the pressure drop in the linearly slowly-varying cylindrical contraction shown in Fig. 9.17.

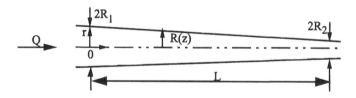

Figure 9.17. *Flow in conical pipe.*

9.2. *Sheet coating* [8]. To apply a thin liquid film on a moving substrate, the arrangement in Fig. 9.18 is used.
(a) Find the pressure difference, $p_0 - p_L$, required to apply a film of final thickness H_f.

(b) Show that the velocity $u_x(x,y)$ under the die wall is given by

$$u_x = \left(1 - \frac{y}{H}\right)\left[1 - 3\left(1 - 2\frac{H_f}{H}\right)\frac{y}{H}\right],$$

where $H=H(x)$. The term in brackets is negative over a portion of the cross section whenever $H > 3H_f$, indicating a negative velocity and a region of backflow near the wall as shown. Find $H_0(x)$ which bounds the recirculating flow.

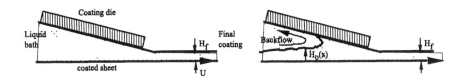

Figure 9.18. *Sheet coating.*

9.3. *Extrusion coating* [11]. The extrusion coating application is shown in Fig. 9.19. A Newtonian liquid of viscosity η and surface tension σ is continuously applied at flow rate Q on a fast-moving substrate at velocity V. The distance between the die and substrate is H_0, and the film thickness far downstream is H_f. Find the film profile $H(x)$ and the ratio $Q/(H_0 V)$.

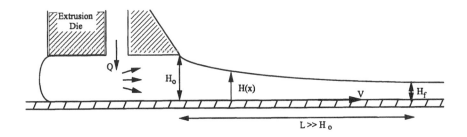

Figure 9.19. *Extrusion coating.*

9.4. *Lubrication equations by perturbation analysis.* Analyze Eqs. (9.9) and (9.10), subjected to appropriate boundary conditions, in the limit of $\alpha=0$ up to the first-order term, i.e., find the solution

$$\mathbf{u}(\alpha, Re, x, y, z) = \mathbf{u}_0(\alpha = 0, Re, x, y, z) + \mathbf{u}_1(Re, x, y, z)\alpha + O(\alpha^2)$$

and

$$p(\alpha, Re, x, y, z) = p_0(\alpha = 0, Re, x, y, z) + p_1(Re, x, y, z)\alpha + O(\alpha^2)$$

for horizontal confined flows. What is the solution in the limit of $\alpha \to \infty$? Does such solution exist?

9.5. Consider the *wire coating* flow depicted in Fig. 9.20. A wire of radius R is advanced at velocity V through a die of radius H_0 by a pulling force. The space between the wire and the die is always filled. Over a distance L, the coating decays from H_0 to the final target film thickness H_∞ under the competing actions of surface tension and velocity V. The physical characteristics of the coating liquid are: density ρ, viscosity η, and surface tension σ. The capillary pressure due to surface tension is $\sigma/H(z)$, where $H(z)$ is the local radius – distance from the z-axis of the free surface.

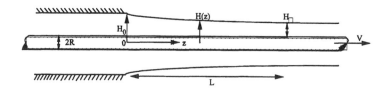

Figure 9.20. *Wire coating.*

(a) What is the velocity profile? What kind of flows does it incorporate? What is the force F required to advance the wire at velocity V?
(b) Derive the Reynolds equation and its appropriate boundary conditions.
(c) Solve the Reynolds equation in the limiting cases of $Ca=0$ and $Ca \to \infty$.
(d) Compare your results with those of Figs. 9.21 and 9.22.

9.6. *Spinning equations by Leibnitz formula.* Show how the one-dimensional fiber-spinning equation at steady state,

$$-\rho u_z \frac{du_z}{dz} = \frac{dp}{dz} + \frac{d\tau_{zz}}{dz} + \rho g = 0 \,,$$

can be transformed to the average spinning equation,

$$\frac{\tau_{zz} - \tau_{rr}}{u} = c \,,$$

by the Leibnitz integration formula.
Hint: let A be the cross-sectional area and V be the volume of the film.

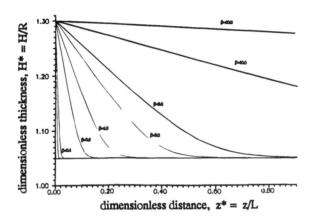

Figure 9.21. *Exact solution of wire coating with $\beta=\alpha/Ca=(R\sigma)/(L\eta V)$. [Provided by D. Hatzinikolaou, MSc. 1990, Univ. of Michigan.]*

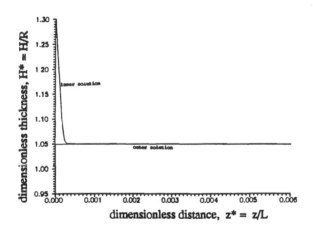

Figure 9.22. *Asymptotic solution in the limit of $\beta=(R\sigma)/(L\eta V)=0.001$. [Provided by D. Hatzinikolaou, MSc. 1990, Univ. of Michigan.]*

9.7. *Uniaxial stretching.* A cylindrical specimen of initial length L_0 and initial radius R_0 is stretched by a constant force F applied at the two edges along its axis of symmetry.

(a) Is the flow an admissible stretching flow? What are the velocity components? What is the pressure if the surface tension of the liquid is σ?

(b) How do the length and the diameter change with time?

(c) Is it a constant extension rate or a constant stress process?

9.8. Consider the *film casting* process depicted in Fig. 9.23. To manufacture plastic films or metal sheets, the hot liquid is forced through a slit die and drawn by a roller applying tension F. Calculate the *production speed* and *volume* for Newtonian liquid supplied at flow rate Q. What is the applied tension by the drawing cylinder? How does this tension propagate upstream? Calculate the film thickness and velocity profiles along the wet film upstream from the drawing cylinder.

[Hint: neglect the extrudate-swell region just after the die exit.]

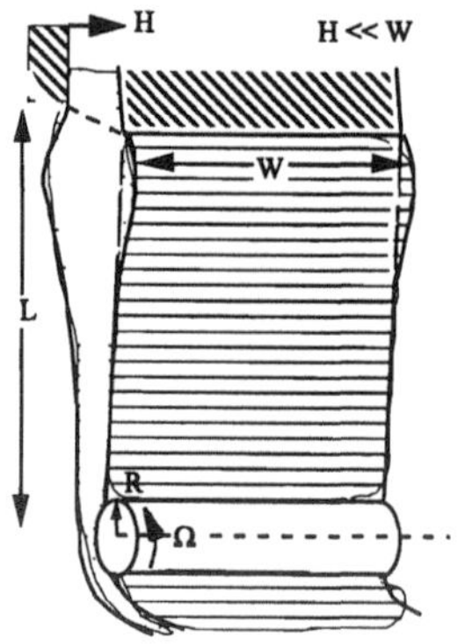

Figure 9.23. *Film casting.*

9.9. *Triaxial stretching.* A cubic specimen of side α is compressed by a squeezing pair-force, F, applied along one of its axes of symmetry

(a) Calculate the resulting three-dimensional, slow stretching flow.

(b) Find the pressure distribution. Is cavitation possible?

(c) How does each side change its length?

(d) What are the resulting extension rates?

9.10. Consider the *film blowing* process [13] illustrated in Fig. 9.24. To manufacture plastic bags for food packaging purposes, the melt is forced through an annulus containing an air supply to keep the walls of the cylindrical film apart. After solidification, the opposite walls are brought together by the tension application system and then cut to bag-pieces. Analyze the axisymmetric angular flow between $0 \leq z \leq L$ for Newtonian liquid of density ρ, viscosity η, and surface tension σ supplied at flow rate Q and average velocity u_0, and drawn by a pair of rotating drums of radius S at Ω revolutions per minute. What is the resulting film thickness

distribution and the required tension?
[Hint: neglect any extrudate-swell effects.]

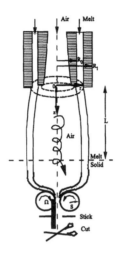

Figure 9.24. *Film blowing.*

9.11. *Jet-stripping coating* [29]. In order to control thickness of coating in dip coating operations, gas knives are often used to strip excessive coating, as shown in Fig. 9.25. Such an arrangement gives rise to external gas pressure and shear stress distributions of the kind shown in the figure.
(a) Sketch representative velocity profiles under the thin film before and after the point of the impingement of the gas knife.
(b) Derive the Reynolds lubrication equation under the combined action of the velocity V, gravity, external shear stress $\tau(x)$, and pressure $p(x)$ due to the gas knife.
(c) Solve the Reynolds equation for limiting cases.

9.12. Analyze the journal-bearing lubrication flow in Fig. 9.26 as a perturbation from the standard circular Couette flow by means of the eccentricity,

$$\epsilon \equiv R_0 - \frac{R_i}{R_0},$$

which is zero for Couette flow between concentric cylinders.

9.13. *Fiber-spinning boundary conditions* [19]. Fiber-spinning of Newtonian liquids under dominant viscous and gravity forces is governed by the one-dimensional

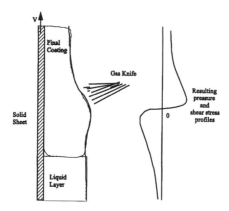

Figure 9.25. *Coating by jet-stripping.*

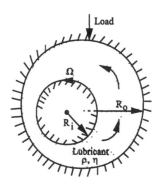

Figure 9.26. *Journal-bearing lubrication.*

equation

$$\frac{d}{dz}\left[3\eta\frac{du_z}{dz}\pi R^2(z)\right] + \rho g\pi R^2(z) = 0 \, ,$$

where $R(z)$ is the radius of the fiber. If V is the inlet velocity, and L the length of the fiber, show that the dimensionless form of the above equation is

$$u_z^*\frac{d}{dz}\left[\frac{1}{u_z^*}\frac{du_z^*}{dz^*}\right] = -\frac{\rho g L^2}{3\eta V} = -St \, .$$

Show, also, how the solution

$$u_z^* = V e^{-c_2 z^*} + \frac{St}{z} z^{*2}$$

may be obtained. What does each of the two terms represent? What are acceptable boundary conditions at the other end of the fiber, and what kind of spinning do they represent (e.g., free falling fiber, drawn fiber)? Justify the limiting value of $u_z^*(z^* \to \infty)$. Show that the fiber will never attain a constant diameter, and explain the physical significance of this fact. What is the solution for fiber drawn by velocity $u_z = V_L$ at $z = L$ (or $z^* = 1$)?

9.14. *Air-sheared film under surface tension* [29]. Figure 9.27 shows a representative configuration of *jet-stripped, continuous coating* of sheet materials, where the final film thickness is controlled by the external pressure, $P(z, t)$, and stress, $T(z, t)$, distributions in the gas jet, in addition to drag by the moving boundary, gravity, and surface tension.

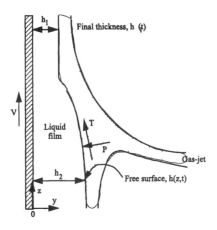

Figure 9.27. *Jet-stripping continuous coating.*

(a) Show that, under lubrication conditions, the dominant velocity component is

$$u_z(z, y, t) = V + \frac{1}{\eta}\left(\rho g + \frac{\partial p}{\partial z} - \sigma \frac{\partial^3 h}{\partial z^3}\right)\left(\frac{y^2}{2} - hy\right) + \frac{Ty}{\eta}, \qquad (9.120)$$

and that the corresponding small vertical velocity component is

$$u_y(z, y, t) = \frac{1}{2\eta}\frac{\partial h}{\partial y}\left(\rho g + \frac{\partial p}{\partial z} - \sigma \frac{\partial^3 h}{\partial z^3}\right) -$$

$$-\frac{1}{\eta}\left(\frac{\partial^2 p}{\partial z^2} - \sigma\frac{\partial^4 h}{\partial z^4}\right)\left(\frac{y^2}{6} - \frac{hy^2}{2}\right) - \frac{\partial T}{\partial y}\frac{y^2}{2\eta}. \qquad (9.121)$$

(b) Show that the profile of the interface, $h(z,t)$, is governed by the kinematic condition,

$$\frac{\partial h}{\partial t} + c\frac{\partial h}{\partial z} = f - \frac{\sigma}{3\eta}\frac{\partial p}{\partial z}\left[h^3\frac{\partial^3 h}{\partial z^3}\right], \qquad (9.122)$$

with

$$c = V + \frac{Th}{\eta} - \frac{1}{2\eta}\left(\rho g + \frac{\partial p}{\partial z}\right)h^2 , \quad f = \frac{h^3}{3\eta}\frac{\partial^2 p}{\partial z^2} - \frac{h^2}{2\eta}\frac{\partial T}{\partial z}. \qquad (9.123)$$

(c) Show that, in case $T = 0$, the resulting steady flow rate expression is

$$Q = Vh - \frac{1}{3\eta}\left(\rho g + \frac{\partial p}{\partial z} - \sigma\frac{\partial^3 h}{\partial z^3}\right)h^3 , \qquad (9.124)$$

with boundary conditions, $\partial h/\partial z = \partial^2 h/\partial z^2 = 0$, as $z \to \pm\infty$.

(d) Show that, under steady conditions, a uniform film is obtained when $T \ll P = c$ and $\sigma = 0$. Show that there are two possible uniform films under these conditions, h_s and h_L, such that $0 < h_s < h_m < h_L$, where $h_m = (V\nu/g)^{1/2}$ is the maximum value of thickness under maximum flow rate $Q_m = (2Vhm)/3$. Show that h_m is the film thickness obtained in the absence of the stripping jet, the presence of which reduces h_m to h_s, which is a solution to Eq. (9.124) in the limit of $\partial p/\partial z = 0$ under boundary conditions $\partial h/\partial z = 0$ and $\partial^2 h/\partial z^2 = 0$ for $z \to \pm\infty$.

(e) Under what combinations of conditions is the steady-state volumetric flow rate given by

$$Q = Vh + \frac{T}{2\eta}h^2 - \frac{1}{3\eta}\left(\rho g + \frac{\partial p}{\partial z} - \sigma\frac{\partial^3 h}{\partial z^3}\right)h^3 ?$$

(In this case a uniform film is possible.)

(f) Solve the transient Eq. (9.122) in the limit of small and large capillary numbers assuming constant external stress, $T = T_0$, and external pressure, $p = -\rho gz$. (Hint: First, nondimensionalize the equation and boundary conditions.)

9.15. *Wetting and contact angles.* In drinking liquids from cylindrical-type caps, it is often observed that an almost circular dry island appears at the bottom, surrounded by a thin film of liquid when the bottom is in some inclination ϕ. Explain this wetting phenomenon. Can the wet island be sustained with horizontal bottom? Make and evaluate any assumptions and approximations.

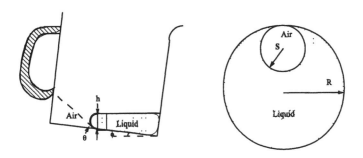

Figure 9.28. *Vertical and horizontal cross section of residual liquid film over an inclined cap bottom with $h \ll S < R$.*

9.4 References

1. O. Reynolds, "Papers on Mechanical and Physical Aspects," *Phil. Trans. Roy. Soc.* **177**, 157 (1886).

2. A. Cameron, *Basic Lubrication Theory*, Longman, 1974.

3. B.V. Deryagin and S.M. Lexi, *Film Coating Theory*, Focal Press, New York, 1964.

4. T.C. Papanastasiou, "Lubrication Flows," *Chem. Eng. Educ.* **24**, 50 (1989).

5. G.K. Batchelor, *An Introduction to Fluid Dynamics*, Cambridge University Press, Cambridge, 1979.

6. N. Tipei, *Theory of Lubrication*, Stanford University Press, 1962.

7. G.K. Miltsios, D.J. Paterson, and T.C. Papanastasiou, "Solution of the lubrication problem and calculation of friction of piston ring," *J. Tribology*, ASME **111**, 635 (1989).

8. M.M. Denn, *Process Fluid Mechanics*, Prentice-Hall, Englewood Cliffs, New Jersey, 1980.

9. D.J. Coyle, *The Fluid Dynamics of Roll Coating: Steady Flows, Stability and Rheology*, Ph.D. Thesis, University of Minnesota, 1984.

10. S. Middleman, *Fundamentals of Polymer Processing*, McGraw-Hill, New York, 1977.

11. N.E. Bixler, *Mechanics and Stability of Extrusion Coating*, Ph.D. Thesis, University of Minnesota, 1983.

12. D.A. Edwards, H. Brenner, and D.T. Wasan, *Interfacial Transport Processes and Rheology*, Butterworth-Heinemann, Boston, 1991.

13. J.R.A. Pearson, *Mechanics of Polymer Processing*, Elsevier Applied Science Publishers, London, 1985.

14. Z. Tadmor and C.G. Gogos, *Principles of Polymer Processing*, Wiley & Sons, New York, 1979.

15. L.E. Scriven and W.J. Suszynski, "Take a closer look at coating problems," *Chem. Eng. Progr.* **24**, September 1990.

16. E.D. Cohen, "Coatings: Going below the surface," *Chem. Eng. Progr.* **19**, September 1990.

17. M. Van Dyke, *Perturbation Methods in Fluid Mechanics*, Academic Press, New York, 1964.

18. N.A. Anturkar, T.C. Papanastasiou, and J.O. Wilkes, "Lubrication theory for n-layer thin-film flow with applications to multilayer extrusion and coating," *Chem. Eng. Sci.* **45**, 3271 (1990).

19. C.J.S. Petrie, *Elongational Flows: Aspects of the Behavior of Model Elastoviscous Fluids*, Pitman, London, 1979.

20. S. Chatraei, C.W. Macosko, and H.H. Winter, "Lubricated squeezing flow: a new biaxial extensional rheometer," *J. Rheology* **34**, 433 (1981).

21. T.C. Papanastasiou, C.W. Macosko, and L.E. Scriven, "Analysis of lubricated squeezing flow," *Int. J. Num. Meth. Fluids* **6**, 819 (1986).

22. D.D. Joseph, "Boundary conditions for thin lubrication layers," *Phys. Fluids* **23**, 2356 (1980).

23. T.M. Geffen, "Oil production to expect from known technology," *Oil Gas J.* **66** (1973).

24. K.J. Mikkelsen, C.W. Macosko, and G.G. Fuller, "Opposed jets: An extensional rheometer for low viscosity liquids," *Proc. Xth Int. Congr. Rheol.*, Sydney, 1989.

25. Z. Chen and T.C. Papanastasiou, "Elongational viscosity by fiber spinning," *Rheol. Acta.* **29**, 385 (1990).

26. L.E. Scriven, *Intermediate Fluid Mechanics*, Lectures, University of Minnesota, 1980.

27. S.M. Alaie and T.C. Papanastasiou, "Film casting of viscoelastic liquid," *Polymer Eng. Sci.* **31**, 67 (1991).

28. E.O. Tuck and J.M. Vanden Broeck, "Influence of surface tension on jet-stripped continuous coating of sheet materials," *AIChE J.* **30**, 808 (1984).

29. P.C. Sukanek, "A model for spin coating with topography," *J. Electrochem. Soc.* **10**, 3019 (1989).

Chapter 10

CREEPING BIDIRECTIONAL FLOWS

Consider the nondimensionalized, steady, Navier–Stokes equation in the absence of body forces,

$$Re\,(\mathbf{u} \cdot \nabla \mathbf{u}) = -\nabla p + \nabla^2 \mathbf{u},\tag{10.1}$$

where

$$Re \equiv \frac{\rho V L}{\eta}\tag{10.2}$$

is the Reynolds number. When the motion of the fluid is "very slow," the Reynolds number is vanishingly small,

$$Re \ll 1,$$

and the flow is said to be *creeping* or *Stokes flow*. In other words, creeping flows are those dominated by viscous forces; the nonlinear inertia term, $Re(\mathbf{u} \cdot \nabla \mathbf{u})$, is negligible compared to the linear viscous term, $\nabla^2 \mathbf{u}$. The Navier-Stokes equation may then be approximated by the *Stokes equation*,

$$-\nabla p + \nabla^2 \mathbf{u} = \mathbf{0}.\tag{10.3}$$

This linear equation, together with the continuity equation,

$$\nabla \cdot \mathbf{u} = 0,\tag{10.4}$$

can be solved analytically for a broad range of problems. Flows at small but nonzero Reynolds numbers are amenable to regular perturbation analysis. The Stokes flow solution can thus be viewed as the zeroth-order approximation to the solution in terms of the Reynolds number.

Note that Eq. (10.3) also holds true for steady, unidirectional flows which are not necessarily creeping; in this case, the inertia term, $\mathbf{u} \cdot \nabla \mathbf{u}$, is identically zero. A similar observation applies to lubrication flows. The inertia term is negligible when

$$\alpha\, Re \ll 1,$$

where α is the inclination of the channel and the Reynolds number is not necessarily vanishingly small.

Reversibility is an important property of Stokes flow. If **u** and p satisfy Eqs. (10.3) and (10.4), then it is evident that the reversed solution, $-$**u** and $-p$, also satisfies the same equations. The reversed flow is obtained by using "reversed" boundary conditions, e.g., **u**$=-f(\mathbf{r})$ instead of **u**$=f(\mathbf{r})$ along a boundary, etc. Reversibility is lost, once the nonlinear convective term, $Re(\mathbf{u}\cdot\nabla\mathbf{u})$, becomes nonzero.

Laminar, slow flow approaching and deflected by a submerged stationary sphere or cylinder or, equivalently, flow induced by a slowly traveling sphere or cylinder in a bath of still liquid, are representative examples of creeping flow. These flows are important in particle mechanics [1] and apply to air cleaning devices from particles [2], to centrifugal or sedimentation separators, to fluidized-bed reactors, and to chemical and physical processes involving gas bubbles or droplets [3]. Slow flows in converging or diverging channels and conical pipes are also examples of important creeping flows [4]. Finally, flows in the vicinity of corners and other geometrical singularities are creeping, being retarded by the encounter with the solid boundaries [5].

This chapter is devoted to creeping, incompressible, bidirectional flows. An excellent analytical tool for solving such flows is the *stream function*. Consider, for example, the creeping bidirectional flow on the xy-plane for which

$$u_x = u_x(x,y)\,, \quad u_y = u_y(x,y) \quad \text{and} \quad u_z = 0\,.$$

For incompressible flow, the continuity equation takes the following form,

$$\frac{\partial u_x}{\partial x} + \frac{\partial u_y}{\partial y} = 0\,; \tag{10.5}$$

the two nonzero components of the Navier–Stokes equation become

$$-\frac{\partial p}{\partial x} + \left(\frac{\partial^2 u_x}{\partial x^2} + \frac{\partial^2 u_x}{\partial y^2}\right) = 0 \tag{10.6}$$

and

$$-\frac{\partial p}{\partial y} + \left(\frac{\partial^2 u_y}{\partial x^2} + \frac{\partial^2 u_y}{\partial y^2}\right) = 0\,. \tag{10.7}$$

Hence, the flow is governed by a system of three PDEs corresponding to the three unknown fields, p, u_x, and u_y.

The continuity equation is automatically satisfied by introducing *Lagrange's stream function*, $\psi(x,y)$, such that

$$u_x = -\frac{\partial \psi}{\partial y} \quad \text{and} \quad u_y = \frac{\partial \psi}{\partial x}\,. \tag{10.8}$$

The pressure, p, can be eliminated by differentiating Eqs. (10.6) and (10.7) with respect to y and x, respectively, and by subtracting one equation from the other. Substituting u_x and u_y in terms of ψ into the resulting equation leads to

$$\frac{\partial^4 \psi}{\partial x^4} + 2 \frac{\partial^4 \psi}{\partial x^2 \partial y^2} + \frac{\partial^4 \psi}{\partial y^4} = 0 \,. \tag{10.9}$$

Recalling that the Laplace operator in Cartesian coordinates is defined by

$$\nabla^2 \equiv \frac{\partial^2}{\partial x^2} + \frac{\partial^2}{\partial y^2} \,, \tag{10.10}$$

Eq. (10.9) can be written in the more concise form

$$\nabla^4 \psi = \nabla^2 \left(\nabla^2 \psi \right) = 0 \,. \tag{10.11}$$

The differential operator ∇^4, defined by

$$\nabla^4 \equiv \nabla^2 \left(\nabla^2 \right) \,, \tag{10.12}$$

is called the *biharmonic operator*. Equation (10.11) is referred to as the *biharmonic* or *Stokes equation*.

The advantage of using the stream function is that, instead of a system of three PDEs for the three unknown fields, u_x, u_y, and p, we have to solve a single PDE for the new dependent variable, ψ. The price we pay is that the highest derivatives of the governing equation are now fourth-order instead of second-order. Once $\psi(x,y)$ is calculated, the velocity components can be obtained by means of Eqs. (10.8). The pressure, p, can then be calculated by integrating the momentum equations (10.6) and (10.7).

In the following three sections, we consider the use of the stream function for three classes of creeping, incompressible, bidirectional flows:

(a) Plane flows in polar coordinates;

(b) Axisymmetric flows in cylindrical coordinates; and

(c) Axisymmetric flows in spherical coordinates.

The various forms of the stream function and the resulting fourth-order PDEs are tabulated in Table 10.1. It should be noted that the use of the stream function is not restricted to creeping flows. The full forms of the momentum equation in terms of the stream function, for all the aforementioned classes of flow, can be found in Reference [6].

Plane flow in Cartesian coordinates

Assumptions: $\quad u_x = u_x(x, y), \quad u_y = u_y(x, y), \quad u_z = 0$

Stream function: $\quad u_x = -\dfrac{\partial \psi}{\partial y}, \quad u_y = \dfrac{\partial \psi}{\partial x}$

Momentum equation: $\quad \nabla^4 \psi = \nabla^2 \left(\nabla^2 \psi \right) = 0$

$$\nabla^2 \equiv \frac{\partial^2}{\partial x^2} + \frac{\partial^2}{\partial y^2}, \qquad \nabla^4 \equiv \frac{\partial^4}{\partial x^4} + 2\frac{\partial^4}{\partial x^2 \partial y^2} + \frac{\partial^4}{\partial y^4}$$

Plane flow in polar coordinates

Assumptions: $\quad u_r = u_r(r, \theta), \quad u_\theta = u_\theta(r, \theta), \quad u_z = 0$

Stream function: $\quad u_r = -\dfrac{1}{r}\dfrac{\partial \psi}{\partial \theta}, \quad u_\theta = \dfrac{\partial \psi}{\partial r}$

Momentum equation: $\quad \nabla^4 \psi = \nabla^2 \left(\nabla^2 \psi \right) = 0$

$$\nabla^2 \equiv \frac{\partial^2}{\partial r^2} + \frac{1}{r}\frac{\partial}{\partial r} + \frac{1}{r^2}\frac{\partial^2}{\partial \theta^2}$$

Axisymmetric flow in cylindrical coordinates

Assumptions: $\quad u_z = u_z(r, z), \quad u_r = u_r(r, z), \quad u_\theta = 0$

Stream function: $\quad u_z = -\dfrac{1}{r}\dfrac{\partial \psi}{\partial r}, \quad u_r = \dfrac{1}{r}\dfrac{\partial \psi}{\partial z}$

Momentum equation: $\quad E^4 \psi = E^2 \left(E^2 \psi \right) = 0$

$$E^2 \equiv \frac{\partial^2}{\partial r^2} - \frac{1}{r}\frac{\partial}{\partial r} + \frac{\partial^2}{\partial z^2}$$

Axisymmetric flow in spherical coordinates

Assumptions: $\quad u_r = u_r(r, \theta), \quad u_\theta = u_\theta(r, \theta), \quad u_\phi = 0$

Stream function: $\quad u_r = -\dfrac{1}{r^2 \sin^2\theta}\dfrac{\partial \psi}{\partial \theta}, \quad u_\theta = \dfrac{1}{r \sin\theta}\dfrac{\partial \psi}{\partial r}$

Momentum equation: $\quad E^4 \psi = E^2 \left(E^2 \psi \right) = 0$

$$E^2 \equiv \frac{\partial^2}{\partial r^2} + \frac{\sin\theta}{r^2}\frac{\partial}{\partial \theta}\left(\frac{1}{\sin\theta}\frac{\partial}{\partial \theta} \right)$$

Table 10.1. *Stokes flow equations in terms of the stream function.*

10.1 Plane Flow in Polar Coordinates

In this section, we consider two-dimensional, creeping, incompressible flows in polar coordinates in which

$$u_r = u_r(r, \theta) \quad \text{and} \quad u_\theta = u_\theta(r, \theta) \,. \tag{10.13}$$

The continuity equation becomes

$$\frac{\partial}{\partial r}(r u_r) + \frac{\partial u_\theta}{\partial \theta} = 0 \tag{10.14}$$

and is automatically satisfied by a Stokes' stream function $\psi(r, \theta)$ defined by

$$u_r = -\frac{1}{r}\frac{\partial \psi}{\partial \theta} \quad \text{and} \quad u_\theta = \frac{\partial \psi}{\partial r} \,. \tag{10.15}$$

Eliminating the pressure from the r- and θ-components of the Navier–Stokes equations, we obtain the biharmonic equation (see Problem 10.1)

$$\nabla^4 \psi = \nabla^2\left(\nabla^2 \psi\right) = 0 \,. \tag{10.16}$$

Recall that, in polar coordinates, the Laplace operator is given by

$$\nabla^2 \equiv \frac{\partial^2}{\partial r^2} + \frac{1}{r}\frac{\partial}{\partial r} + \frac{1}{r^2}\frac{\partial^2}{\partial \theta^2} \,. \tag{10.17}$$

As demonstrated by Lugt and Schwiderski [7], Eq. (10.16) admits separated solutions of the form

$$\psi(r, \theta) = r^{\lambda+1} f_\lambda(\theta) \tag{10.18}$$

where the exponent λ may be complex. For the Laplacian of ψ, we get

$$\nabla^2 \psi = r^{\lambda-1}\left[f_\lambda''(\theta) + (\lambda+1)^2 f_\lambda(\theta)\right] \,,$$

where the primes designate differentiation with respect to θ. Another application of the Laplace operator yields

$$\nabla^4 \psi = r^{\lambda-3}\left\{f_\lambda''''(\theta) + \left[(\lambda-1)^2 + (\lambda+1)^2\right]f_\lambda''(\theta) + (\lambda-1)^2(\lambda+1)^2 f_\lambda(\theta)\right\} \,.$$

Due to Eq. (10.16),

$$f_\lambda''''(\theta) + \left[(\lambda-1)^2 + (\lambda+1)^2\right]f_\lambda''(\theta) + (\lambda-1)^2(\lambda+1)^2 f_\lambda(\theta) = 0 \,. \tag{10.19}$$

This is a linear, homogeneous, fourth-order ordinary differential equation, the characteristic equation of which is

$$\left[m^2 + (\lambda + 1)^2\right]\left[m^2 + (\lambda - 1)^2\right] = 0.$$

Hence, the general solution for $f_\lambda(\theta)$ is

$$f_\lambda(\theta) = A_\lambda \cos(\lambda+1)\theta + B_\lambda \sin(\lambda+1)\theta + C_\lambda \cos(\lambda-1)\theta + D_\lambda \sin(\lambda-1)\theta, \quad (10.20)$$

where the constants A_λ, B_λ, C_λ, and D_λ may be complex.

Therefore, the general solution of Eq. (10.16) is

$$\psi(r,\theta) = r^{\lambda+1}\left[A_\lambda \cos(\lambda + 1)\theta + B_\lambda \sin(\lambda + 1)\theta + C_\lambda \cos(\lambda - 1)\theta + D_\lambda \sin(\lambda - 1)\theta\right].$$
$$(10.21)$$

The two velocity components are now easily obtained:

$$u_r(r,\theta) = -r^\lambda\left[-A_\lambda\,(\lambda + 1)\,\sin(\lambda + 1)\theta \;+\; B_\lambda\,(\lambda + 1)\,\cos(\lambda + 1)\theta\right.$$

$$\left.-\; C_\lambda\,(\lambda - 1)\,\sin(\lambda - 1)\theta \;+\; D_\lambda\,(\lambda - 1)\cos(\lambda - 1)\theta\right], \quad (10.22)$$

$$u_\theta(r,\theta) = (\lambda+1)r^\lambda\left[A_\lambda\,\cos(\lambda + 1)\theta \;+\; B_\lambda\,\sin(\lambda + 1)\theta\right.$$

$$\left.+\; C_\lambda\,\cos(\lambda - 1)\theta \;+\; D_\lambda\,\sin(\lambda - 1)\theta\right]. \quad (10.23)$$

The pressure p is calculated by integrating the r- and θ-momentum equations (see Problem 10.2):

$$p = -4\eta\,r^{\lambda-1}\left[C_\lambda\,\sin(\lambda - 1)\theta \;-\; D_\lambda\,\cos(\lambda - 1)\theta\right]. \quad (10.24)$$

In general, there is an infinite number of admissible values of λ which depend on the geometry and the boundary conditions. Since the problem is linear, these solutions may be superimposed. The particular solutions to Eq. (10.16) for $\lambda = -1$, 0, and 1 are considered in Problem 10.3. A particular solution independent of θ is given in Problem 10.4.

Example 10.1.1. Flow near a corner

Consider flow of a viscous liquid between two rigid boundaries fixed at an angle 2α (Fig. 10.1). Since the velocity on the two walls is zero, inertia terms are negligible near the neighborhood of the corner. Therefore, the flow can be assumed to be *locally* creeping. The solution to this flow problem was determined by Dean and Montagnon [8]. The stream function is expanded in a series of the form

$$\psi(r,\theta) = \sum_{k=1}^{\infty} \psi_{\lambda_k} = \sum_{k=1}^{\infty} a_{\lambda_k}\, r^{\lambda_k+1}\, f_{\lambda_k}(\theta), \quad (10.25)$$

where the polar coordinates (r, θ) are centered at the vertex. The exponents λ_k are suitably ordered so that

$$0 < \mathcal{R}e(\lambda_1) < \mathcal{R}e(\lambda_2) < \cdots .$$

The first of the inequalities ensures that the velocity vanishes at the corner. The second one ensures that the first term in the summation will dominate unless $a_{\lambda_1}=0$.

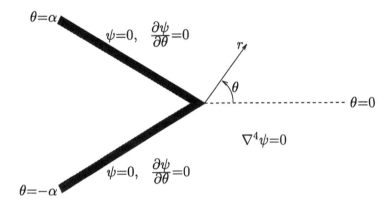

Figure 10.1. *Creeping flow near a corner.*

As pointed out by Dean and Montagnon [8], a disturbance far from the corner can generate either an *antisymmetric* or a *symmetric* flow pattern near the corner, and the corresponding stream function is an even or odd function of θ, respectively. Taking advantage of the linearity of the Stokes equation, we study the two types of flow separately.

Antisymmetric flow near a corner
For this type of flow, $f_\lambda(\theta)$ is even $(B_\lambda=D_\lambda=0)$ and

$$f_\lambda(\theta) = A_\lambda \, \cos(\lambda + 1)\theta \, + \, C_\lambda \, \cos(\lambda - 1)\theta \,, \tag{10.26}$$

The boundary conditions $u_r=u_\theta=0$ on $\theta=\pm\alpha$ demand that

$$f_\lambda(\theta) = f'_\lambda(\theta) = 0 \quad \text{on} \quad \theta = \pm\alpha \,,$$

which gives the following two equations:

$$A_\lambda \, \cos(\lambda + 1)\alpha \, + \, C_\lambda \, \cos(\lambda - 1)\alpha \, = \, 0$$

$$A_\lambda \, (\lambda + 1) \, \sin(\lambda + 1)\alpha \, + \, C_\lambda \, (\lambda - 1) \, \sin(\lambda - 1)\alpha \, = \, 0$$

For a nontrivial solution for A_λ and C_λ to exist, the determinant of the coefficient matrix must be zero,

$$\begin{vmatrix} \cos(\lambda+1)\alpha & \cos(\lambda-1)\alpha \\ (\lambda+1)\,\sin(\lambda+1)\alpha & (\lambda-1)\,\sin(\lambda-1)\alpha \end{vmatrix} = 0 \,.$$

With a little manipulation, we get the following eigenvalue equation

$$\sin 2\lambda\alpha \;=\; -\lambda\,\sin 2\alpha\,. \tag{10.27}$$

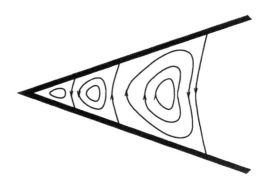

Figure 10.2. *Sketch of Moffatt's vortices in a sharp corner.*

With the obvious exception of the trivial solution $\lambda=0$, the eigenvalues are necessarily complex when 2α is less than approximately 146.4°. This implies the existence of an infinite sequence of eddies near the corner. These were predicted analytically by Moffatt [9]. A sketch of Moffatt's vortices is shown in Fig. 10.2. The strength of these vortices decays exponentially as the corner is approached. The ratio of the distance of the centers of successive vortices is given by

$$\frac{r_i}{r_{i+1}} \;=\; e^{\pi/q_1}\,,$$

where q_1 is the imaginary part of the leading eigenvalue, $\lambda_1=p_1+iq_1$. Table 10.2 provides the real and imaginary parts of λ_1 for various values of 2α.

For values of 2α greater than 146.4°, all the solutions of Eq. (10.27) are real. As shown in Table 10.2, the value of the leading exponent λ_1 decreases with the angle 2α. When $2\alpha=180^\circ$, $\lambda_1=1$, which corresponds to simple shear flow. For values of 2α greater than 180°, λ_1 is less than unity. From Eqs. (10.22) and (10.23) we deduce that, in such a case, the velocity derivatives and, consequently, the pressure and

2α (in ^{o})	p_1	q_1
30.0	8.0630	4.2029
60.0	4.0593	1.9520
90.0	2.7396	1.1190
120.0	2.0941	0.6046
146.4	1.7892	0.0000
150.0	1.9130	
180.0	1.0000	
210.0	0.7520	
240.0	0.6157	
270.0	0.5445	
300.0	0.5122	
330.0	0.5015	
360.0	0.5000	

Table 10.2. *Real and imaginary parts of the leading exponent,* $\lambda_1 = p_1 + iq_1$, *for antisymmetric flow near a corner.*

the stress components go to infinity at the corner. This is an example of a *stress singularity* that is caused by the nonsmoothness of the boundary. The strongest singularity appears at $2\alpha = 360^o$; in this case, $\lambda_1 = 0.5$, which corresponds to flow around a semi-infinite flat plate.

Symmetric flow near a corner

For this type of flow, $f_\lambda(\theta)$ is odd ($A_\lambda = C_\lambda = 0$) and

$$f_\lambda(\theta) \; = \; B_\lambda \, \sin(\lambda + 1)\theta \, + \, D_\lambda \, \sin(\lambda - 1)\theta \,. \tag{10.28}$$

In this case, the eigenvalue equation is

$$\sin 2\lambda\alpha \; = \; \lambda \, \sin 2\alpha \,. \tag{10.29}$$

The real and imaginary parts of the leading exponent, $\lambda_1 = p_1 + iq_1$, are tabulated in Table 9.3 for various values of the angle 2α. (The trivial solutions $\lambda_1 = 0$ and 1 are not taken into account.) For 2α greater than approximately 159.2^o, Eq. (10.29) admits only real solutions. For $2\alpha = 180^o$, $\lambda_1 = 2$, which corresponds to orthogonal stagnation-point flow.   □

2α (in ^{o})	p_1	q_1
30.0	14.3303	5.1964
60.0	7.1820	2.4557
90.0	4.8083	1.4639
120.0	3.6307	0.8812
150.0	2.9367	0.3637
159.2	2.8144	0.0000
180.0	2.0000	
210.0	1.4858	
240.0	1.1489	
270.0	0.9085	
300.0	0.7309	
330.0	0.5982	
360.0	0.5000	

Table 10.3. *Real and imaginary parts of the leading exponent, $\lambda_1 = p_1 + iq_1$, for symmetric flow near a corner.*

Example 10.1.2. Intersection of a wall and a free surface

Due to symmetry, the preceding analysis of creeping flow near a corner also holds for flow near the intersection of a rigid boundary and a free surface positioned at $\theta = 0$ (Fig. 10.3). Along the free surface,

$$u_\theta = 0 \quad \text{and} \quad \tau_{r\theta} = \eta \left[r \frac{\partial}{\partial r} \left(\frac{u_\theta}{r} \right) + \frac{1}{r} \frac{\partial u_r}{\partial \theta} \right] = 0 \, ;$$

consequently,

$$u_\theta = \frac{\partial u_r}{\partial \theta} = 0 \quad \Longrightarrow \quad \psi = \frac{\partial^2 \psi}{\partial \theta^2} = 0 \, .$$

Using physical arguments, Michael [10] showed that the angle of separation, α, cannot take arbitrary values. He showed that, when the surface tension is zero, α mus be equal to π. Therefore, we focus on the case of flow near a wall and a free surface meeting at an angle π, as in Fig. 10.4.

Even set of solutions

Even solutions,

$$\psi_\lambda = r^{\lambda+1} [A_\lambda \cos(\lambda+1)\theta + C_\lambda \cos(\lambda-1)\theta] \, ,$$

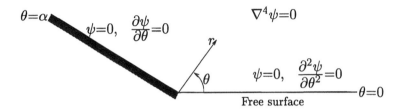

Figure 10.3. *Creeping flow near the intersection of a wall and a free surface.*

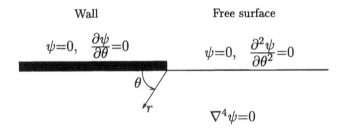

Figure 10.4. *Creeping flow near a wall and a free surface meeting at an angle π.*

satisfy automatically the conditions $u_r=0$ at $\theta=0$ and $u_\theta=0$ at $\theta=\pi$ (see Fig. 10.4, for the definition of θ). The condition $u_\theta=0$ at $\theta=0$ demands that

$$A_\lambda \cos(\lambda+1)\theta + C_\lambda \cos(\lambda-1)\theta = 0 \quad \text{at} \quad \theta = 0 \,,$$

which yields $C_\lambda=-A_\lambda$. Finally, the condition $\tau_{r\theta}=0$ at $\theta=\pi$ leads to the following equation

$$(\lambda+1)^2 \cos(\lambda+1)\pi - (\lambda-1)^2 \cos(\lambda-1)\pi = 0 \,.$$

From standard trigonometric identities, we get

$$2(\lambda^2+1) \sin\lambda\pi \sin\pi + 4\lambda\cos\lambda\pi \cos\pi = 0 \quad \Longrightarrow$$

$$\cos\lambda\pi = 0 \quad \Longrightarrow \quad \lambda = \frac{1}{2}, \frac{3}{2}, \cdots$$

Therefore, even solutions are given by

$$\psi_\lambda = a_\lambda \, r^{\lambda+1} [\cos(\lambda+1)\theta - \cos(\lambda-1)\theta] \,, \quad \lambda = \frac{1}{2}, \frac{3}{2}, \cdots \tag{10.30}$$

where $a_\lambda = A_\lambda = -C_\lambda$. The corresponding velocity components and the pressure are:

$$u_r \;=\; -a_\lambda\, r^\lambda\, \left[-(\lambda+1)\,\sin(\lambda+1)\theta + (\lambda-1)\,\sin(\lambda-1)\theta\right], \qquad (10.31)$$

$$u_\theta \;=\; a_\lambda\,(\lambda+1)\,r^\lambda\,\left[\cos(\lambda+1)\theta - \cos(\lambda-1)\theta\right], \qquad (10.32)$$

$$p \;=\; 4a_\lambda \eta \lambda\, r^{\lambda-1}\,\sin(\lambda-1)\theta. \qquad (10.33)$$

Note that the leading term of the pressure is characterized by an inverse-square-root singularity:

$$p \sim 2a_{1/2}\,\eta\,\frac{1}{\sqrt{r}}\,\sin\frac{\theta}{2}.$$

This is also true for the velocity derivatives and the stress components. This is an example of a stress singularity caused by the sudden change of the boundary condition along a smooth boundary.

Odd set of solutions
Odd solutions,

$$\psi_\lambda \;=\; r^{\lambda+1}\,\left[B_\lambda\,\sin(\lambda+1)\theta + D_\lambda\,\sin(\lambda-1)\theta\right],$$

satisfy automatically the conditions $u_\theta = 0$ at $\theta = 0$ and $\tau_{r\theta} = 0$ at $\theta = \pi$. The condition $u_r = 0$ at $\theta = 0$ requires that

$$(\lambda+1)B_\lambda + (\lambda-1)D_\lambda \;=\; 0.$$

The remaining condition $u_\theta = 0$ at $\theta = \pi$ leads to

$$(\lambda-1)\,\sin(\lambda+1)\pi - (\lambda+1)\,\sin(\lambda+1)\pi \;=\; 0 \qquad \Longrightarrow$$

$$2\lambda\,\cos\pi\,\sin\lambda\pi - 2\lambda\,\sin\pi\,\sin\lambda\pi \;=\; 0 \qquad \Longrightarrow$$

$$\sin\lambda\pi \;=\; 0 \qquad \Longrightarrow \qquad \lambda = 2, 3, \cdots$$

Note that the trivial solution $\lambda = 1$ has been omitted. Therefore, odd solutions are of the form

$$\psi_\lambda \;=\; a_\lambda\, r^{\lambda+1}\,\left[(\lambda-1)\,\sin(\lambda+1)\theta + (\lambda+1)\,\sin(\lambda-1)\theta\right], \qquad \lambda = 2, 3, \cdots \quad (10.34)$$

where $(\lambda^2 - 1)a_\lambda = (\lambda + 1)B_\lambda = -(\lambda - 1)D_\lambda$. The corresponding solutions for u_r, u_θ, and p are:

$$u_r = -a_\lambda (\lambda^2 - 1) r^\lambda [\cos(\lambda + 1)\theta - \cos(\lambda - 1)\theta] , \qquad (10.35)$$

$$u_\theta = a_\lambda (\lambda + 1) r^\lambda [(\lambda - 1) \sin(\lambda + 1)\theta + (\lambda + 1) \sin(\lambda - 1)\theta] , \quad (10.36)$$

$$p = -4a_\lambda \eta \lambda (\lambda + 1) r^{\lambda - 1} \cos(\lambda - 1)\theta . \qquad (10.37)$$

$\square$

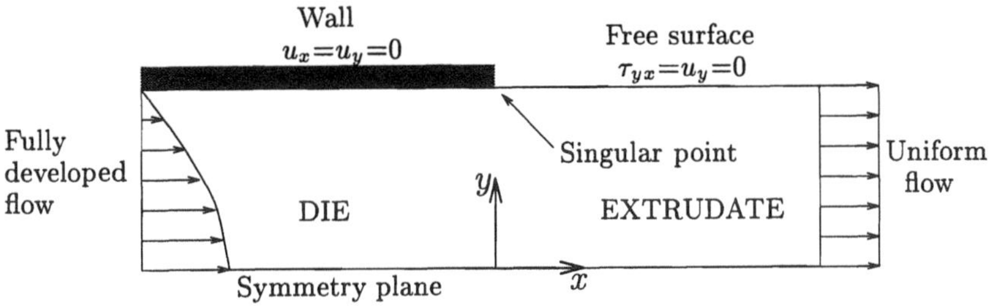

Figure 10.5. *The plane stick-slip problem..*

It should be noted that the solutions discussed in the previous two examples hold only *locally*. The constants a_λ and b_λ are determined from the boundary conditions of the global problem. Consider, for example, the so-called plane *stick-slip* problem, illustrated in Fig. 10.5. This problem owes its name to the fact that the boundary conditions change suddenly at the exit of the die, from no slip to slip. The stick-slip problem is the special case of the *extrudate-swell problem* in the limit of infinite surface tension which causes the free surface to be flat. The singular solution obtained in Example 10.1.2 holds in the neighborhood of the exit of the die. The leading term of the local solution is

$$\psi_{1/2} = a_{1/2} r^{3/2} \left(\cos \frac{3\theta}{2} - \cos \frac{\theta}{2} \right) ,$$

where the polar coordinates (r, θ) are centered at the exit of the die. The plane stick-slip problem was solved analytically by Richardson [11]. It turns out that $a_{1/2} = \sqrt{3/2\pi} = 0.690988$.

As already mentioned, the velocity derivatives and the stresses corresponding to the leading term of the local solution are characterized by an inverse-square-root singularity. This has a negative effect on the performance of standard numerical methods used to model the stick-slip (or the extrudate-swell) flow. The rate of convergence with mesh refinement and the accuracy are, in general, poor in the neighborhood of stress singularities. The strength of the singularity may be alleviated by using slip along the wall which leads to nonsingular finite stresses [12,13]. Alternatively, special numerical techniques, such as singular finite elements [14,15] or special mesh refinement methods [16], must be employed, in order to get accurate results in the neighborhood of the singularity.

10.2 Axisymmetric Flow in Cylindrical Coordinates

Consider a creeping, axisymmetric, incompressible flow in cylindrical coordinates such that

$$u_z = u_z(r, z), \quad u_r = u_r(r, z), \quad \text{and} \quad u_\theta = 0. \tag{10.38}$$

It is easily shown that the stream function $\psi(r, z)$, defined by

$$u_z = -\frac{1}{r}\frac{\partial \psi}{\partial r} \quad \text{and} \quad u_r = \frac{1}{r}\frac{\partial \psi}{\partial z}, \tag{10.39}$$

satisfies the continuity equation identically. Substituting u_z and u_r into the z- and r-components of the Navier–Stokes equation, and eliminating the pressure lead to the following equation (Problem 10.7):

$$E^4\psi = E^2\left(E^2\psi\right) = 0, \tag{10.40}$$

where the differential operator E^2 is defined by

$$E^2 \equiv \frac{\partial^2}{\partial r^2} - \frac{1}{r}\frac{\partial}{\partial r} + \frac{\partial^2}{\partial z^2}. \tag{10.41}$$

Separating the axial from the radial dependence and stipulating a power-law functional dependence on r, we seek a solution to Eq. (10.40) of the form

$$\psi = r^\lambda f(z). \tag{10.42}$$

Applying the operator E^2 to the above solution yields

$$E^2\psi = \lambda(\lambda - 2) r^{\lambda-2} f(z) + r^\lambda f''(z),$$

where the primes denote differentiation with respect to z. Applying the operator E^2 once again, we get

$$E^4\psi = \lambda(\lambda-2)^2(\lambda-4)r^{\lambda-4}f(z) + 2\lambda(\lambda-2)r^{\lambda-2}f''(z) + r^\lambda f''''(z).$$

Due to the Stokes flow Eq. (10.40), the only admissible values of λ are 0 and 2. For both values we get the simple fourth-order ODE

$$f''''(z) = 0,$$

the general solution of which is

$$f(z) = c_0 + c_1 z + c_2 z^2 + c_3 z^3. \tag{10.43}$$

The value $\lambda=0$ corresponds to the solution $\psi=f(z)$ which is independent of r. Let us, however, focus on the more interesting case of $\lambda=2$ in which we have

$$\psi(r,z) = r^2\left(c_0 + c_1 z + c_2 z^2 + c_3 z^3\right). \tag{10.44}$$

The values of the constants c_0, c_1, c_2, and c_3 are determined from the boundary conditions. For the velocity components we get:

$$u_z(r,z) = -\frac{1}{r}\frac{\partial\psi}{\partial r} = -2f(z) = -2\left(c_0 + c_1 z + c_2 z^2 + c_3 z^3\right) \tag{10.45}$$

$$u_r(r,z) = \frac{1}{r}\frac{\partial\psi}{\partial z} = rf'(z) = r\left(c_1 + 2c_2 z + 3c_3 z^2\right). \tag{10.46}$$

It can be shown that the z- and r-components of the Navier–Stokes become

$$-\frac{\partial p}{\partial z} + \eta\frac{\partial^2 u_z}{\partial z^2} = 0 \quad\text{and}\quad -\frac{\partial p}{\partial r} + \eta\frac{\partial^2 u_r}{\partial z^2} - 0$$

or

$$\frac{\partial p}{\partial z} = -2\eta\, f''(z) \quad\text{and}\quad \frac{\partial p}{\partial r} = \eta\, r\, f'''(z),$$

respectively. Integration of the above two equations yields

$$p(r,z) = -3\eta\, c_3\left(2z^2 - r^2\right) - 4\eta\, c_2 z + c, \tag{10.47}$$

where c is a constant.

Example 10.2.1. Axisymmetric squeezing flow

Squeezing flows are induced by externally applied normal stresses or vertical velocities by means of a mobile boundary. The induced normal velocity propagates within the liquid due to incompressibility, and changes direction due to obstacles to normal penetration. The most characteristic example is *Stefan's squeezing flow* [17], illustrated in Fig. 10.6. The vertically moving fronts meet the resistance of the inner liquid layers and are deflected radially. For small values of the velocity V of the two plates, the gap $2H$ changes slowly with time and can be assumed to be constant, that is, the flow can be assumed to be *quasi-steady*. If, in addition, the fluid is highly viscous, then the creeping flow approximation is a valid assumption.

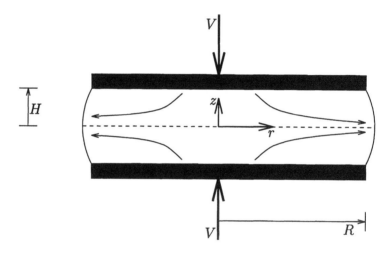

Figure 10.6. *Squeezing flow.*

Introducing the cylindrical coordinates shown in Fig. 10.6 and employing the stream function defined in Eq. (10.39), we can make use of the previous analysis. The stream function is thus given by Eq. (10.44):

$$\psi(r, z) = r^2 f(z) = r^2 \left(c_0 + c_1 z + c_2 z^2 + c_3 z^3 \right) .$$

The four constants are determined from the boundary conditions. At $z=0$, symmetry requires that $u_z = \partial u_r / \partial z = 0$, and therefore

$$f(z) = f''(z) = 0 \quad \text{at} \quad z = 0 ;$$

consequently, $c_0 = c_1 = 0$. At $z = H$, $u_z = V$ and $u_r = 0$, which gives

$$c_1 H + c_3 H^3 = \frac{V}{2} \quad \text{and} \quad c_1 + 3 c_3 H^2 = 0 .$$

Solving for c_1 and c_3, we get:

$$c_1 = \frac{3V}{4H} \quad \text{and} \quad c_3 = -\frac{V}{4H^3}.$$

Therefore,

$$f(z) = \frac{V}{4}\left[3\left(\frac{z}{H}\right) - \left(\frac{z}{H}\right)^3\right].$$

The stream function and the two velocity components are given by

$$\psi = \frac{V}{4}r^2\left[3\left(\frac{z}{H}\right) - \left(\frac{z}{H}\right)^3\right]; \tag{10.48}$$

$$u_z = -\frac{V}{2}\left[3\left(\frac{z}{H}\right) - \left(\frac{z}{H}\right)^3\right]; \tag{10.49}$$

$$u_r = -\frac{3V}{4H}r\left[1 - \left(\frac{z}{H}\right)^2\right]. \tag{10.50}$$

Finally, from Eq. (10.47), the pressure distribution is

$$p(r,z) = \frac{3\eta V}{4H^3}\left(2z^2 - r^2\right) + c.$$

Assuming that $p=p_0$ at $r=R$ and $z=0$, we find that

$$p(r,z) = \frac{3\eta V}{4H^3}\left(2z^2 + R^2 - r^2\right) + p_0. \tag{10.51}$$

□

10.3 Axisymmetric Flow in Spherical Coordinates

In this section, we consider the case of axisymmetric flow in spherical coordinates such that

$$u_r = u_r(r,\theta), \quad u_\theta = u_\theta(r,\theta), \quad \text{and} \quad u_\phi = 0. \tag{10.52}$$

The Stokes stream function, defined by

$$u_r = -\frac{1}{r^2\sin\theta}\frac{\partial\psi}{\partial\theta} \quad \text{and} \quad u_\theta = \frac{1}{r\sin\theta}\frac{\partial\psi}{\partial r}, \tag{10.53}$$

satisfies the continuity equation identically. Substituting Eqs. (10.53) into the r- and θ-momentum equations and eliminating the pressure, we obtain (Problem 10.8)

$$E^4\psi = E^2\left(E^2\psi\right) = 0\,, \tag{10.54}$$

where the differential operator E^2 is defined by

$$E^2 \equiv \frac{\partial^2}{\partial r^2} + \frac{\sin\theta}{r^2}\frac{\partial}{\partial\theta}\left(\frac{1}{\sin\theta}\frac{\partial}{\partial\theta}\right)\,. \tag{10.55}$$

Example 10.3.1. Creeping flow past a fixed sphere

As already mentioned, flows around submerged bodies are of great importance in a plethora of applications. The most important example of axisymmetric flow is the very slow flow past a fixed sphere, illustrated in Fig. 10.7. A practically unbounded viscous, incompressible fluid approaches, with uniform speed U, a sphere of radius R. The sphere is held stationary by some applied external force. Clearly, the resulting flow is axisymmetric with $u_\phi=0$.

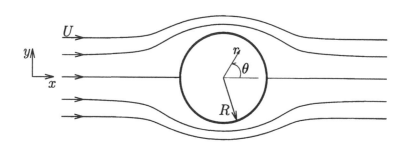

Figure 10.7. *Creeping flow past a sphere.*

The boundary conditions for this flow are as follows.
(a) On $r=R$, $u_r=u_\theta=0$. In terms of the stream function defined in Eq. (10.53), we get

$$\frac{\partial\psi}{\partial\theta} = \frac{\partial\psi}{\partial r} = 0 \quad \text{on} \quad r = R\,. \tag{10.56}$$

(b) As $r \to \infty$,

$$\mathbf{u} = U\,\mathbf{i} = U\left(\cos\theta\,\mathbf{e}_r - \sin\theta\,\mathbf{e}_\theta\right)$$

which gives

$$u_r = U\cos\theta \quad \text{and} \quad u_\theta = -U\sin\theta \quad \text{as} \quad r \to \infty\,.$$

In terms of ψ, we get

$$\frac{\partial \psi}{\partial \theta} = -Ur^2 \sin\theta \cos\theta \quad \text{and} \quad \frac{\partial \psi}{\partial r} = -Ur \sin^2\theta \quad \text{as} \quad r \to \infty \,.$$

Integrating the above two equations, we get

$$\psi = -\frac{U}{2} r^2 \sin^2\theta \quad \text{as} \quad r \to \infty \,. \tag{10.57}$$

The above condition suggests seeking a separated solution to Eq. (10.54) of the form

$$\psi(r,\theta) = U f(r) \sin^2\theta \,. \tag{10.58}$$

In terms of $f(r)$, the boundary conditions (10.56) and (10.57) become:

$$f(r) = f'(r) = 0 \quad \text{at} \quad r = R \tag{10.59}$$

and

$$f(r) = -\frac{1}{2} r^2 \quad \text{as} \quad r \to \infty \,. \tag{10.60}$$

Applying the operator E^2 to the separated solution (10.58) yields

$$E^2 f(r) \sin^2\theta = \sin^2\theta \left(\frac{d^2}{dr^2} - \frac{2}{r^2} \right) f(r) \,,$$

and thus

$$E^4 f(r) \sin^2\theta = \sin^2\theta \left(\frac{d^2}{dr^2} - \frac{2}{r^2} \right)^2 f(r) \,.$$

From Eq. (10.54), we get

$$\left(\frac{d^2}{dr^2} - \frac{2}{r^2} \right)^2 f(r) = 0 \,. \tag{10.61}$$

This equation is homogeneous in r and is known to have solutions of the form $f(r) = r^\lambda$. Substituting into Eq. (10.61), we get

$$[\lambda (\lambda - 1) - 2] [(\lambda - 2) (\lambda - 3) - 2] r^{\lambda - 4} = 0 \,.$$

The admissible values of λ are, therefore, the roots of the equation

$$[\lambda (\lambda - 1) - 2] [(\lambda - 2) (\lambda - 3) - 2] = 0 \,, \tag{10.62}$$

i.e., $\lambda=-1$, 1, 2, and 4. The general solution for f is then

$$f(r) = \frac{A}{r} + Br + Cr^2 + Dr^4 . \tag{10.63}$$

For the boundary condition (10.60) to be satisfied, we must have

$$C = -\frac{1}{2} \quad \text{and} \quad D = 0 .$$

From the boundary conditions (10.59), we then get

$$\left.\begin{array}{l} \frac{A}{R} + BR = \frac{1}{2}R^2 \\[2mm] -\frac{A}{R^2} + B = R \end{array}\right\} \quad \Longrightarrow \quad A = -\frac{1}{4} \quad \text{and} \quad B = \frac{3}{4}R .$$

Therefore,

$$f(r) = -\frac{UR^2}{4}\left[2\left(\frac{r}{R}\right)^2 - 3\left(\frac{r}{R}\right) + \left(\frac{R}{r}\right)\right] \tag{10.64}$$

and

$$\psi(r,\theta) = -\frac{UR^2}{4}\left[2\left(\frac{r}{R}\right)^2 - 3\left(\frac{r}{R}\right) + \left(\frac{R}{r}\right)\right]\sin^2\theta . \tag{10.65}$$

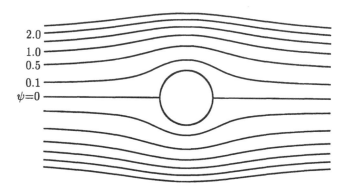

Figure 10.8. *Calculated streamlines of creeping flow past a sphere.*

The two velocity components become:

$$u_r = \frac{U}{2}\left[2 - 3\left(\frac{R}{r}\right) + \left(\frac{R}{r}\right)^3\right]\cos\theta , \tag{10.66}$$

$$u_\theta = -\frac{U}{4} \left[4 - 3 \left(\frac{R}{r} \right) - \left(\frac{R}{r} \right)^3 \right] \sin\theta \,. \tag{10.67}$$

Note that reversing the direction of the flow leads to a change of the sign of **u** everywhere. In Fig. 10.8 we show streamlines as predicted by Eq. (10.65). These are symmetric fore and aft of the sphere.

A quantity of major interest is the *drag force*, F_D, on the sphere. For its calculation we need to know the pressure and the stress components. To obtain the pressure we first substitute u_r and u_θ into the r- and θ-components of the Navier–Stokes equation which yields

$$\frac{\partial p}{\partial r} = 3\eta \, RU \, \frac{\cos\theta}{r^3} \quad \text{and} \quad \frac{\partial p}{\partial \theta} = \frac{3}{2}\eta \, RU \, \frac{\sin\theta}{r^2} \,.$$

We then integrate the above equations getting

$$p(r,\theta) = p_\infty - \frac{3}{2}\eta \, RU \, \frac{\cos\theta}{r^2} \,, \tag{10.68}$$

where p_∞ is the uniform pressure at infinity. Therefore, on the sphere,

$$p(R,\theta) = p_\infty - \frac{3}{2}\frac{\eta U}{R} \cos\theta \,. \tag{10.69}$$

The rr- and $r\theta$-components of the total stress tensor on the surface of the sphere are:

$$T_{rr} = -p + \tau_{rr} = -p + 2\eta \, \frac{\partial u_r}{\partial r} = -p_\infty + \frac{3}{2}\frac{\eta U}{R} \cos\theta + 0 = -p_\infty + \frac{3}{2}\frac{\eta U}{R} \cos\theta$$

$[\tau_{rr}=0$ due to Eq. (10.59)] and

$$T_{r\theta} = \tau_{r\theta} = \eta \left[r\frac{\partial}{\partial r} \left(\frac{u_\theta}{r} \right) + \frac{1}{r}\frac{\partial u_r}{\partial \theta} \right] = -\frac{3}{2}\frac{\eta U}{R} \sin\theta \,.$$

The force per unit area exerted on the sphere is given by

$$\mathbf{f} = \mathbf{e}_r \cdot \mathbf{T} = T_{rr}\,\mathbf{e}_r + T_{r\theta}\,\mathbf{e}_\theta = T_{rr}\,(\cos\theta\,\mathbf{i} + \sin\theta\,\mathbf{j}) + T_{r\theta}\,(-\sin\theta\,\mathbf{i} + \cos\theta\,\mathbf{j}) \implies$$

$$\mathbf{f} = (T_{rr}\cos\theta - T_{r\theta}\sin\theta)\,\mathbf{i} + (T_{rr}\sin\theta + T_{r\theta}\cos\theta)\,\mathbf{j} \,.$$

Due to symmetry, the net force f on the sphere is in the direction **i** of the uniform flow, i.e.,

$$f = \mathbf{i} \cdot \mathbf{f} = T_{rr}\cos\theta - T_{r\theta}\sin\theta \,.$$

Substitution of T_{rr} and $T_{r\theta}$ leads to

$$f = -p_\infty \cos\theta + \frac{3}{2}\frac{\eta U}{R} . \qquad (10.70)$$

To obtain the drag force, we integrate f over the sphere surface:

$$F_D = \int_0^{2\pi} \int_0^\pi f\, R^2 \sin\theta\, d\theta d\phi = 2\pi R^2 \int_0^\pi \left(-p_\infty \cos\theta + \frac{3}{2}\frac{\eta U}{R}\right) \sin\theta\, d\theta \quad \Longrightarrow$$

$$F_D = 6\pi\, \eta R U . \qquad (10.71)$$

Note that, due to symmetry, the term $-p_\infty \cos\theta$ does not contribute to the drag force. Equation (10.71) is the famous *Stokes law* for creeping flow past a sphere [18].

Equation (10.71) can also be cast in the general form

$$\mathbf{F}_D = \eta\, \mathbf{R}_S \cdot (U\, \mathbf{i}) , \qquad (10.72)$$

where $\mathbf{R}_S$ denotes the *shape tensor*. In the case of an isotropic sphere, the shape tensor is obviously given by

$$\mathbf{R}_S = 6\pi R\, \mathbf{I} . \qquad (10.73)$$

Shape tensors for several bodies are given in Reference [19].

The *drag coefficient*, C_D, is generally obtained by dividing the drag force by $1/2\rho U^2$, and by the area of the body projected on a plane normal to the direction of the uniform velocity field. Therefore, in the present case,

$$C_D \equiv \frac{F_D}{\frac{1}{2}\rho U^2\, (\pi R^2)} , \qquad (10.74)$$

which takes the form

$$C_D = \frac{24}{Re} , \qquad (10.75)$$

where the Reynolds number, Re, is defined by

$$Re \equiv \frac{2\rho U R}{\eta} . \qquad (10.76)$$

It should be noted that the creeping flow assumption,

$$|\mathbf{u} \cdot \nabla \mathbf{u}| \ll \left|\nu\nabla^2\mathbf{u}\right| ,$$

is not valid far from the sphere where the velocity gradients are vanishing and, consequently, inertia forces become comparable to viscous forces [5]. The failure of Stokes flow is more striking in the case of two-dimensional flow past a circular cylinder. In this flow problem, the assumption of a separated solution of the form $\psi(r,\theta)=Uf(r)\sin\theta$ leads to (Problem 10.8)

$$\psi(r,\theta) = U\left(\frac{A}{r} + Br + Cr^3 + Dr\ln r\right)\sin\theta.$$

The trouble with the above solution is that there is no choice of the arbitrary constants with which all the boundary conditions are satisfied. Historically, this failure is known as *Stokes' paradox*.

To overcome the failure of Stokes flow far from the sphere, Oseen [20] used the substitution

$$\mathbf{u} = U\mathbf{i} + \mathbf{u}', \tag{10.77}$$

with which the Navier–Stokes equation becomes

$$U\mathbf{i}\cdot\nabla\mathbf{u}' + \mathbf{u}'\cdot\nabla\mathbf{u}' = -\frac{1}{\rho}\nabla p + \nu\nabla^2\mathbf{u}'. \tag{10.78}$$

The nonlinear inertia term, $\mathbf{u}'\cdot\nabla\mathbf{u}'$, is vanishingly small and can be neglected. Therefore,

$$U\mathbf{i}\cdot\nabla\mathbf{u}' = -\frac{1}{\rho}\nabla p + \nu\nabla^2\mathbf{u}'. \tag{10.79}$$

Equation (10.79) is known as *Oseen's equation*, and its solution is called *Oseen's approximation*. Lamb [21] obtained an approximate solution to Eq. (10.79) for the sphere problem which yields

$$F_D = 6\pi\,\eta RU\left[1 + \frac{3}{16}Re + O\left(Re^2\right)\right]. \tag{10.80}$$

Proudman and Pearson [22] solved the full Navier–Stokes equation at small Reynolds number using a singular perturbation method and obtained the following expression for the drag force:

$$F_D = 6\pi\,\eta RU\left[1 + \frac{3}{16}Re + \frac{9}{160}Re^2\ln Re + O\left(Re^2\right)\right]. \tag{10.81}$$

$\square$

Example 10.3.2. Creeping flow around a translating sphere

Consider creeping flow around a sphere translating steadily with velocity $U\mathbf{i}$ through an incompressible, Newtonian liquid which is otherwise undisturbed. Setting the origin of the spherical coordinate system at the instantaneous position of the center of the sphere, we obtain the velocity of the liquid by adding

$$-U\mathbf{i} = -U\cos\theta\,\mathbf{e}_r + U\sin\theta\,\mathbf{e}_\theta$$

to the velocity vector found in the previous example. We thus get

$$u_r = \frac{U}{2}\left[-3\left(\frac{R}{r}\right) + \left(\frac{R}{r}\right)^3\right]\cos\theta \tag{10.82}$$

and

$$u_\theta = \frac{U}{4}\left[3\left(\frac{R}{r}\right) + \left(\frac{R}{r}\right)^3\right]\sin\theta\,. \tag{10.83}$$

The corresponding stream function is given by

$$\psi(r,\theta) = \frac{UR^2}{4}\left[3\left(\frac{r}{R}\right) - \left(\frac{R}{r}\right)\right]\sin^2\theta\,. \tag{10.84}$$

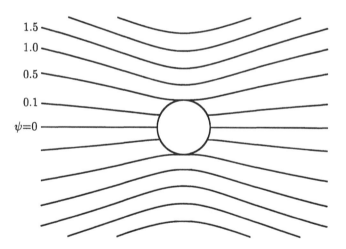

Figure 10.9. *Calculated streamlines of creeping flow around a translating sphere.*

In Fig. 10.9, we show streamlines predicted by Eq. (10.84). Note that the disturbance due to the motion of the sphere propagates to a considerable distance from the sphere. □

Example 10.3.3. Creeping flow around bubbles and droplets

The analysis for flow around a gas bubble of zero viscosity is the same as that for flow past a solid sphere studied in Example 10.3.1, except that the boundary conditions at the liquid-gas interface become:

$$u_r = 0 \quad \text{at} \quad r = R \quad \text{(no penetration in gas volume)} \, ;$$
$$\tau_{r\theta} = 0 \quad \text{at} \quad r = R \quad \text{(no traction on the free surface)} \, .$$

The second condition implies that u_θ is, in general, nonzero on the interface. The drag force turns out to be (Problem 10.9)

$$F_D = 4\pi \, \eta R U \, . \tag{10.85}$$

Hence, the corresponding shape tensor is $\mathbf{R}_S = 4\pi R \mathbf{I}$.

In the case of creeping flow of a Newtonian liquid of viscosity η_o past a spherical droplet of another Newtonian liquid of viscosity η_i, the boundary conditions on the interface are:

$$\mathbf{u}^o = \mathbf{u}^i \quad \text{at} \quad r = R \quad \text{(continuity of velocity)} \, ;$$
$$\tau_{r\theta}^o = \tau_{r\theta}^i \quad \text{at} \quad r = R \quad \text{(continuity of shear stress)} \, .$$

The drag force is given by (Problem 10.10)

$$F_D = 4\pi \, \eta_o \, \frac{\eta_o + \frac{3}{2}\eta_i}{\eta_o + \eta_i} \, RU \, . \tag{10.86}$$

Equation (10.86) contains the case of creeping flow past a solid sphere in the limit $\eta_i \to \infty$, and the case of creeping flow past a gas bubble in the limit $\eta_i \to 0$.

The preceding analyses apply to bubbles and droplets of small size so that surface tension forces are sufficiently strong to suppress the deforming effect of viscous forces, and to keep the bubbles or droplets approximately spherical [5]. □

Example 10.3.4. Terminal velocity

Consider a solid spherical particle of radius R and density ρ_p falling under gravity in a bath of a Newtonian fluid of density ρ and viscosity η. The sphere attains a constant velocity U_t, called the *terminal velocity*, once the gravitational force is counterbalanced by the hydrodynamic forces exerted on the sphere, i.e., the buoyancy and drag forces:

$$\frac{4}{3}\pi \, R^3 \rho_p g \, - \, \frac{4}{3}\pi \, R^3 \rho g \, - \, 6\pi \, \eta R \, U_t = 0 \, . \tag{10.87}$$

Solving for U_t yields

$$U_t = \frac{2R^2 \, (\rho_p - \rho) \, g}{9\eta} \, . \tag{10.88}$$

Note that when the particle is less dense than the fluid, $\rho_p - \rho < 0$, the terminal velocity is negative, which obviously means that the particle would be rising in the surrounding fluid.

From Eq. (10.88) we deduce that Stokes' law holds when

$$Re \equiv \frac{2\rho U R}{\eta} = \frac{4R^3 (\rho_p - \rho)\rho\, g}{9\eta^2} \ll 1\,,$$

i.e, when

$$R \ll \left[\frac{9\eta^2}{4(\rho_p - \rho)\rho\, g} \right]^{1/3} \tag{10.89}$$

$\square$

10.4   Problems

10.1. Show that the stream function, ψ, defined in Eq. (10.15) satisfies the biharmonic equation, Eq. (10.16), in polar coordinates for creeping plane incompressible flow.

10.2. By integrating the r- and θ-momentum equations, show that the general form of the pressure p, in creeping plane incompressible flow in polar coordinates, is given by Eq. (10.24).

10.3. Show that, in the particular cases $\lambda=-1$, 0, and 1, the function $f_\lambda(\theta)$ in Eq. (10.18) degenerates to the following forms:

$$
\begin{aligned}
f_{-1}(\theta) &= A\cos2\theta + B\sin2\theta + C\theta + D & (10.90)\\
f_0(\theta) &= A\cos\theta + B\sin\theta + C\theta\cos\theta + D\theta\sin\theta & (10.91)\\
f_1(\theta) &= A\cos2\theta + B\sin2\theta + C\theta + D & (10.92)
\end{aligned}
$$

10.4. Show that Eq. (10.16) has the following particular solution which is independent of θ:

$$\psi(r) = A r^2 \ln r + B \ln r + C r^2 + D\,. \tag{10.93}$$

Show that, in this case, the pressure p is given by

$$p = 4A\,\eta\,\theta + c\,, \tag{10.94}$$

where c is the integration constant.

10.5. Consider the creeping flow of a Newtonian liquid in a corner formed by two plates, one of which is sliding on the other with constant speed U, as shown in Fig. 10.10. The angle α between the two plates is constant.

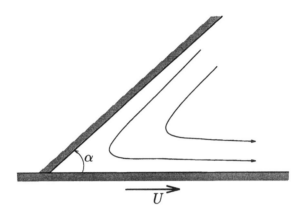

Figure 10.10. *Creeping flow near a corner with one sliding plate.*

(a) Introducing polar coordinates centered at the corner, write down the governing equation and the boundary conditions in terms of the stream function $\psi(r, \theta)$.

(b) Show that the particular solution

$$\psi(r, \theta) = Ur f_0(\theta) = Ur (A \cos\theta + B \sin\theta + C\theta \cos\theta + D\theta \sin\theta) \qquad (10.95)$$

(found in Problem 10.2), satisfies all the boundary conditions.

(c) Show that the stream function is given by

$$\psi(r, \theta) = \frac{Ur}{\alpha^2 - \sin^2\alpha} \left[\alpha(\theta - \alpha) \sin\theta - \theta \sin\alpha \sin(\theta - \alpha) \right]. \qquad (10.96)$$

(d) Calculate the velocity components.

(e) Determine the shear stress at $\theta = 0$ and show that it exhibits a $1/r$ singularity which suggests that an infinite force is required in order to maintain the motion of the sliding plate. What is the origin of this nonphysical result?

10.6. Show that the stream function ψ defined in Eq. (10.39) satisfies Eq. (10.40) for creeping, axisymmetric, incompressible flow in cylindrical coordinates.

10.7. Show that the stream function ψ defined in Eq. (10.53) satisfies Eq. (10.54) for creeping, axisymmetric, incompressible flow in spherical coordinates.

10.8. Consider the creeping flow of a Newtonian liquid past a fixed circular cylinder of radius R assuming that, far from the cylinder, the flow is uniform with speed U.

(a) Introducing polar coordinates centered at the axis of symmetry, write down the governing equation and the boundary conditions for this flow in terms of the stream function $\psi(r, \theta)$.

(b) In view of the boundary condition at $r \to \infty$, assume a solution of the form

$$\psi(r, \theta) = U f(r) \sin\theta, \tag{10.97}$$

and show that this leads to

$$\psi(r, \theta) = U \left(\frac{A}{r} + Br + Cr^3 + Dr \ln r \right) \sin\theta. \tag{10.98}$$

(c) Show that there is no choice of the constants A, B, C, and D for which all the boundary conditions are satisfied (*Stokes' paradox*). Why does the Stokes flow assumption fail? Explain how a well-posed problem can be obtained.

10.9. Consider the creeping flow of an incompressible Newtonian liquid approaching, with uniform speed U, a fixed spherical bubble of radius R.

(a) Introducing spherical coordinates with the origin at the center of the bubble, write down the governing equation and the boundary conditions for this flow in terms of the stream function $\psi(r, \theta)$.

(b) Show that the stream function is given by

$$\psi = -\frac{UR^2}{2} \left[\left(\frac{r}{R} \right)^2 - \frac{r}{R} \right] \sin^2\theta, \quad r \geq R. \tag{10.99}$$

(c) Calculate the two nonzero velocity components and the pressure.

(d) Show that the drag force exerted on the bubble is given by Eq. (10.85).

10.10. Consider the creeping flow of an incompressible Newtonian liquid of viscosity η_o approaching, with uniform speed U, a fixed spherical droplet of viscosity η_i and radius R.

(a) Introducing spherical coordinates with the origin at the center of the bubble, write down the governing equation and the boundary conditions for this flow in terms of the stream function $\psi(r, \theta)$.

(b) Show that the stream function is given by

$$\psi_o = -\frac{UR^2}{4} \left[2 \left(\frac{r}{R} \right)^2 - \frac{2\eta_o + 3\eta_i}{\eta_o + \eta_i} \frac{r}{R} + \frac{2\eta_i}{\eta_o + \eta_i} \frac{R}{r} \right] \sin^2\theta, \quad r \geq R \tag{10.100}$$

outside the droplet, and by

$$\psi_i = \frac{UR^2}{4} \frac{\eta_o}{\eta_o + \eta_i} \left[\left(\frac{r}{R} \right)^2 - \left(\frac{r}{R} \right)^4 \right] \sin^2\theta, \quad r \leq R \tag{10.101}$$

inside the droplet.

(c) Calculate the two nonzero velocity components and the pressure.

(d) Show that the drag force exerted on the droplet is given by Eq. (10.86).

10.11. Calculate the *terminal velocity* of

(a) a spherical bubble rising under gravity in a pool of a Newtonian liquid, and

(b) a spherical droplet of density ρ_i and viscosity η_i falling under gravity in a Newtonian liquid of density ρ_o and viscosity η_o.

10.12. Consider the creeping flow of an incompressible Newtonian liquid of viscosity η and density ρ approaching, with uniform speed U, a fixed solid sphere of radius R, and introduce spherical coordinates with the origin at the center of the sphere. Assume that slip occurs on the sphere surface according to

$$\tau_{r\theta} = \beta\, u_\theta \quad \text{at} \quad r = R\,, \tag{10.102}$$

where β is a slip parameter.

(a) Show that the drag force exerted on the sphere by the liquid is given by

$$F_D = 6\pi\,\eta RU\,\frac{2\eta + \beta R}{3\eta + \beta R}\,, \tag{10.103}$$

and identify limiting cases of the above result.

(b) Calculate the terminal velocity of a solid sphere in a pool of Newtonian liquid when slip occurs on the sphere surface according to Eq. (10.102).

10.5   References

1. S. Kim and S.J. Karrila, *Microhydrodynamics: Principles and Selected Applications*, Butterworth-Heinemann, Boston, 1991.

2. K. Wark and C.F. Warner, *Air Pollution: Its Origin and Control*, Harper & Row, New York, 1975.

3. J.H. Seinfeld, *Atmospheric Chemistry and Physics of Air Pollution*, Wiley & Sons, New York, 1985.

4. S. Middleman, *Fundamentals of Polymer Processing*, McGraw-Hill, New York, 1980.

5. G.K. Batchelor, *An Introduction to Fluid Dynamics*, Cambridge University Press, Cambridge, 1967.

6. R.B. Bird, R.C. Armstrong, and O. Hassager, *Dynamics of Polymeric Liquids: Volume 1, Fluid Mechanics*, Wiley & Sons, New York, 1977.

7. H.J. Lugt and E.W. Schwiderski, "Flows around dihedral angles, I. Eigenmotion analysis," *Proc. Roy. Soc. London* **A285**, 382-399 (1965).

8. W.R. Dean and P.E. Montagnon, "On the steady motion of viscous liquid in a corner," *Proc. Camb. Phil. Soc.* **45**, 389-394 (1949).

9. H.K. Moffatt, "Viscous and resistive eddies near a sharp corner," *J. Fluid Mech.* **18**, 1-18 (1964).

10. D.H. Michael, "The separation of a viscous liquid at a straight edge," *Mathematica* **5**, 82-84 (1958).

11. S. Richardson, "A 'stick-slip' problem related to the motion of a free jet at low Reynolds numbers," *Proc. Camb. Phil. Soc.* **67**, 477-489 (1970).

12. W.J. Silliman and L.E. Scriven, "Separating flow near a static contact line: Slip at the wall and shape of a free surface," *J. Comp. Phys.* **34**, 287-313 (1980).

13. T.R. Salamon, D.E. Bornside, R.C. Armstrong, and R.A. Brown, "Local similarity solutions in the presence of a slip boundary condition," *Physics of Fluids* **9**, 1235-1247 (1997).

14. G.C. Georgiou, L.G. Olson, W.W. Schultz, and S. Sagan, "A singular finite element for Stokes flow: the stick-slip problem," *Int. J. Numer. Methods Fluids* **9**, 1353-1367 (1989).

15. G.C. Georgiou and A. Boudouvis, "Converged solutions of the Newtonian extrudate swell problem," *Int. J. Numer. Methods Fluids* **29**, 363-371 (1999).

16. T.R. Salamon, D.E. Bornside, R.C. Armstrong, and R.A. Brown, "The role of surface tension in the dominant balance in the die well singularity," *Phys. Fluids* **7**, 2328 (1995).

17. M.J. Stefan, "Versuch über die Scheinbare Adhäsion," *Akad. Wissensch. Wien, Math.-Natur.* **69**, 713 (1874).

18. G.G. Stokes, "On the effect of the internal friction of fluids on the motion of pendulums," *Trans. Camb. Phil. Soc.* **9**, 8 (1851).

19. H. Brenner and R.G. Cox, "The resistance to a particle of arbitrary shape in translational motion at small Reynolds numbers," *J. Fluid Mech.* **17**, 561 (1963).

20. C.W. Oseen, "Über die Stokessche formel und über die verwandte aufgabe in der hydrodynamik," *Arkiv Mat. Astron. Fysik* **6**, 29 (1910).

21. H. Lamb, "On the uniform motion of a sphere through a viscous fluid," *Phil. Mag.* **21**, 112 (1911).

22. I. Proudman and J.R.A. Pearson, "Expansions at small Reynolds number for the flow past a sphere and a circular cylinder," *J. Fluid Mech.* **2**, 237 (1957).

List of Symbols

The most frequently used symbols are listed below. Note that some of them are used in multiple contexts. Symbols not listed here are defined at their first place of use.

a	Distance between parallel plates; dimension
$\mathbf{a}$	Acceleration vector; vector
b	Width; dimension
$\mathbf{B}$	Vector potential; Finger strain tensor
c	Integration constant; height; dimension; concentration
c_i	Arbitrary constant
C	Curve
$\mathbf{C}$	Cauchy strain tensor
Ca	Capillary number, $Ca \equiv \frac{\eta \bar{u}}{\sigma}$
C_D	Drag coefficient
C_p	Specific heat at constant pressure
C_v	Specific heat at constant volume
d	Diameter; distance
$d\ell$	Differential arc length
dS	Differential surface
$d\mathbf{S}$	Directed differential surface, $d\mathbf{S} \equiv \mathbf{n}dS$
ds	Differential length
dV	Differential volume
D	Diameter
$\mathbf{D}$	Rate-of-strain tensor, $\mathbf{D} \equiv \frac{1}{2}\left[\nabla u + (\nabla u)^T\right]$
$\frac{D}{Dt}$	Substantial derivative operator
$\mathbf{e}_i$	Unit vector in the x_i-direction
E	energy

$\dot{E}$	Rate of energy conversion
E^2	Stokes stream function operator
E^4	Stokes stream function operator, $E^4 \equiv E^2(E^2)$
Eu	Euler number, $Eu \equiv \dfrac{2\,\Delta p}{\rho V^2}$
$\mathbf{f}$	Traction force
$\mathbf{F}$	Force
F_D	Drag force
Fr	Froude number, $Fr \equiv \dfrac{V^2}{gL}$
g	Gravitational acceleration
$\mathbf{g}$	Gravitational acceleration vector
$\mathbf{G}$	Green strain tensor
h	Height; elevation
H	Distance between parallel plates; thermal energy; enthalpy
$\dot{H}$	rate of production of thermal energy
i	Imaginary unit, $i \equiv \sqrt{-1}$; index
$\mathbf{i}$	Cartesian unit vector in the x-direction
I	First invariant of a tensor
$\mathbf{I}$	Unit tensor
II	Second invariant of a tensor
III	Third invariant of a tensor
$\mathbf{j}$	Cartesian unit vector in the y-direction
J_n	nth-order Bessel function of the first kind
$\mathbf{J}$	Linear momentum, $\mathbf{J} \equiv m\mathbf{u}$
$\dot{\mathbf{J}}$	Rate of momentum convection
$\mathbf{J}_\theta$	Angular momentum, $\mathbf{J}_\theta \equiv \mathbf{r} \times \mathbf{J}$
k	Thermal conductivity; diffusion coefficient; Boltzman constant; index
$\mathbf{k}$	Cartesian unit vector in the z-direction
L	Length; characteristic length
m	Mass; meter (unit of length)
$\dot{m}$	Mass flow rate
M	Molecular weight
$\mathbf{M}$	Moment
$\mathbf{n}$	Unit normal vector
N	Newton (unit of force)
O	Order of
p	Pressure
p_0	Reference pressure

p_∞	Pressure at infinity
P	Equilibrium pressure
Q	Volumetric flow rate
r	Radial coordinate; radial distance
$\mathbf{r}$	Position vector
R	Radius; ideal gas constant
Re	Reynolds number, $Re \equiv \dfrac{L\bar{u}\rho}{\eta}$
$\mathcal{R}e$	Real part of
s	Length; second (time unit)
S	Surface; surface area
$\mathbf{S}$	Vorticity tensor, $\mathbf{S} \equiv \frac{1}{2}\left[\nabla u - (\nabla u)^T\right]$
St	Stokes number, $St \equiv \dfrac{\rho g L^2}{\eta u}$
t	Time
$\mathbf{t}$	Unit tangent vector
T	Absolute temperature
T_0	Reference temperature
$\mathbf{T}$	Total stress tensor
T_{ij}	ij-component of the total stress tensor
$\mathbf{u}$	Vector; velocity vector
$\bar{u}$	Mean velocity
u_r	Radial velocity component
u_w	Slip velocity (at a wall)
u_x	x-velocity component
u_y	y-velocity component
u_z	z-velocity component
u_θ	azimuthal velocity component
u_ϕ	ϕ-velocity component
U	Velocity (magnitude of); internal energy per unit mass, $dU \equiv C_v dT$
U_t	Terminal velocity
$\mathbf{v}$	Vector
V	Volume; velocity (magnitude of); characteristic velocity
$\hat{V}$	Specific volume
V_M	Molecular volume
W	Width; work; weight
$\dot{W}$	Rate of production of work
We	Weber number, $We \equiv \dfrac{\rho V^2 L}{\sigma}$
x	Cartesian coordinate

x_i	Cartesian coordinate
y	Cartesian coordinate
Y_n	nth-order Bessel function of the second kind
z	Cartesian or cylindrical or spherical coordinate

Greek letters

α	Inclination; angle; dimension; coefficient of thermal expansion
β	Isothermal compressibility; slip coefficient
Γ	Circulation
δ	Film thickness; boundary layer thickness
δ_{ij}	Kronecker's delta
Δ	Difference; local rate of expansion
Δp	Pressure drop
$\Delta p / \Delta L$	Constant pressure gradient
$\Delta \mathbf{r}$	Separation vector
ϵ	Aspect ratio, e.g., $\epsilon \equiv \dfrac{H}{L}$; perturbation parameter
ϵ_{ijk}	Permutation symbol
η	Viscosity; similarity variable
η_v	Bulk viscosity
θ	Cylindrical or spherical coordinate; angle
λ	Second viscosity coefficient
ν	Kinematic viscosity, $\nu \equiv \dfrac{\eta}{\rho}$
Π	Dimensionless number
ξ	Stretching coordinate; similarity variable
ρ	Density
σ	Surface tension
$\boldsymbol{\sigma}$	Tensor
σ_{ij}	ij-component of $\boldsymbol{\sigma}$
$\boldsymbol{\tau}$	Viscous stress tensor; tensor
τ_{ij}	ij viscous stress component
τ_w	Wall shear stress
ϕ	Spherical coordinate; angle; scalar function
ψ	Stream function
ω	Vorticity; angular frequency
$\boldsymbol{\omega}$	Vorticity vector
Ω	Angular velocity
$\boldsymbol{\Omega}$	Angular velocity vector

Other symbols

∇	Nabla operator
∇_{II}	Nabla operator in natural coordinates $(\mathbf{t}, \mathbf{n})$, $\nabla_{II} \equiv \frac{\partial}{\partial t}\,\mathbf{t} + \frac{\partial}{\partial n}\,\mathbf{n}$
$\nabla\mathbf{u}$	Velocity gradient tensor
∇^2	Laplace operator
∇^4	Biharmonic operator, $\nabla^4 \equiv \nabla^2(\nabla^2)$
$\cdot$	Dot product
:	Double dot product
$\times$	Cross product

Superscripts

T	Transpose (of a matrix or a tensor)
-1	Inverse (of a matrix or a tensor)
$*$	Dimensionless variable

Abbreviations

1D	One-dimensional
2D	Two-dimensional
3D	Three-dimensional
CFD	Computational Fluid Dynamics
ODE(s)	Ordinary differential equation(s)
PDE(s)	Partial differential equation(s)

Index

Milton Keynes UK
Ingram Content Group UK Ltd.
UKHW051943071024
449327UK00026B/2140